T0321799

Control and Signal Processing Applications for Mobile and Aerial Robotic Systems

Oleg Sergiyenko
Universidad Autónoma de Baja California, Mexico

Moises Rivas–Lopez
Universidad Autónoma de Baja California, Mexico

Wendy Flores–Fuentes
Universidad Autónoma de Baja California, Mexico

Julio Cesar Rodríguez–Quiñonez
Universidad Autónoma de Baja California, Mexico

Lars Lindner
Universidad Autónoma de Baja California, Mexico

A volume in the Advances in
Computational Intelligence and
Robotics (ACIR) Book Series

Published in the United States of America by
IGI Global
Engineering Science Reference (an imprint of IGI Global)
701 E. Chocolate Avenue
Hershey PA, USA 17033
Tel: 717-533-8845
Fax: 717-533-8661
E-mail: cust@igi-global.com
Web site: http://www.igi-global.com

Library of Congress Cataloging-in-Publication Data

Names: Sergiyenko, Oleg, 1969- editor. | Rivas-Lopez, Moises, 1960- editor. |
  Flores-Fuentes, Wendy, 1978- editor. | Rodríguez-Quiñonez, Julio C.,
  1985- editor. | Lindner, Lars, editor.
Title: Control and signal processing applications for mobile and aerial
  robotic systems / Oleg Sergiyenko, Moises Rivas-Lopez, Wendy
  Flores-Fuentes, Julio C. Rodriguez-Quinonez, and Lars Lindner, editors.
Description: Hershey, PA : Engineering Science Reference, an imprint of IGI
  Global, [2019] | Includes bibliographical references and index.
Identifiers: LCCN 2019013250| ISBN 9781522599241 (hardcover) | ISBN
  9781522599265 (ebook) | ISBN 9781522599258 (softcover)
Subjects: LCSH: Drone aircraft--Control systems. | Mobile robots. | Signal
  processing.
Classification: LCC TL589.4 .C65 2019 | DDC 629.8/93--dc23 LC record available at https://lccn.
loc.gov/2019013250

This book is published in the IGI Global book series Advances in Computational Intelligence and Robotics (ACIR) (ISSN: 2327-0411; eISSN: 2327-042X)

British Cataloguing in Publication Data
A Cataloguing in Publication record for this book is available from the British Library.

All work contributed to this book is new, previously-unpublished material.
The views expressed in this book are those of the authors, but not necessarily of the publisher.

For electronic access to this publication, please contact: eresources@igi-global.com.

Advances in Computational Intelligence and Robotics (ACIR) Book Series

Ivan Giannoccaro
University of Salento, Italy

ISSN:2327-0411
EISSN:2327-042X

MISSION

While intelligence is traditionally a term applied to humans and human cognition, technology has progressed in such a way to allow for the development of intelligent systems able to simulate many human traits. With this new era of simulated and artificial intelligence, much research is needed in order to continue to advance the field and also to evaluate the ethical and societal concerns of the existence of artificial life and machine learning.

The **Advances in Computational Intelligence and Robotics (ACIR) Book Series** encourages scholarly discourse on all topics pertaining to evolutionary computing, artificial life, computational intelligence, machine learning, and robotics. ACIR presents the latest research being conducted on diverse topics in intelligence technologies with the goal of advancing knowledge and applications in this rapidly evolving field.

COVERAGE

- Pattern Recognition
- Artificial Intelligence
- Adaptive and Complex Systems
- Computational Intelligence
- Evolutionary Computing
- Fuzzy Systems
- Brain Simulation
- Cyborgs
- Computational Logic
- Machine Learning

IGI Global is currently accepting manuscripts for publication within this series. To submit a proposal for a volume in this series, please contact our Acquisition Editors at Acquisitions@igi-global.com or visit: http://www.igi-global.com/publish/.

Titles in this Series

For a list of additional titles in this series, please visit:
https://www.igi-global.com/book-series/advances-computational-intelligence-robotics/73674

Examining Fractal Image Processing and Analysis
Soumya Ranjan Nayak (Chitkara University, India) and Jibitesh Mishra (College of Engineering and Technology, India)
Engineering Science Reference • © 2020 • 305pp • H/C (ISBN: 9781799800668) • US $245.00

AI and Big Data's Potential for Disruptive Innovation
Moses Strydom (Emeritus, France) and Sheryl Buckley (University of South Africa, South Africa)
Engineering Science Reference • © 2020 • 405pp • H/C (ISBN: 9781522596875) • US $225.00

Handbook of Research on the Internet of Things Applications in Robotics and Automation
Rajesh Singh (Lovely Professional University, India) Anita Gehlot (Lovely Professional University, India) Vishal Jain (Bharati Vidyapeeth's Institute of Computer Applications and Management (BVICAM), New Delhi, India) and Praveen Kumar Malik (Lovely Professional University, India)
Engineering Science Reference • © 2020 • 433pp • H/C (ISBN: 9781522595748) • US $295.00

Handbook of Research on Applications and Implementations of Machine Learning Techniques
Sathiyamoorthi Velayutham (Sona College of Technology, India)
Engineering Science Reference • © 2020 • 461pp • H/C (ISBN: 9781522599029) • US $295.00

Handbook of Research on Advanced Mechatronic Systems and Intelligent Robotics
Maki K. Habib (The American University in Cairo, Egypt)
Engineering Science Reference • © 2020 • 466pp • H/C (ISBN: 9781799801375) • US $295.00

701 East Chocolate Avenue, Hershey, PA 17033, USA
Tel: 717-533-8845 x100 • Fax: 717-533-8661
E-Mail: cust@igi-global.com • www.igi-global.com

Table of Contents

Detailed Table of Contents

    Viktor Melnyk, Kharkiv Petro Vasylenko National Technical University
        of Agriculture, Ukraine
    Roman Antoshchenkov, Kharkiv Petro Vasylenko National Technical
        University of Agriculture, Ukraine
    Viktor Antoshchenkov, Kharkiv Petro Vasylenko National Technical
        University of Agriculture, Ukraine

To determine the wheel dynamics of the mobile machine, a sensor has been developed, which consists of a three-axis accelerometer, a three-axis gyroscope, and a three-axis magnetometer. The sensor is connected to the microcontroller, which transmits the data received via the 2.4 GHz channel. It is attached coaxially to the center of the wheel. In the first step of processing data from the accelerometer and gyroscope, their values are corrected. The corrected acceleration signal and angular velocities are processed using a Butterworth filter. Madgwick filter determines the angles of orientation of the sensor in space. In the next step, the authors deduct the centrifugal component from accelerations. Further, the gravitational component is subtracted from the accelerations to get its real value. The wheel speed is obtained by integrating accelerations. Angular wheel speed and accelerations are processed with a Kalman filter. The outcomes of experimental studies of the proposed method and sensor to determine the wheel dynamics on the tractors with 4x4 wheel formula have been analyzed.

Chapter 2

Fabian N. Murrieta-Rico, Universidad Autónoma de Baja California,
Mexico
Vitalii Petranovskii, Universidad Nacional Autónoma de Mexico, Mexico
Juan de Dios Sanchez-Lopez, Universidad Autónoma de Baja
California, Mexico
Juan Ivan Nieto-Hipolito, Universidad Autónoma de Baja California,
Mexico
Mabel Vazquez-Briseño, Universidad Autónoma de Baja California,
Mexico
Joel Antúnez-García, Universidad Nacional Autónoma de Mexico,
Mexico
Rosario I. Yocupicio-Gaxiola, Universidad Nacional Autónoma de
Mexico, Mexico
Vera Tyrsa, Universidad Autónoma de Baja California, Mexico

In most aerial vehicles, accurate information about critical parameters like position, velocity, and altitude is critical. In these systems, such information is acquired through an inertial measurement unit. Parameters like acceleration, velocity, and position are obtained after processing data from sensors; some of them are the accelerometers. In this case, the signal generated by the accelerometer has a frequency that depends from the acceleration experienced by the sensor. Since the time available for frequency estimation is critical in an aerial device, the frequency measurement algorithm is critical. This chapter proposes the principle of rational approximations for measuring the frequency from accelerometer-generated signals. In addition, the effect of different measurement parameters is shown, discussed, and evaluated.

Chapter 3

Lyudmila Yurievna Vorochaeva, Southwest State University, Russia
Sergey Igorevich Savin, Innopolis University, Russia
Andrei Vasilievich Malchikov, Southwest State University, Russia
Andres Santiago Martinez Leon, Southwest State University, Russia

This chapter is dedicated to tackling the issues related to the design and locomotion control of a hybrid wheeled jumping monitoring platform. The studied robot consists of a body mounted on a wheeled platform and of a jump acceleration module. An approach to making design decisions regarding the structure of the investigated robot is proposed. To select the kinematic structure of the robot, classifications of possible variants of hybrid jumping platforms and accelerating modules are presented. Methods for controlling the function of the accelerating modules and the analysis of their work is carried out. Various implementations of jumping motion are discussed; these implementations are characterized by different combinations of relative links movements during various stages of motion. Each of the proposed jump motion types requires the development of a control system, which is also discussed in this chapter.

Chapter 4

The Arduino is programmed to control the robot navigation. The Garbage Collection Robot is designed to collect solid waste at public places (schools, workplaces, and parks) and residential areas. The design of the robot is such that when it starts, it maneuvers as per programmed route. The Garbage Collector can sense by means of capacitive proximity sensors if the obstacle is living (for example, a human being) or non-living (for example, vehicle) and then gives appropriate warning signals like flashing light, hoot, or voice commands. The robot is equipped with vision capabilities in order for it to detect colors, namely green, red, yellow, blue, and black for organics, plastic, metal, paper, and glass, respectively. When the GCR sees a particular color code on garbage container, it picks up the bin, carries it in its carriage, then offloads it at a desired station to wait for recycling or final dumping.

Chapter 5

In this chapter, the authors consider the need and relevance of cryptographic transformation of images and video files that are transmitted from unmanned aircraft, airborne robots. The authors propose and consider new multifunctional matrix-algebraic models of cryptographic image transformations, the variety of matrix models, including block parametrical and matrix affine permutation ciphers. The authors show the advantages of the cryptographic models, such as adaptability to various formats, multi-functionality, ease of implementation on matrix parallel structures, interchangeability of iterative procedures and matrix exponentiation modulo, ease of selection, and control of cryptographic transformation parameters. The simulation results of the proposed algorithms and procedures for the direct and inverse transformation of images with the aim of masking them during transmission are demonstrated and discussed in this chapter. The authors evaluate the effectiveness and implementation reliability of matrix-algebraic models of cryptographic image transformations.

Chapter 6
    Miguel Reyes-Garcia, Universidad Autónoma de Baja California, Mexico

    Cesar Sepulveda-Valdez, Universidad Autónoma de Baja California, Mexico

    Oleg Sergiyenko, Universidad Autónoma de Baja California, Mexico

    Moisés Rivas-López, Universidad Autónoma de Baja California, Mexico

    Julio Rodríguez-Quiñonez, Universidad Autónoma de Baja California, Mexico

    Wendy Flores-Fuentes, Universidad Autónoma de Baja California, Mexico

    Daniel Hernandez-Balbuena, Universidad Autónoma de Baja California, Mexico

    Juan-Ivan Nieto-Hipolito, Universidad Autónoma de Baja California, Mexico

    Fabian N. Murrieta-Rico, Universidad Autónoma de Baja California, Mexico

    Lars Lindner, Universidad Autónoma de Baja California, Mexico

    Mykhailo Ivanov, Universidad Autónoma de Baja California, Mexico

Positioning technologies are useful in a great number of applications, which are oriented for pick and place robots, manipulation of machine tools, especially on machines oriented for artificial vision and detection systems, such as vision-guided robotic systems and object existence in a limited environment. Due to the high demand of those applications, this chapter presents digital control theory application using a laser positioner, which obtains 3D coordinates in a defined field of view. Using the LM629N-8 motion controller, representing the main digital controller for the motion task of the laser positioner as an active element, which is analyzed via modeling and simulation using Matlab-Simulink. Additionally, this chapter focuses on some of the principal sources of uncertainties that exist in a laser scanning system and mainly on the receptive part of such system, which is driven by a brushed DC motor. The processed signal will be analyzed in different environmental conditions to analyze how it is affected by the instability characteristics of this main actuator.

In this chapter, the problem of trajectory generation for bipedal walking robots is considered. A number of modern techniques are discussed, and their limitations are shown. The chapter focuses on zero-moment point methods for trajectory generation, where the desired trajectory of that point can be used to allow the robot to keep vertical stability if followed, and presents an instrument to calculate the desired trajectory for the center of mass for the robot. The chapter presents an algorithm based on quadratic programming, with an introduction of a slack variable to make the problem feasible and a change of variables to improve the numeric properties of the resulting optimization problem. Modern optimization tools allow one to solve such problems in real time, making it a viable solution for trajectory planning for the walking robots. The chapter shows a few results from the numerical simulation made for the algorithm, demonstrating its properties.

Navigation of mobile autonomous robots on unknown terrain in the absence of GPS is extremely difficult. The general aim of the chapter is to analyze the possibilities of reliable detecting landmarks and determining their coordinates for navigation purposes. It is shown that the method of solving such a problem is the complex use of the meters operating on different physical principles. The main attention is paid to the radar method of measuring the angular coordinates of the landmarks by an antenna with a small size. For a radar with a wide antenna pattern, the possibility of angular resolving of two or more closely spaced landmarks is estimated. The most reliable method providing angular resolution is the creation of a synthesized aperture of antennae in the process of linear movement of a robot. The possibilities of such antennae are analyzed, considering random phase distortions and errors.

In this chapter, the almost periodicity of the first principal components is used to carry out the reconstruction of images. First, the principal component analysis technique is applied to an image. Then, the periodicity of its principal components is analyzed. Next, this periodicity is used to build periodic vectors of the same length of the original principal components, and finally, these periodic vectors are used to reconstruct the original image. The proposed method was compared against the JPEG (Joint Photographic Experts Group) compression technique. The mean square error and peak signal-to-noise ratio were used to perform the above-mentioned comparison. The experimental results showed that the proposed method performed better than JPEG, when the original image was reconstructed using the principal components modified by periodicity.

Chapter 10
*Danilo Caceres Hernandez, Universidad Tecnológica de Panamá,
Panama*
Laksono Kurnianggoro, University of Ulsan, South Korea
Alexander Filonenko, ABBYY, Russia
Kang-Hyun Jo, University of Ulsan, South Korea

In the field of advanced driver-assistance and autonomous vehicle systems, understanding the surrounding vehicles plays a vital role to ensure a robust and safe navigation. To solve detection and classification problem, an obstacle classification strategy based on laser sensor is presented. Objects are classified according the geometry, distance range, reflectance, and disorder of each of the detected object. In order to define the best number of features that allows the algorithm to classify these objects, a feature analysis is performed. To do this, the set of features were divided into four groups based on the characteristic, distance, reflectance, and the entropy of the object. Finally, the classification task is performed using the support vector machines (SVM) and adaptive boosting (AdaBoost) algorithms. The evaluation indicates that the method proposes a feasible solution for intelligent vehicle applications, achieving a detection rate of 87.96% at 48.32 ms for the SVM and 98.19% at 79.18ms for the AdaBoost.

Foreword

Today we are witnessing how mobile robotics is firmly entering our daily lives. A few decades ago, mobile robots had mainly military, rescue and research applications, now unmanned vehicles are appearing on our highways. Robotic assistants help us in airports, libraries, hospitals, and autonomous harvesters work on agricultural fields while UAVs scan crops to determine a vegetation index. UAVs will soon be delivering parcels, carrying heavy loads and passengers. In other words, the range of usage of mobile robotics is huge. It has found application in a variety of industries, including manufacturing, construction, agriculture, aerospace, mining, medicine, etc.

The complexity of robots and their application requires the consolidation of knowledge of mathematics and mechanical engineering, electrical and control engineering, computer science, and signal processing, pattern recognition and machine learning. Despite the efforts of a large number of scientists and engineers that have led to the solution of many problems, many things remain to be improved and refined, and some problems are still awaiting solutions. This book aims to familiarize the reader with some solutions of such important robotics tasks as localization and mapping, trajectory planning, reliability and security of transmitting information from a mobile robot, etc.

An active area of research in mobile robotics is Simultaneous Localization and Mapping (SLAM), where a map of the environment is autonomously constructed and at the same time, the location of a robot is estimated. The instability of receiving the GPS signal forced researchers to return to onboard autonomous navigation systems using inertial navigation and odometry. However, the use of an Inertial Measurement Unit (IMU) to determine the movement parameters and the position of a robot is also associated with certain difficulties. Chapters 1 and 2 of this book deal with overcoming these issues.

The improvement of Machine Vision Systems (MVS) to assist in the solution of SLAM problems is still a current research topic. MVS can be based on different sensors, like camera, laser, radar, sonar or lidar sensors, as well as a combination of these sensors and is for studying the environment, detecting various landmarks, identifying them, determining their size and shape, as well as the position and orientation of the robot in relation to these landmarks. Chapter 5 discusses the reliability and security of the image transmission received from the cameras of a robot MVS. Thereby, Chapter 6 of the book is devoted to improving the efficiency of using lasers in a robot MVS.

When searching for a landmark and its location using electromagnetic or ultrasonic waves, echo signals can arrive from the surrounding environment. These signals are easily confused with signals that are reflected from a landmark. Chapter 8 proposes an approach that always makes it possible to identify landmarks using a radar.

Robots can move in many ways: there are flying and swimming robots, wheeled and legged robots, and even crawling or jumping robots. A jumping robot can turn in a limited space where other robots cannot. Such a robot can also move on challenging terrains with obstructions and obstacles. If we add wheels to the jumping robot, its manoeuvrability and speed will increase due to the combination of two modes of movement. Chapter 3 is about the issues related to the design and motion control of a hybrid wheeled jumping robots. At the same time, Chapter 7 describes the solution of the trajectory-planning task of a walking robot.

One useful application of mobile robotics is a garbage collection robot. The famous robotic vacuum cleaner is the most successful household robot. Further development of the robotic garbage collector will be a robot that identifies and sorts the garbage. A step towards creating such a cleaner robot is the robot described in Chapter 4 of present book.

Thus, this book covers the essential issues of mobile and aerial robotics and makes a significant contribution to their development. This book is of great value for advanced undergraduate readers as well as for researchers and designers who deal with improving the existing and creating new robots.

Alexander Gurko
Kharkiv National Automobile and Highway University, Ukraine
August 2019

Preface

Mobile robotic systems are utilizing a lot of complex combined electronic devices, which are applying a wide range of control and signal processing techniques to detect and capture their surrounding environment in real-time. Different kinds of sensory parts are used, like for example CCD cameras, laser scanners, infrared and ultrasonic sensors, IMU and GPS sensors. Also, a combination of previously mentioned sensory parts is used, which measure actual physical parameters of the environment, as well as the current global position and speed of the robotic system, as also the precise time traveling along the trajectory. Using the mixed measurements of the actual physical parameters, different hardware processing units are utilized, implementing the suitable control algorithms to determine the control signals for the robot actuators and take the decisions based on external conditions of the robotic system. Mobile robots therefore play a crucial role in modern industrial applications, as well as its research and development.

DEFINITIONS

Robotics can be defined as an interface between engineering and science, based on principles of electrical engineering, mechanical engineering, information technology and computer science to develop and research machines that mimic human movements. These machines, also called *robots*, can perform various handling or manufacturing tasks, using sensors, grippers, tools and free programming of the movement sequences. Robotics is concerned with the design, control, production and operation of robots. Robotics must work closely with human-machine interaction, psychology, sociology (social robotics) and philosophy (machine ethics). The results of robotics research are important in many fields, such as industry (industrial, agricultural and service robots), science (research and experimental robots), society (service robots and assistance systems), health care (care and therapy robots), transport (robotic cars) and military (combat robots). All the types of robots can be mainly classified in 5 types: Industrial robots, Service robots, Mobile robots, Micro and Nano robots and

Humanoid robots. Today, the industrial robots are the furthest in use. The industrial robots are characterized by very high precision and repeatability of their performed tasks, and requires a very robust design. Service robots are available for the private (lawn mower or vacuum cleaner) and the professional sector (agricultural robots, cleaning robots, diving robots). Mobile robots are available in various designs: driverless transport system or exploration robot. Micro robots are sometimes only a few millimeters in size and can act in swarms. Humanoid robots try to simulate the human completely, which often places high demands on sensor and motor technology.

Signals are physical quantities, which transfer information's and which are variable in time. The main task of *Signal Processing* is to extract this information from a given signal, whereby the information content should not be lost and the information of the signal is encoded in the temporal or spatial change of the signal. The received signal is usually converted into an electrical signal for further processing. To convert the signal, microphones, cameras or radio antennas are used, to name only few. Signal processing and its application has relationships with many fields of knowledge. The most important applications of signal processing are communications engineering, control engineering, acoustics and metrology.

Control Technology is an engineering branch, which designs, constructs and analyses automatically operating machines and devices, which are often combined into complex and industrial systems. In order to control dynamic systems, the control technology uses the closed-loop method, which is defined by DIN 19226 as follows: "Control is a process in which a variable to be controlled (controlled variable) is continuously recorded, compared with another variable (reference variable) and influenced in terms of an approximation to the reference variable. Characteristic for control is the closed-loop principle of operation, in which the controlled variable continuously influences itself over the closed-loop." The use of control engineering methods is largely independent of the respective application. The problems to be solved have mostly similarities and come from technical, biological, economic and sociological areas.

HISTORY OF ROBOTICS

The idea of mechanical robots goes back more than 2,000 years. At 60 B.C. Heron of Alexandria constructed a three-wheeled car that could follow a pre-programmed route. The car was powered by a falling weight that pulled on a rope wound around two axles. The direction of the car could be changed with the help of pins. This primitive mechanism is similar to today's binary code. In 1478, Leonardo da Vinci constructed the first self-propelled trolley operated by clock springs. He also created the first historically documented humanoid with the drawing of his knight robot. This idea of interaction with humans is used today in industry. The two-armed, humanoid robot Baxter has an LCD display as face and can respond to human interactions. Robots can now work alongside people, which has led to new applications. Around 1740 Jacques de Vaucanson constructed an automatic duck and a machine that could play flute. He also build a fully automatic loom. Towards the end of the 19th century, the military increasingly searched for machines and remote-controlled technology. The term "robot" was first used in 1920 by the Czech writer Karel Capek in his drama RUR. RUR means "Rossum's Universal Robots", where the word Robots is derived from the Czech word "robota", which means forced labor or conscription.

Other innovations in the field of complex mechanics and the introduction of electricity were important impulses for robotics. The development of modern robotics has also depended on several important inventions, such as the electric motor for providing the drive and electronic devices for automation and control. In the 20th century, new concepts from cybernetics and the development of William Grey Walter's "Machina Speculatrix" were added. These machines were two simple cell systems that could do different things, such as avoid obstacles and turn light sources on and off. They were commonly considered the first stand-alone electronic robots. The Russian-American science-fiction author Isaac Asimov has taken up the term "robot" and applied it to machines. In his short story "Runaround" from 1942, for the first time the term "robotics" arose and also for the first time the "Three Laws of Robotics" were defined.

The first large-scale introduction of robots was in factories, which is still their biggest field of applications today. The first programmable robot arm capable of performing tasks was developed by George Devol and Joe Engelberger. He later became known as UNIMATE as the first industrial robot. In 1961 he was installed in a production line of General Motors and carried out primarily repetitive and dangerous tasks. The so-called Selective Compliance Articulated Robot Arm (SCARA) was introduced in 1978. This robot type can perform a unique 4-axis motion and was ideal for pick-and-place applications. Manipulating industrial robots are generally not operated intelligently, but carry out multi-purpose work and are reprogrammable. Industrial robots were originally used for hot, heavy and dangerous tasks and are now used for a wide variety of applications. Industrial robots have been a true success story since their introduction in 1961. The latest generation of industrial robots are referred to as collaborative robots. Sensors have been developed, which shut down the robot when it gets too close to a human. Many collaborative robots do not need complex programming.

Intelligent machines are increasingly available. Robots are expected to become ubiquitous in the years to come, in areas such as industry, military, search and rescue, and research. In addition, robots are expected to become increasingly popular in the home environment as well.

SIGNAL PROCESSING AND ROBOTICS

The use of sensors is becoming increasingly important in the development of autonomous and intelligent robot systems. Here, the environment is perceived via sensors and adaptively changed. A perception-action cycle is performed, which includes the following steps. The received signals are first preprocessed (filtered, normalized, etc.) and then fused with other sensor data. Therefore, redundant or multi-dimensional sensor data are combined in order to obtain more robust measurement data. After, some features are extracted from the received data and certain patterns are identified. For the technical realization of human perception, features are calculated, which describe the perception mathematically. The identified patterns are used, to determine the subsequent action of the robot, for further interaction with the changeable environment.

CONTROL AND ROBOTICS

Today, robots cannot adapt flexibly to new tasks and unexpected situations yet, and this is the subject to failure. Thereby, the control technology is widely used in robotics and it needed in some cases to analyze the complicated motion sequences and to design the suitable controllers for them. Since the physical relationships of a robot are non-linear partially, the non-linear controllers must be in use. Machine learning is already being used in robotics to adapt the actions of the robot to local environmental conditions. For example, machine learning can be used to allow a robot quickly locate the items in an unknown environment, safely capture items that they do not know, and independently learn appropriate rules for new tasks. In order to make robots more stable and safer, the control algorithms should be programmed so that machines can continually optimize their actions and react to disturbances.

IMPACT AND CONTRIBUTION OF THE BOOK

Present book "Control and Signal Processing Applications for Mobile and Aerial Robotic Systems" is the third title, published by the same research team and part of the international book series "Advances in Computational Intelligence and Robotics" (ACIR, ISSN 2327-0411). The previous two published books are titled "Developing and Applying Optoelectronics in Machine Vision" (ISBN 1522506322) and "Optoelectronics in Machine Vision-Based Theories and Applications" (ISBN 1522557512). This book fits in seamlessly with the major topics machine vision and optoelectronics, with a special focus on control theory, signal processing and mobile robot applications.

Today, mobile robots play a big role in helping people with different tasks and in different areas. They will play an even bigger role in the future, with major advances in intelligent signal processing, integration of microcontrollers in low-cost hardware, machine learning and intelligent algorithms (AI) over the last decade. Above all, the aerial robots (drones) will play a big role in the future in the implementation of transport tasks, photo and film recordings, measurement of building technology and inspection tasks of installations that are not (or only with difficulty) reachable by humans. Mobile robots are therefore a very current topic in research and development. In order to make mobile robots even smarter and more flexible, novel algorithms for control and signal processing have to be developed, implemented and tested.

This book is intended to address this very specific interface between robotics and control and signal processing. It presents current algorithms for control and signal processing implemented for mobile and aerial robots. Thereby, "Control and Signal Processing Applications for Mobile and Aerial Robotic Systems" shows the current state of the art in control and signal processing of mobile robotic systems. This book serves as a compilation of these topics and as a reference book for researchers, engineers or specialists. Original research contributions, tutorials and review chapters are sought in education, research, development and industrial applications. Furthermore, present book serves as a reference work for students and researchers of control and signal processing methods and applications, as well as a reference book for engineers and technicians in today's industry of mobile robotic systems. It gives a clear vision of perspective of this area development for researchers, whose contributions will signify in the future.

MAIN TOPICS OF THE BOOK

In the following list, all the topics covered in this book are presented, this list is not exhaustive:

- The wheel dynamics of mobile machines, using a novel developed sensor, which consists of a three-axis accelerometer, a three-axis gyroscope and a three-axis magnetometer. Various filters are used to process the obtained signal information.
- The measurement principle of the rational approximation, which is based on the treatment of integers and rational numbers, to perform high-precision frequency measurements from accelerometers generated from an inertial measurement system.

- The design and analysis of a hybrid wheeled jumping monitoring platform is presented extensively. The mathematical fundamentals of the jump motion are presented, as well as the technical implementation of the control algorithms.
- A garbage collection by robot, which collects solid waste at public places and which is controlled using an Arduino single-board microcontroller, with the extensive design and development of such garbage robot.
- Cryptographic transformations of image and video files using different cryptography techniques. Thereby, new multifunctional matrix-algebraic models of cryptographic image transformations and algorithms for generating the required matrix keys are presented.
- Digital control theory and signal processing is applied in a laser scanning system used by a mobile robot, to measure and capture his surrounding areas. The implementation of the digital controller LM629, to reduce the final angular error position of a DC-motor shaft is described and the sources of uncertainties for the measuring part of the laser scanning system are discussed.
- The zero-moment point method for trajectory generation of a bipedal walking robot is presented. This method is used to stabilize a walking robot about its vertical axis and thus to simulate the human physics of walking.
- Navigation of mobile autonomous robots on unknown terrain in the absence of GPS is presented, using radar signals and there reflection of landmarks. To increase the angular resolution of landmarks detection, antennas with synthesized aperture are proposed.

ORGANIZATION OF THE BOOK

The book contains ten chapters, which are briefly described in the following:

Chapter 1 describes the development of a measuring system for mobile wheeled platforms on the example of a tractor, which allows to obtain information about the actual dynamics of the wheels and the use of this information to control them. The chapter proposes a system for controlling the wheels of the device, using a sensor consisting of an accelerometer, a gyroscope and a magnetometer and describes an algorithm for processing the information obtained by the sensor. Different filters (Butterworth, Madgwick and Kalman) are used to process the obtained signals information. Furthermore, the outcomes of experimental studies of the proposed method and sensor to determine the wheel dynamics on the tractors with 4x4 wheel formula will be analyzed.

Chapter 2 presents the principle of rational approximations for measuring the frequency from accelerometers generated from an inertial measurement system, used by aerial vehicles for example. The chapter presents the simulation and analysis of the measurement process during aircraft landing. Different types of accelerometers with three frequency signals are compared, as well as different measurement parameters are shown, discussed, and evaluated.

Chapter 3 introduces the design and analyzes the movement of a wheeled jumping robot. It describes the basic principles of constructive and creative development of the jumping robot. It introduces the determination of the jump technique and the assignment of its phases and stages, as well as the design of a control system, which provides the implementation of different variants of the jump. The chapter introduces the basic concepts of jump movements, describes the properties of the jump and introduces its phases. This chapter also introduces tools for selecting a design for a jumping robot.

Chapter 4 is devoted to the development of a robot for collecting and transporting garbage. The chapter proposes a device design, perform a detailed calculation of all components, perform test strength calculations of parts and develop a robot control system. In addition, the text presents a kinematic analysis of the device and programs to simulate its functioning. The introduction of the chapter gives an overview of a large number of literature sources and explains the topics covered in various sources.

Chapter 5 provides background information about different cryptography techniques for image and video files processing, which are transmitted from unmanned aircraft or airborne robots. Thereby, new multifunctional matrix-algebraic models of cryptographic image transformations are analyzed. The algorithms and protocols for generating the required matrix keys are presented in the chapter. Also, the simulation results of the proposed algorithms and methods for direct and inverse transformation of images are discussed in this chapter.

Chapter 6 offers the application of digital control theory used for the implementation of a laser scanning system, which obtains 3D coordinates of any object under observation in a defined field of view. This chapter also focuses on some of the major sources of uncertainty that can occur in a laser scanning system. The received and processed light signal is examined under various environmental conditions, to analyze how it is affected by the instability characteristics of the main actuator of the laser scanning receiving unit.

Chapter 7 studies the problem of zero moment point (ZMP) - based trajectory generation applied for bipedal walking robots. An overview of the current methods for planning, design and control of current running robots is discussed. A detailed explanation of ZMP-based methods for generating trajectories for the center of mass (CoM) of walking robots is given. Furthermore, a quadratic programming-based method for simultaneous ZMP-CoM trajectory planning is presented.

Chapter 8 is dedicated to specific questions of landmark detection for improvement of mobile robot navigation. The chapter also describes the initial state of the problems of landmarks detection and estimation of their coordinates and the improvements of obtained results by increasing the angular resolution of landmarks using antennas with synthesized aperture. Simulation results are presented, which indicate the possibility of implementation of such synthesized apertures into practice of mobile robots.

Chapter 9 describes an image compression technique based on some principal components periodicity. Images are reconstructed, using the principal component analysis technique and the almost periodicity of the first principal components. The periodicity is used to build periodic vectors, and finally these periodic vectors are used to reconstruct the original image. The chapter also compares this method against the widely known JPEG format.

Chapter 10 describes an image compression technique based on some principal components periodicity. Images are reconstructed, using the principal component analysis technique and the almost periodicity of the first principal components. The periodicity is used to build periodic vectors, and finally these periodic vectors are used to reconstruct the original image. The chapter also compares this method against the widely known JPEG format.

Oleg Sergiyenko
Universidad Autónoma de Baja California, Mexico

Moisés Rivas-López
Universidad Autónoma de Baja California, Mexico

Julio C. Rodríguez-Quiñonez
Universidad Autónoma de Baja California, Mexico

Wendy Flores-Fuentes
Universidad Autónoma de Baja California, Mexico

Lars Lindner
Universidad Autónoma de Baja California, Mexico

Acknowledgment

In the name of all editors, we would like to thank all the people involved in this project and, more specifically, to the authors and reviewers that took part in the reviewing process. Without their support, this book would not have been completed.

First, the editors would also like to thank each one of the authors for their contributions. Our sincere gratitude goes to the chapter's authors who contributed with their time and expertise to this book.

Second, the editors wish to acknowledge the valuable contributions of the reviewers regarding the improvement of quality, coherence, and content presentation of chapters. Most of the authors also served as referees; we highly appreciate their double task.

Oleg Sergiyenko
Universidad Autónoma de Baja California, Mexico

Moisés Rivas-López
Universidad Autónoma de Baja California, Mexico

Wendy Flores-Fuentes
Universidad Autónoma de Baja California, Mexico

Julio C. Rodríguez-Quiñonez
Universidad Autónoma de Baja California, Mexico

Lars Lindner
Universidad Autónoma de Baja California, Mexico

Chapter 1
Determination of Mobile Machine Wheel Dynamics

Viktor Melnyk
https://orcid.org/0000-0002-1176-2831
Kharkiv Petro Vasylenko National Technical University of Agriculture, Ukraine

Roman Antoshchenkov
https://orcid.org/0000-0003-0769-7464
Kharkiv Petro Vasylenko National Technical University of Agriculture, Ukraine

Viktor Antoshchenkov
https://orcid.org/0000-0002-1136-5430
Kharkiv Petro Vasylenko National Technical University of Agriculture, Ukraine

ABSTRACT

To determine the wheel dynamics of the mobile machine, a sensor has been developed, which consists of a three-axis accelerometer, a three-axis gyroscope, and a three-axis magnetometer. The sensor is connected to the microcontroller, which transmits the data received via the 2.4 GHz channel. It is attached coaxially to the center of the wheel. In the first step of processing data from the accelerometer and gyroscope, their values are corrected. The corrected acceleration signal and angular velocities are processed using a Butterworth filter. Madgwick filter determines the angles of orientation of the sensor in space. In the next step, the authors deduct the centrifugal component from accelerations. Further, the gravitational component is subtracted from the accelerations to get its real value. The wheel speed is obtained by integrating accelerations. Angular wheel speed and accelerations are processed with a Kalman filter. The outcomes of experimental studies of the proposed method and sensor to determine the wheel dynamics on the tractors with 4x4 wheel formula have been analyzed.

DOI: 10.4018/978-1-5225-9924-1.ch001

INTRODUCTION

The study of the dynamics of mobile machines functioning requires determining the parameters of its state both as a whole and its individual elements.

The issue of creating measurement systems for the functioning parameters of mobile agricultural machines appeared almost simultaneously with the creation of the tractor. At first, measuring systems were designed to improve the performance of machine-tractor units and were based mainly on mechanical computing elements. The lack of accuracy of these devices makes it impossible to use them effectively on tractors at elevated speeds. Previously (Kodenko M. N., Lebedev, A. T., 1969), the prospects of research on the optimal control of a tractor unit using self-adjusting systems have been proved. In this paper, the dynamics of traction efficiency of the tractor has been first evaluated. This made it possible to optimize its operation modes under unsteady load.

For experimental research of mobile machines, the scientists developed measuring instruments and systems (Debain, C. A, Chateau, T., Berducat, M., Martinet, P., & Bonton, P., 2000; Mojtaba, N., Alimardani, R., Sharifi, A., & Tabatabaeefar, A., 2009; Serrano, J. M., Peça, J. O., Shahidian, S., Nunes, M. C., Ribeiro, L., & Santos F., 2011). Some of them are universal (Drenkow, G., 2006), others are specific (Eremenko, A. V., Maloletov, A. V., & Skakunov, V. N., 2010; Kuvachev, V. P., Ayubov, A.M., & Kotov, O. G., 2007). There are systems that read operating parameters from diagnostic interfaces (Čupera, J., & Sedlak, P., 2011), as well as virtual systems (Kring, J., & Travis, J., 2006).

A group of scientists developed the measuring information system IP-256M for measuring and calculating the data of energy and traction tests of tractors and agricultural machines (Kadochnikov, G. N., 2006). The measuring system has a limited number of analogue (specialized) channels (6), to which temperature sensors, strain-gauge links and 8 discrete inputs can be connected.

Nowadays, manufacturers of semiconductor devices have created a sufficient number of sensors of physical quantities, therefore, the information system IP-256M needs only an analogue-to-digital converter and a microcontroller to collect data from sensors.

A group of scientists has developed a data collection and processing system based on the Dewe-2010 PC industrial data collection system (A., Yahya, M., Zohadie, A. F., Kheiralla, S. K., Gew, B.S., Wee, &. E. B., Ng, 2004). The developed data acquisition system includes a global positioning system with differential correction (DGPS-RTK). In this case, the processing and storage of data in this system occurs in real time. The tractor Massey Ferguson 3060 was equipped with such a measuring system. Additional features of this system are the design of the traction force sensor and the torque sensor on the tractor wheels, which require intervention into the tractor design (Yahya, A., 2000).

A sufficient number of sensors installed on the tractor made it possible to determine: the geographical position of the machine, terrain parameters, tillage quality and the traction characteristics of the tractor.

To determine the longitudinal, lateral and vertical accelerations in the process of taking off and the acceleration of an agricultural unit, a mobile measuring complex was developed which consists of Freescale Semiconductor MMA7260QT acceleration sensors and a laptop for processing and storing the data obtained during the experiment (Artemov, N. P., Lebedev, A. T., Podrygalo, M. A., Polyansky, A. S., & Klets, D. M., 2012).

MEMS accelerometers are used as sensors in the mobile measurement system. The popularity of MEMS accelerometers and gyroscopes is caused by their great potential for use in both domestic and industrial equipment (Sysoev, S., 2009).

MEMS sensors are widely used in the automotive industry to control airbags, burglar alarms and navigation systems to calculate the distance traveled or determine the route.

The combination of an accelerometer and a gyroscope allows you to create integrated inertial measuring systems (IMU) or inertial navigation systems (Gebre-Egziabher, D., Hayward, R. C., & Powell, J. D., 1998), used in studies of the movement of mobile machines.

When carrying out dynamic traction testing of wheeled mobile vehicles, it is necessary to determine the rotational speeds of the driving and driven wheels. For steering wheels it's necessary to know them to assess the stability and controllability of mobile machines. Slipping is one of the indicators of the efficiency of the movement of mobile machines. Slipping the drive wheels of the machine reduces the dynamics during acceleration. The highest traction characteristics of wheeled tractors are achieved with 10-15% slipping of the drive wheels. These parameters must be determined in experimental studies and tests of mobile machines. Therefore, the goal of this study is the scientific and applied task of determining the wheel dynamics of a mobile vehicle.

BACKGROUND

There are universal measuring systems to be used in various branches of engineering. The versatility of such systems consists in the ability to connect a variety of analogue or digital inputs of the measuring system.

The determination of the position and speed of rotation of the wheel is associated with certain technical issues that are aggravated with increasing measurement accuracy.

The slipping of the wheels of mobile vehicles is determined through the actual speed of the vehicle and the speed of rotation. The speed of rotation of the machine wheels is determined in two ways:

- with an external encoder;
- with a speed sensor based on the Hall effect.

The encoder is widely used in robotics to control the speed of movement and the position of the robot (I., Zunaidi, K., Norihiko, N., Yoshihiko & M., Hirokazu., 2019).

The external encoder has difficulties with the installation and requires an external arm for fixing the sensor body in a stationary position relative to the machine body. The wheel speed sensor based on the Hall effect requires the intervention into the design of the machine, i.e. manufacturing technical holes, fasteners. Modern tractors and cars have installed sensors necessary for the operation of the ABS\ESP system.

When calculating the slipping, it is necessary to take into account the dynamic radius of the wheels, which is assumed to be constant for all modes of movement of the machine.

The wheel slip measurement method, which requires a machine vision system, is rather demanding on the performance of a digital signal processing system (X., Song, L., Seneviratne & K. Althoefer, 2009) and does not work well enough in dusty conditions.

In navigation systems associated with an automatic steering system, the signal from the steering wheel position sensor is used to assess the motion path. The automatic steering system is designed to actuate the steering rod.

The feedback sensor of the control loop must provide an estimate of the position of the wheel being turned. This wheel will usually be one of two front wheels on tractors, sprayers and trucks (Ackerman steering) or one of the rear wheels on combine harvesters (Inverse Ackerman). There are three ways to determine the value of the angle of rotation of the controlled wheel (F. R., Más, Q., Zhang & A. C., Hansen, 2011):

- direct measurement of the angle of rotation with an encoder;
- indirect measurement of the angle by estimating the displacement of the hydraulic cylinder, which actuates the steering gear;
- indirect measurement of the angle of rotation of the wheel by monitoring the flow of hydraulic fluid passing through the steering cylinder.

The estimation of the angle of rotation by means of linear movement of the steering rod cylinder requires calibration of the sensor in order to associate linear movements with angles. The steering, which contains the steering linkage, turns the wheels at a different angle. Thus, it is necessary to take into account the non-linear relationship between the two wheels, since the sensor usually estimates the angle of rotation of one of them.

Linear potentiometers are used as such sensors. Optical sensors are an alternative to linear potentiometers. Such rotary encoders are mounted on a wheel axle, the angle of which is measured. In this case, the assembly is difficult because it is necessary to attach either the sensor body or the sensor shaft to the chassis of the machine so that relative movements can be monitored and the angles of rotation of the wheels measured.

The studies of the dynamics of machines, which include determining the position in space, the speed of movement, acceleration, angular velocity of the machine, are performed by inertial measuring devices. The algorithms for data processing of incoming from the sensors are proposed in (R. Dorobantu & B. Zebhauser, 1999). The determination of the angles of orientation of the object in space according to the accelerometer, gyroscope and magnetometer is described in detail in (S. O. H. Madgwick, 2010).

To prevent random errors that occur due to the lateral entry of wheels, skidding, irregularities of the support surface on which the machine moves, filtering methods are used. In addition, the elimination of random errors contributes to the integration of information from various sensors located on the platform of the machine. Therefore, the best way to obtain accurate parameters of the current location is Kalman filtering, which gives the minimum-standard root-mean-square error of the position estimate. The advantages of Kalman filtering are as follows (R. Dorobantu & B. Zebhauser, 1999):

- relative simplicity and availability for engineering developments of non-stationary filters in various technical applications;
- the possibility of analytical proof and confirmation of the optimality of filtering in versions of the design of filters of various complexity;
- the visualization of the analytical apparatus based on ordinary differential equations, or difference equations;
- the ability to assess the state of the system in the time domain based on statistical data on the sources and types of errors;
- the possibility of building filters for multidimensional dynamic systems based on the Hilbert representation of the state space;
- the possibility of obtaining a recurrent system of algorithms and recursive optimal filtering procedures, which is much more convenient when using modern computers.

The use of various options for constructing the Kalman filter (linear or adaptive) in the problem of the relative localization of mobile machines is described in (Negenborn, R., 2003). Also, this filter is used in experimental studies of automobiles (Klets, D. M., 2012) and agricultural units (Artyomov, N. P., Shuliak, M. L., 2015).

MAIN FOCUS OF THE CHAPTER

Measuring system of dynamics and power of mobile machines

For experimental studies has been developed measuring system for the dynamics and energy of mobile machines (Antoshchenkov, R. V., 2017). The measuring system relates to the technical means of diagnosis and operational control and can be used in agriculture and engineering industry. The measuring system is designed to determine kinematic, dynamic, power and energy characteristics of mobile machines and their elements during road, field and bench tests.

The main components of the measuring system are (Fig. 1):

- computation module;
- sensors;
- power supply.

The computing module is designed for processing, visualization and storage of the data coming from the sensors. The power supply allows the measuring system to operate autonomously or to receive power from the on-board system of the mobile machine. Sensors determine the measured parameters.

It is used to determine the dynamic and energy properties of trucks and cars, buses and road trains, tractors, military wheeled and tracked vehicles, as well as their elements while in service, during the autotechnical examination and in other cases requiring operational monitoring of the state of the machine.

The system defines linear accelerations and angular rotational speeds around the axes of symmetry of the mobile machine, translational speed, geographic location, fuel consumption, tractive effort, wheel rotation speeds and other parameters.

The system provides additional opportunities:

- the measurement results are displayed in real time on the screen of the computation unit;
- saving measurement results on an external storage medium (USB Flash drive, USB HDD drive);
- review of the saved measurement results.

Figure 1. The block diagram of the measuring system of dynamics and power of mobile machines

1 – measuring\main unit; 2 – keyboard; 3 – GPS receiver; 4 – power supply; 5 – CAN data bus; 6 – inertial measuring device; 7 – rotation angle sensor; 8 – fuel flow meter; 9 – rpm sensor (rotation speed); 10 – electronic dynamometer; 11 – analogue inputs; 12 – discrete inputs; 13 – computer; 14 – wheel dynamics sensor

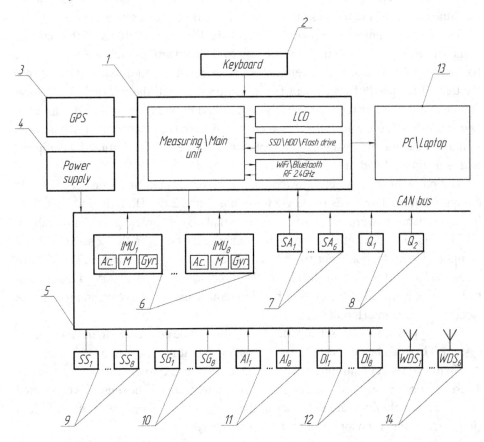

The measuring system works as follows: The sensors necessary for the study are attached to the computation module (for example, IMU, traction force sensor, gyroscope, and others). The signals from the sensors are sent through the signal cable to the computation module, where they are converted (digitally processed), saved or later sent to the computer. Voltage comes from the power supply to the computation module through the cables.

The number and types of sensors the machine is equipped with during the tests depend on its type and parameters that must be determined. The measuring system can be equipped with inertial measuring devices 6, consisting of a gyroscope, accelerometer and magnetometer, the number of which depends on the number of elements of the mobile machine. Also, this sensor allows to determine the vibration and the actual trajectory of movement of the elements of the mobile machine in three planes. The system can process the data from eight inertial measuring devices.

Obtaining navigation information, trajectories, speed and altitude above sea level is done using a GPS navigation receiver *3*. To determine the rotation angles of the machine elements, the angle sensors *7* which are potentiometers are used.

The fuel consumption sensors *8* are installed in the fuel lines of the internal combustion engine of the mobile machine in the forward and reverse direction of the fuel supply, that is, the system takes into account the fuel that is drained into the tank. The speed of rotation of the transmission shaft, the internal combustion engine and the power take-off shaft is determined by the rotation speed sensors 9. Dynamometers *10* are installed between the elements of a mobile car, a car and a trailer or a tractor and an agricultural machine. The device is additionally equipped with analogue *11* and discrete inputs *12*.

The communication between the sensors, inertial measuring devices and the computation module takes place via CAN bus 5 and 2.4 GHz radio channel. This bus has several protection degrees, including that against breakage of signal cables.

The maximum length of the signal cables that connect the sensors and the computing module should not exceed 40 m. The maximum number of sensors that are connected to the CAN-bus is limited to the standard (Lapin, A., 2005). Termination resistors are installed at the ends of bus *5* to reduce signal reflection and reduce the electromagnetic interference effect.

Other types of sensors serially manufactured can be connected to the measuring system. They can be connected to analogue or digital inputs or to a CAN-bus.

Up to two fuel consumption sensors can be connected to the measuring system. These sensors are installed in the engine compartment. One of the sensors is connected to the fuel supply line, and the other to the drain line. Sensors are volumetric fuel flow meters. Each sensor has a fuel temperature sensor.

The power of the measuring system is supplied from the gel lead-acid battery AGM which is located in the power supply or on-board network of the machine.

Wheel Dynamics Sensor

A wheel sensor for a mobile machine wheel, which consists of a three axial accelerometer and a magnetometer (LSM303DLHC), as well as a three-axial gyroscope (L3GD20) (Fig. 2) has been developed. The sensor is an inertial measuring device, which is assembled on the basis of STM32F3DISCOVERY. The microcontroller (STM32F303VCT6) transmits data over the 2.4 GHz channel through a radio module (nRF24L01). The sensor contains a 3.7V Li-pol battery.

The sensor detects the wheel rotation speed of the mobile machine (tractor or car) or the drive sprocket of a caterpillar tractor. It allows determining the actual wheel rotation speed and the orientation in space, without intervention into the design.

The accelerometer mounted in the sensor measures the acceleration a_x, a_y, a_z in three orthogonal axes (Fig. 2). Similarly, the angular velocity ω_x, ω_y, ω_z and the intensity of the magnetic field m_x, m_y, m_z are measured in three axes.

The wheel dynamics sensor is mounted coaxially and in the center of the mobile machine wheel. The installation diagram of the wheel dynamics sensor is shown in Fig. 3, and the installed sensor on the on-board wheel tractor gearbox is shown on Fig. 4.

The developed sensor is mounted in the center of the wheel *1* of the mobile machine (Fig. 3). The sensor and the wheel are connected rigidly. The center of the sensor, in which the accelerometer, gyroscope and magnetometer are located, should coincide with the center of the mobile machine wheel. The mounting surfaces of the sensor and the wheel should be parallel. The *z*-axis of the sensor should match the axis of the wheel rotation.

Figure 2. The diagram of the parameters measured by the wheel dynamics sensor
a_x, a_y, a_z – *acceleration sensors (accelerometers) along the axes x, y, z;* ω_x, ω_y, ω_z – *angular velocity sensors (gyroscopes) along the axes x, y, z;* m_x, m_y, m_z – *magnetic field sensors (magnetometers) along the axes x, y, z*

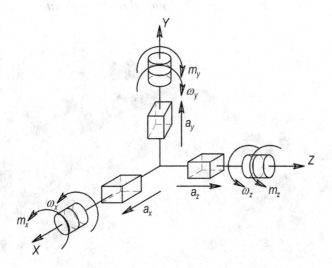

Figure 3. The scheme of installation of the wheel dynamics sensor
1 – wheel; 2 – sensor

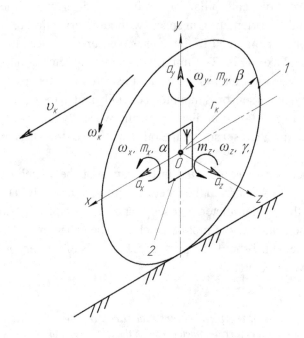

Figure 4. The wheel dynamics sensor mounted on the onboard gear wheel of the wheeled tractor

Sensor Signal Processing

Mobile machines such as tractors and automobiles operate in the conditions of vibrations and noise generated by the environment and technological equipment. Analytically, the IMU sensor acceleration consists of four components:

- the actual acceleration (this is the acceleration acting along the corresponding axis of the mobile machine);
- the angle of the machine relative to the horizon;
- the vibrations created by the environment when the machine is moving or generated by the technological machine in a stationary position;
- the inherent noise of the sensor.

The elimination of the ambient noise and unwanted data in the sensor signals requires mathematical data processing. The data processing method of the wheel dynamics sensor is shown below and consists of six steps.

Step 1: The first step in processing data from the accelerometer, gyroscope and magnetometer consists in adjusting their values:

$$
\mathbf{a_c} = \begin{bmatrix} a_x c \\ a_y c \\ a_z c \end{bmatrix}; \begin{bmatrix} a_x c \\ a_y c \\ a_z c \\ 1 \end{bmatrix} = \begin{bmatrix} \mathbf{a_r} \\ 1 \end{bmatrix} \cdot \mathbf{A} = \begin{bmatrix} a_x r \\ a_y r \\ a_z r \\ 1 \end{bmatrix} \begin{bmatrix} a_{11} & a_{12} & a_{13} & b_1 \\ a_{21} & a_{22} & a_{23} & b_2 \\ a_{31} & a_{32} & a_{33} & b_3 \\ 0 & 0 & 0 & 1 \end{bmatrix};
$$

$$
\boldsymbol{\omega_c} = \begin{bmatrix} \omega_x c \\ \omega_y c \\ \omega_z c \end{bmatrix}; \begin{bmatrix} \omega_x c \\ \omega_y c \\ \omega_z c \\ 1 \end{bmatrix} = \begin{bmatrix} \boldsymbol{\omega_r} \\ 1 \end{bmatrix} \cdot \mathbf{G} = \begin{bmatrix} \omega_x r \\ \omega_y r \\ \omega_z r \\ 1 \end{bmatrix} \begin{bmatrix} g_{11} & g_{12} & g_{13} & c_1 \\ g_{21} & g_{22} & g_{23} & c_2 \\ g_{31} & g_{32} & g_{33} & c_3 \\ 0 & 0 & 0 & 1 \end{bmatrix}; \qquad (1)
$$

$$
\mathbf{m_c} = \begin{bmatrix} m_x c \\ m_y c \\ m_z c \end{bmatrix}; \begin{bmatrix} m_x c \\ m_y c \\ m_z c \\ 1 \end{bmatrix} = \begin{bmatrix} \mathbf{m_r} \\ 1 \end{bmatrix} \cdot \mathbf{M} = \begin{bmatrix} m_x r \\ m_y r \\ m_z r \\ 1 \end{bmatrix} \begin{bmatrix} m_{11} & m_{12} & m_{13} & d_1 \\ m_{21} & m_{22} & m_{23} & d_2 \\ m_{31} & m_{32} & m_{33} & d_3 \\ 0 & 0 & 0 & 1 \end{bmatrix},
$$

where $\mathbf{a_r} = [a_x r \ a_y r \ a_z r]^T$ – the accelerometer input signal; $\boldsymbol{\omega_r} = [\omega x_r \ \omega y_r \ \omega z_r]^T$ – the gyroscope input signal; $\mathbf{m_r} = [mx_r my_r mz_r]^T$ – the magnetometer input signal; $a_c = [ax_c ay_c az_c]^T$ – the adjusted accelerometer signal; $\boldsymbol{\omega_c} = [\omega_{xc} \ \omega_{yc} \ \omega_{zc}]^T$ – the adjusted gyroscope signal; A, G, M – the accelerometer, gyroscope and magnetometer adjustment matrixes; $a_{ij}, b_i, g_{ij}, c_i, m_{ij}, d_i$ – the adjustment matrix coefficients

At this step, the installation errors of the accelerometer, gyroscope and magnetometer in the sensor housing are compensated. The coefficients a_{ij}, b_i, g_{ij}, c_i, m_{ij}, di for the corresponding adjustment matrices are obtained when calibrating the instrument on the bench.

Step 2: The adjusted acceleration signal $\boldsymbol{\alpha}_c$ and angular velocities signal $\boldsymbol{\omega}_c$ are processed using a Butterworth filter of the 3rd order with a cut-off frequency of 3 Hz and a sampling rate of 100 Hz:

$$\mathbf{a_f} = filter\_butter\left(\mathbf{a_c}\right);$$
$$\omega = filter\_butter\left(\omega_c\right),$$

(2)

where $\mathbf{a}_f=[a_x f \, a_y f \, a_z f]^T$ – the matrix-vector of the filtered accelerometer signal; $\boldsymbol{\omega}=[\omega_x, \omega_y, \omega_z]^T$ – the matrix-vector of the filtered gyroscope signal.

Step 3: There is no need to process the magnetometer data with the Butterworth filter, since the Madgwick filter is a low-pass filter. The matrix-vector $\boldsymbol{\theta}$ contains the orientation angles of the wheel dynamics sensor α, β, γ around the axes x, y, z, which is calculated with the Madgwick filter (S. O.H. Madgwick, 2010):

$$\theta = \begin{bmatrix} \alpha \\ \beta \\ \gamma \end{bmatrix} = f_{\text{Madgwick}}\left(\mathbf{a_f}, \omega, \mathbf{m_c}\right),$$

(3)

where f_{Madgwick} – Madgwick filter function; $\mathbf{m}_c=[m_x c \, m_y c \, m_z c]^T$ – magnetometer sensor data

The Madgwick filter settings for accelerometer and magnetometer (LSM303DLHC) and also for three-axis gyroscope (L3GD20) is $\beta=0.6$, $f_s=100$Hz.

Step 4: Deduct the centrifugal force from the acceleration:

$$\mathbf{a_b} = \begin{bmatrix} a_x b \\ a_y b \\ a_z b \end{bmatrix} = \mathbf{a_f} - \omega \cdot \upsilon = \begin{bmatrix} a_x f \\ a_y f \\ a_z f \end{bmatrix} - \begin{bmatrix} \omega_x \\ \omega_y \\ \omega_z \end{bmatrix} \cdot \begin{bmatrix} \upsilon_x \\ \upsilon_y \\ \upsilon_z \end{bmatrix},$$

(4)

where $\upsilon=[\upsilon_x \, \upsilon_y \, \upsilon_z]^T$ represents the speed of motion along the axes x, y, z

After subtracting the centrifugal force from the acceleration, it is necessary to deduct the gravitational component and thus obtain its real value:

$$
\mathbf{a} = \begin{bmatrix} a_x \\ a_y \\ a_z \end{bmatrix} = \mathbf{a_b} + g \cdot \begin{bmatrix} -\sin\beta \\ \cos\beta \cdot \sin\alpha \\ \cos\beta \cdot \cos\alpha \end{bmatrix} = \begin{bmatrix} a_x b \\ a_y b \\ a_z b \end{bmatrix} + g \cdot \begin{bmatrix} -\sin\beta \\ \cos\beta \cdot \sin\alpha \\ \cos\beta \cdot \cos\alpha \end{bmatrix},
\tag{5}
$$

where g – acceleration of gravity

Step 5: Calculate the speed of motion, which is necessary for the account of the centrifugal forces, through the integration of accelerations:

$$
\upsilon = \begin{bmatrix} \upsilon_x \\ \upsilon_y \\ \upsilon_z \end{bmatrix} = \begin{bmatrix} \int_0^t a_x dt + \upsilon_x\big|_{t=0} \\ \int_0^t a_y dt + \upsilon_y\big|_{t=0} \\ \int_0^t a_z dt + \upsilon_z\big|_{t=0} \end{bmatrix},
\tag{6}
$$

where $\upsilon_x\big|_{t=0}$, $\upsilon_y\big|_{t=0}$, $\upsilon_z\big|_{t=0}$ – initial values of the speeds

Step 6: The final step in processing wheel dynamics sensor data is the processing of the data of acceleration, angular velocities and orientation angles using a Kalman filter.

The purpose of the Kalman filter is to minimize the variance of the estimate of a vector random process $\mathbf{x}(k)$ that changes over time:

$$
\mathbf{x}(k+1) = \mathbf{A}(k)\mathbf{x}(k) + \mathbf{u}(k),
\tag{7}
$$

where $\mathbf{A}(k)$ – transition matrix; $\mathbf{u}(k)$ – control vector, or process noise, which has a normal distribution over the correlation matrix $\mathbf{Q}_p(k)$

A linear process $\mathbf{y}(k)$, on which noise is imposed, is available for observation:

$$
\mathbf{y}(k) = \mathbf{H}(k)\mathbf{x}(k) + \mathbf{w}(k),
\tag{8}
$$

13

where $\mathbf{H}(k)$ – observation matrix; $\mathbf{w}(k)$ – observation noise, which is a random vector that has a normal distribution over the covariance matrix $\mathbf{Q}_{\mathbf{M}}(k)$

The recursive recovery process evaluation algorithm $\bar{\mathbf{x}}(k)$ has the following sequence:

Step 6.1: predicted value of the observed signal:

$$\bar{\mathbf{y}}(k) = \mathbf{C}(k)\mathbf{A}(k)\bar{\mathbf{x}}(k-1), \tag{9}$$

where $\mathbf{C}(k)$ – measurement matrix

Step 6.2: determining the difference between the predicted and observed values:

$$\mathbf{e}(k) = \mathbf{y}(k) - \bar{\mathbf{y}}(k); \tag{10}$$

Step 6.3: filter gain calculation:

$$\mathbf{K}(k) = \mathbf{P}(k-1)\mathbf{C}^{T}(k) \times \left(\mathbf{C}(k)\mathbf{P}(k-1)\mathbf{C}^{T}(k) + \mathbf{Q}_{\mathbf{M}}(k)\right)^{-1}; \tag{11}$$

where $\mathbf{P}(k)$ – estimation of the correlation matrix

Step 6.4: process assessment recovery $\mathbf{x}(k)$:

$$\bar{\mathbf{x}}(k) = \mathbf{A}(k)\bar{\mathbf{x}}(k-1) + \mathbf{K}(k)\mathbf{e}(k); \tag{12}$$

where $\mathbf{e}(k)$ – interference vector matrix
Step 6.5: recovery of the evaluation of the correlation matrix of filtering errors:

$$\mathbf{P}(k) = \mathbf{A}(k)\left[\mathbf{P}(k-1) - \mathbf{K}(k)\mathbf{C}(k)\mathbf{P}(k-1)\right]\Phi^{T}(k) + \mathbf{Q}_{M}(k). \tag{13}$$

where $\mathbf{\Phi T}(k)$ – error correlation matrix
The Kalman filter is executed in two stages: the first is the prediction (7), (13); the second is the correction (11), (12).

The state vector **x** of the wheel dynamics sensor, which contains acceleration sensors, a gyroscope and a magnetometer, has the form:

$$\mathbf{x} = \begin{bmatrix} \theta & \omega & \mathbf{a} \end{bmatrix}^T = \begin{bmatrix} \alpha & \beta & \gamma & \omega_x & \omega_y & \omega_z & a_x & a_y & a_z \end{bmatrix}^T. \qquad (14)$$

The transition matrix **A** is calculated from the dynamic sensor system, which is constructed according to the classical laws of the mechanics of motion in the three-dimensional Cartesian coordinate system.

In the absence of the control vector **u**, the equation of the state vector (7) estimation is:

$$\mathbf{x}_{k+1} = \mathbf{A} \cdot \mathbf{x}_k, \qquad (15)$$

accordingly, in expanded form, it is:

$$\mathbf{x}_{k+1} = \begin{bmatrix} 1 & 0 & 0 & \Delta t & 0 & 0 & \frac{1}{2}\Delta t^2 & 0 & 0 \\ 0 & 1 & 0 & 0 & \Delta t & 0 & 0 & \frac{1}{2}\Delta t^2 & 0 \\ 0 & 0 & 1 & 0 & 0 & \Delta t & 0 & 0 & \frac{1}{2}\Delta t^2 \\ 0 & 0 & 0 & 1 & 0 & 0 & \Delta t & 0 & 0 \\ 0 & 0 & 0 & 0 & 1 & 0 & 0 & \Delta t & 0 \\ 0 & 0 & 0 & 0 & 0 & 1 & 0 & 0 & \Delta t \\ 0 & 0 & 0 & 0 & 0 & 0 & 1 & 0 & 0 \\ 0 & 0 & 0 & 0 & 0 & 0 & 0 & 1 & 0 \\ 0 & 0 & 0 & 0 & 0 & 0 & 0 & 0 & 1 \end{bmatrix} \cdot \begin{bmatrix} \alpha \\ \beta \\ \gamma \\ \omega_x \\ \omega_y \\ \omega_z \\ a_x \\ a_y \\ a_z \end{bmatrix}_k^T \qquad (16)$$

where $\Delta t = 0.001$ s is sampling time

The observation equations for a given dynamic system, under the condition of simultaneous measurement of all elements of the vector **x** (14), will be:

$$\mathbf{y}_k = \mathbf{H}_k \cdot \mathbf{x}_k, \qquad (17)$$

accordingly, the observation matrix is:

$$
\mathbf{H}_k =
\begin{bmatrix}
1 & 0 & 0 & 0 & 0 & 0 & 0 & 0 & 0 \\
0 & 1 & 0 & 0 & 0 & 0 & 0 & 0 & 0 \\
0 & 0 & 1 & 0 & 0 & 0 & 0 & 0 & 0 \\
0 & 0 & 0 & 1 & 0 & 0 & 0 & 0 & 0 \\
0 & 0 & 0 & 0 & 1 & 0 & 0 & 0 & 0 \\
0 & 0 & 0 & 0 & 0 & 1 & 0 & 0 & 0 \\
0 & 0 & 0 & 0 & 0 & 0 & 1 & 0 & 0 \\
0 & 0 & 0 & 0 & 0 & 0 & 0 & 1 & 0 \\
0 & 0 & 0 & 0 & 0 & 0 & 0 & 0 & 1
\end{bmatrix}.
$$

According to the results of previous studies and technical characteristics of the sensors, we obtain the value of the standard deviation of the error σ in measuring physical quantities (J., Gomez-Gil, S., Alonso-Garcia, F. J., Gómez-Gil & T., Stombaugh, 2011; O., Maklouf, A., Ghila, A., Abdulla & A. Yousef., 2013):

$$
\sigma_{a_x} = \sigma_{a_y} = \sigma_{a_z} = 0.01; \quad \sigma_{\omega_x} = \sigma_{\omega_y} = \sigma_{\omega_z} = 0.1; \quad \sigma_\alpha = \sigma_\beta = \sigma_\gamma = 0.05. \tag{18}
$$

Considering the measurement errors of physical quantities by sensors (18), we calculate the covariance matrix of the measurement noise $\mathbf{R}$ and the covariance matrix of the estimation process $\mathbf{Q}$. The covariance matrix of the measurement noise is:

$$
\mathbf{R} =
\begin{bmatrix}
\Delta t & 0 & 0 & 0 & 0 & 0 & 0 & 0 & 0 \\
0 & \Delta t & 0 & 0 & 0 & 0 & 0 & 0 & 0 \\
0 & 0 & \Delta t & 0 & 0 & 0 & 0 & 0 & 0 \\
0 & 0 & 0 & 0.25\Delta t & 0 & 0 & 0 & 0 & 0 \\
0 & 0 & 0 & 0 & 0.25\Delta t & 0 & 0 & 0 & 0 \\
0 & 0 & 0 & 0 & 0 & 0.25\Delta t & 0 & 0 & 0 \\
0 & 0 & 0 & 0 & 0 & 0 & 36\Delta t & 0 & 0 \\
0 & 0 & 0 & 0 & 0 & 0 & 0 & 36\Delta t & 0 \\
0 & 0 & 0 & 0 & 0 & 0 & 0 & 0 & 36\Delta t
\end{bmatrix}.
$$

Covariance matrix of the assessment process:

$$\mathbf{Q} = \begin{bmatrix} 0.01\Delta t^2 & 0 & 0 & 0 & 0 & 0 & 0 & 0 & 0 \\ 0 & 0.01\Delta t^2 & 0 & 0 & 0 & 0 & 0 & 0 & 0 \\ 0 & 0 & 0.01\Delta t^2 & 0 & 0 & 0 & 0 & 0 & 0 \\ 0 & 0 & 0 & 0.05\Delta t^3 & 0 & 0 & 0 & 0 & 0 \\ 0 & 0 & 0 & 0 & 0.05\Delta t^3 & 0 & 0 & 0 & 0 \\ 0 & 0 & 0 & 0 & 0 & 0.05\Delta t^3 & 0 & 0 & 0 \\ 0 & 0 & 0 & 0 & 0 & 0 & 0.6\Delta t^4 & 0 & 0 \\ 0 & 0 & 0 & 0 & 0 & 0 & 0 & 0.6\Delta t^4 & 0 \\ 0 & 0 & 0 & 0 & 0 & 0 & 0 & 0 & 0.6\Delta t^4 \end{bmatrix}.$$

The matrices $\mathbf{R}$ and $\mathbf{Q}$ obtained are substituted into the Kalman filter (7)–(17).

A diagram of the wheel dynamics sensor data processing, which includes processing with the Butterworth filter, coriolis acceleration compensation, inclination (orientation) angles, gravity acceleration and Kalman filter processing, is shown in Fig. 5.

The study of the dynamics of the tractor requires the determination of the wheel slipping. The method for determining tractor wheel slipping, described in ISO 789-9:2018 and OECD Code 2, allows to get the value of the slipping after running the measured section. This method makes it possible to determine the slipping of the mobile machine wheels in statics. With the help of the developed sensor and measuring system, it is possible to determine the wheel slippage in the dynamics, i.e. its instant value. The wheel dynamics sensor determines the angular speed of the wheel rotation, which is converted into the slippage, taking into account the speed of the machine:

$$\delta = \frac{\pi r_{\mathrm{w}} \dfrac{\omega_z}{180} - \upsilon_{\mathrm{GPS}}}{\upsilon_{\mathrm{GPS}}} \cdot 100\%, \tag{19}$$

where ω_z – angular speed of the wheel rotation which is determined by the developed wheel dynamics sensor, $°/_s$; r_{w} – dynamic radius of the wheel which is determined according to ISO 4251-1:2017, m; υ_{GPS} – actual speed of motion determined by the GPS receiver, m/s

Determining the speed of the mobile machine wheel rotation is based on determining the angular velocity of rotation which is measured with a gyroscope and an accelerometer located in the center of the mobile machine wheel. The angular

Figure 5. Block diagram of the data processing algorithm of the measuring system

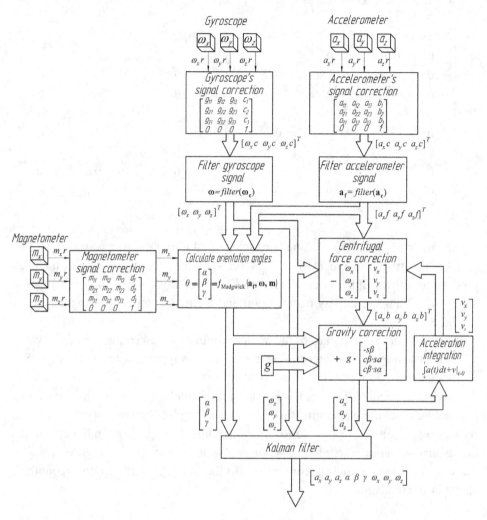

velocity measured by the gyroscope is equal to the angular velocity of the wheel rotation. When the wheel rotates, the accelerometer signal (along the *x* or *y* axis) varies sinusoidally, and its signal frequency is equal to the speed of the wheel.

RESULTS

With the help of the developed measuring system of the dynamics and power of mobile machines, experimental studies of the KhTP-243K.20 tractor (produced by Kharkiv Tractor Plant, Ukraine) were carried out. The tractor is an all-wheel drive with a 4x4 wheel formula and an articulated frame. During experimental studies, the wheel dynamics sensors were installed on the on-board gearboxes of the tractor. The actual speed of the motion was measured by a GPS receiver. The acceleration data, the angular velocities and magnetic field strengths have been obtained and processed with the proposed data processing method. According to this data, the orientation angles of the mobile machine wheels have been calculated. The orientation angles of the rear right (Fig. 6) and rear left (Fig. 7) wheels of the tractor have been calculated.

Figure 6. The orientation angles of the rear right wheel of the tractor
α, β, γ – the orientation angles of the tractor wheels; $α_{Kf}$, $β_{Kf}$, $γ_{Kf}$ – the orientation angles of the tractor wheels, which are treated with a Kalman filter

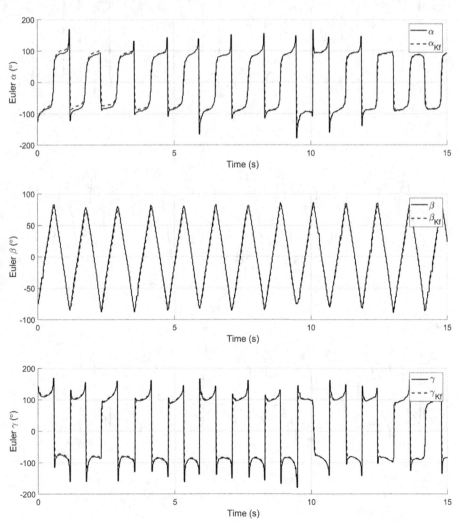

The orientation angle β has the shape of a sawtooth harmonic signal, whose frequency is proportional to the wheel rotation speed. The amplitude of the signal is equal to 90°. The sensor installation error (not parallel to the sensor and wheel planes) results in signals of orientation angles α and γ of a specific shape (Fig. 6, 7).

The slipping of the tractor's drive wheels has been determined (Fig. 8).

The mean value of the tractor driving wheels slipping is at the level of δ=5%, i.e. it is low enough for this tractor.

Figure 7. The orientation angles of the rear left wheel of the tractor

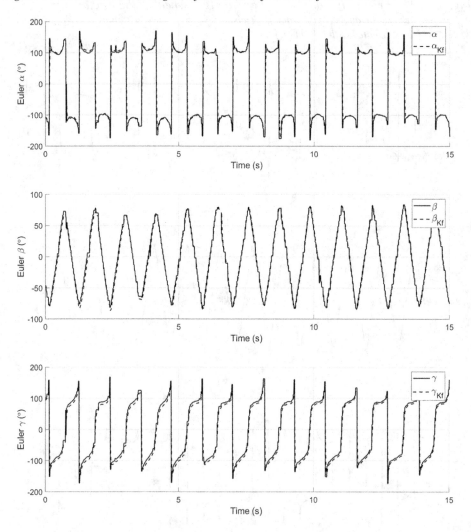

Figure 8. Slipping of the tractor's drive wheels

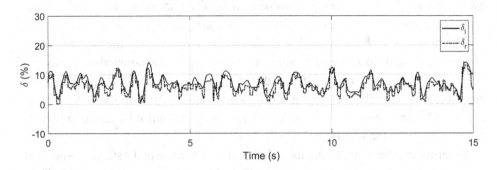

Designed wheel dynamics sensor suitable for industrial use in the manufacture and testing of mobile machines such as tractors and automobiles. The data processing algorithm proposed by the authors for this wheel dynamics sensor is used for the first time.

SOLUTIONS AND RECOMMENDATIONS

To track the traction-dynamic and power characteristics of mobile machines in real time, including automobiles and agricultural units, a measuring system of the dynamics and energy of mobile machines has been developed and manufactured.

The developed sensor is recommended to be installed coaxially with the center of rotation of the mobile machine wheel. Improving the accuracy of sensor signals data processing requires the clarification of the coefficient β of the Madgwick filter. This coefficient depends on the error in measuring the angular velocity with the gyroscope. It is also necessary to set the sampling (discretization) rate correctly in the same filter.

When using other models (types) of accelerometers, gyroscopes and magnetometers in the wheel dynamics sensor, it is necessary to clarify the Kalman filter coefficients such as R and Q, and the transition matrix A.

FUTURE RESEARCH DIRECTIONS

The developed measuring system and wheel dynamics sensor enables to carry out experimental studies of mobile machines effectively. The mathematical processing unit needs to be improved and transferred to the microcontroller, which is located in the sensor itself. To do this it is necessary to solve the problem of computational efficiency of the developed mathematical apparatus.

In subsequent research, it is necessary to study the effect of the Butterworth filter overlay on accelerometer and gyroscope signals. The parameters of the filter itself, such as the cut-off frequency and order, also require further elaboration.

The developed wheel dynamics sensor can be used to study the steering wheel dynamics of mobile machines. During such studies, the sensor mounted on the steering wheel, will evaluate its rotational speed and angular acceleration. According to the data obtained from the accelerometer, it is possible to estimate the vibro-tension of the mobile machine operator, that is, the level of vibrations transmitted by the steering wheel to the hands.

Determining the angular velocities of rotation and position of the steering wheel will allow, in further research, to study the stability of the mobile machine motion. These parameters can be recalculated in the angles of rotation of the steering gear elements.

One of the promising studies in the field of agricultural tractor dynamics is connected with the determination of the dynamic radius of the wheel. This parameter can be determined with the developed wheel dynamics sensor.

CONCLUSION

To track in real time the traction-dynamic and power characteristics of mobile machines, including cars and agricultural machines, a measuring system for the dynamics and energy of mobile machines has been developed and manufactured. It has been designed to determine the kinematic, dynamic, power and energy characteristics of mobile machines and their elements during road, field and bench tests. The modularity of the measuring system allows adapting it to measure only the necessary parameters of the dynamics of mobile machines.

The measuring system can include such sensors as: an inertial measuring device consisting of a microcontroller, an accelerometer, a gyroscope and a magnetometer, which is designed to determine vibrations, the actual trajectory of motion, acceleration and angular velocities of the unit element in three planes, orientation in space; a shaft speed sensor; a wheel dynamics sensor, which allows, without interfering with the design to determine the speed of rotation and skidding; a fuel consumption sensor; a GPS navigation receiver. Other types of sensors that are commercially manufactured in industry can be connected to the measuring system. They can be connected to the analogue, digital inputs of the measuring system or via the CAN-bus.

It has been established that the acceleration signal consists of four components: acceleration itself (this acceleration acts along the corresponding axis), the angle of inclination relative to the horizon; vibrations created by the environment during movement, and, in stationary mode, a technological machine; own noise sensor (that of a sensitive element).

An algorithm for processing data by a measuring system has been developed, eliminating the influence of indicators of the location of the sensors, the acceleration of gravity, the irregularities of the reference surface and its inclination. To combat the random errors that occur during research Madgwick and Kalman filters were used.

The developed wheel dynamics sensor allows determining the slippage of the mobile machine propulsor. This method of determining the slippage is fairly accurate and does not require any intervention in the design of the machine. Determining the speed of rotation of the mobile machine wheels is based on determining the angular velocity of rotation, which is measured with a gyroscope and an accelerometer located in the center of the mobile machine wheel. The angular velocity measured with the gyroscope is equal to the angular velocity of the wheel rotation. When the wheel rotates, the accelerometer signal (along the x or y axis) varies sinusoidally, and its signal frequency is equal to the speed of the wheel rotation.

The developed wheel dynamics sensor is suitable for industrial production and use when testing mobile vehicles such as tractors and automobiles. The data processing algorithm proposed by the authors for this wheel dynamics sensor is used for the first time to determine wheel parameters.

REFERENCES

Antoshchenkov, R. V. (2017). *Dynamics and energy of motion of multi-element machine-tractor aggregates: a monograph*. Kharkiv: KhNTUA.

Artemov, N. P., Lebedev, A. T., Podrygalo, M. A., Polyansky, A. S., & Klets, D. M. (2012). The method of partial accelerations and its applications in the dynamics of mobile machines: a monograph. Kharkiv: Urban Press.

Artyomov, N. P., & Shuliak, M. L. (2015). Use of a filter to increase the accuracy of the study of the dynamics of mobile machines. *Scientific Bulletin of NUBIP of Ukraine*, 226, 290–295.

Čupera, J., & Sedlak, P. (2011). The use of CAN-BUS messages of an agricultural tractor for monitoring its operation. *Research in Agricultural Engineering*, 57(4), 117–127. doi:10.17221/20/2011-RAE

Debain, C. A., Chateau, T., Berducat, M., Martinet, P., & Bonton, P. (2000). Guidance-assistance system for agricultural vehicles. *Computers and Electronics in Agriculture*, 25(1-2), 29–51. doi:10.1016/S0168-1699(99)00054-X

Dorobantu, R., & Zebhauser, B. (1999). *Field evaluation of a low-coststrapdown IMU by means GPS. In Ortung und Navigation* (pp. 51–65). Bonn: DGON.

Drenkow, G. (2006). LXI – A New Generation of Measuring Systems. *Electronics NTB - Scientific and Technical Journal. Control and Measurement, 6*, 13–16.

Eremenko, A. V., Maloletov, A. V., & Skakunov, V. N. (2010). Microprocessor control system of robotic manipulators. News of VolgGTU, 3, 88–94.

Gebre-Egziabher, D., Hayward, R. C., & Powell, J. D. (1998). A Low-Cost GPS/Inertial Attitude Heading Reference System (AHRS) for General Aviation Application. *Proc. of IEEE Position Location and Navigation Symp.,* 518–525. 10.1109/PLANS.1998.670207

Gomez-Gil, J., Alonso-Garcia, S., Gómez-Gil, F. J., & Stombaugh, T. (2011). A Simple Method to Improve Autonomous GPS Positioning for Tractors. *Sensors (Basel), 11*(6), 5630–5644. doi:10.3390110605630 PMID:22163917

Kadochnikov, G. N. (2006). *Test report No. 07-06-2006 (1200012). Information measuring system IP-256M.* FGU: Kuban MIS.

Klets, D. M. (2012). Improving the accuracy of the experimental evaluation of the performance properties of wheeled vehicles using Kalman filter. *Technological and technological aspects of development and development of new technical and technological technologies for the Ukrainian government: Zbirnik nauk.pr. DNU UkrNDIPVT im. L. Pogoriloy,* Doslidnitske. *UkrNDIPVT, 16*(30), 467–484.

Kodenko, M. N., & Lebedev, A. T. (1969). *Automation of tractor units.* Moskow: Mashinostroenie.

Kring, J., & Travis, J. (2006). LabVIEW for Everyone: Graphical Programming Made Easy and Fun (3rd ed.). Prentice Hall.

Kuvachev, V. P., Ayubov, A. M., & Kotov, O. G. (2007). Improvement of the method of registration of vertical vibrations of mobile energy means. *Proceedings of Tavria DATE. Melitopol. TDAATA, 7*(1), 139–145.

Lapin, A. (2005). New generation of Texas Instruments products for a controlled electric drive. *Electronics: Science, Technology Business, 7,* 56–59.

Madgwick. (2010). *An effcient orientation filter for inertial and inertial/magnetic sensor arrays.* Academic Press.

Maklouf, O., Ghila, A., Abdulla, A., & Yousef, A. (2013). Low Cost IMU\GPS Integration Using Kalman Filtering for Land Vehicle Navigation Application. *International Journal of Electrical, Computer, Energetic Electronic and Communication Engineering, 7*(2), 184–190.

Más, F. R., Zhang, Q., & Hansen, A. C. (2011). *Mechatronics and Intelligent Systems for Off-road Vehicles*. London: Springer-Verlag London.

Mojtaba, N., Alimardani, R., Sharifi, A., & Tabatabaeefar, A. (2009). A Microcontroller-Based Data Logging System for Cone. *Tarim Makinalaji Bilimi Dergisi, 5*(4), 379–384.

Negenborn, R. (2003). *Robot Localization and Kalman Filters: On finding your position in a noisy world* (Thesis). Utrecht University.

Serrano, J. M., Peça, J. O., Shahidian, S., Nunes, M. C., Ribeiro, L., & Santos, F. (2011). Development of a Data Acquisition System to optimizing the Agricultural Tractor Performance. *Journal of Agricultural Science and Technology*, 756–766.

Song, X., Seneviratne, L., & Althoefer, K. (2009). *A Vision Based Wheel Slip Estimation Technique for Mining Vehicles*. IFACMMM.

Sysoev, S. (2009). Magnetically controlled, MEMS and multisensory motion sensors of 2009 are more functional, more precise, miniature predecessors. *Component Technology, 97*, 54–63.

Yahya, A. (2000). *Tractor with Built-in DGPS for Mapping Power and Energy Demand of Agricultural Field Operations in Malaysia*. UPM Research Report 1997-2000, II/2, 129–131.

Yahya, A., Zohadie, M., Kheiralla, A. F., Gew, S. K., Wee, B. S., & Ng, E. B. (2004). Dewe-2000. Precision system for mapping terrain trafficability, tractor-implement performance and tillage quality. *Proceedings of the 7th International Conference on Precision Agriculture and Other Precision Resources Management*, 23–41.

Zunaidi, I., Norihiko, K., Yoshihiko, N., & Hirokazu, M. (2019). Positioning System for 4 Wheel Mobile Robot: Encoder, Gyro and Accelerometer Data Fusion with Error Model Method. *CMU. Journal, 5*(1), 1–14.

Chapter 2
Application of the Principle of Rational Approximations for Measuring Dynamic Frequency Values Generated by an IMU

Fabian N. Murrieta-Rico
https://orcid.org/0000-0001-9829-3013
Universidad Autónoma de Baja California, Mexico

Vitalii Petranovskii
Universidad Nacional Autónoma de Mexico, Mexico

Juan de Dios Sanchez-Lopez
Universidad Autónoma de Baja California, Mexico

Juan Ivan Nieto-Hipolito
https://orcid.org/0000-0003-0105-6789
Universidad Autónoma de Baja California, Mexico

Mabel Vazquez-Briseño
Universidad Autónoma de Baja California, Mexico

Joel Antúnez-García
https://orcid.org/0000-0003-3668-1701
Universidad Nacional Autónoma de Mexico, Mexico

Rosario I. Yocupicio-Gaxiola
Universidad Nacional Autónoma de Mexico, Mexico

Vera Tyrsa
Universidad Autónoma de Baja California, Mexico

DOI: 10.4018/978-1-5225-9924-1.ch002

ABSTRACT

In most aerial vehicles, accurate information about critical parameters like position, velocity, and altitude is critical. In these systems, such information is acquired through an inertial measurement unit. Parameters like acceleration, velocity, and position are obtained after processing data from sensors; some of them are the accelerometers. In this case, the signal generated by the accelerometer has a frequency that depends from the acceleration experienced by the sensor. Since the time available for frequency estimation is critical in an aerial device, the frequency measurement algorithm is critical. This chapter proposes the principle of rational approximations for measuring the frequency from accelerometer-generated signals. In addition, the effect of different measurement parameters is shown, discussed, and evaluated.

INTRODUCTION

For aerial vehicles, like aircrafts, unmanned aerial vehicles (UAV) or drones (Sabatini et al., 2015), the estimation of position, also known as georeferencing (Schwarz, 1996), and velocity are some of the most important tasks that are done during "the fly". For most flying systems, such parameters are calculated using combined data from Global Positioning System (GPS) and the Inertial Navigation System (INS). Since the INS is located inside the vehicle, navigation information can be generated *in situ* without external references. At the same time, within the INS, the desired parameters are calculated by the Inertial Measurement Unit (IMU), as it is illustrated in Fig. 1. In the case of GPS outage, aircraft positioning depends from INS. In such a case, positional and attitude precision degrades rapidly due to INS sensor errors (Zhao et al., 2011). For this reason, the functioning of sensors incorporated in IMU can be improved by means of sensors construction, operation and signal processing. Accelerometers and gyroscopes are the primary source of information for an IMU (Ahmad et al., 2013). As stated by Sankar et al. (2009), there are many types of accelerometers, which include piezoelectric, piezoresistive, capacitive, tunneling, vibrating/resonating beam. This work focus on the accelerometers that generate a frequency output signal, this value is a function of the experienced acceleration. For some applications it is enough if the frequency output is interpreted as voltage or electrical current, which is easily achievable using well known frequency to voltage converters. Nevertheless, an accelerometer with a frequency output offers desirable characteristics such as quasi-digital signals, high sensitivity, high resolution, wide dynamic range, anti-interference capacity and good stability (Huang et al., 2013). For these reasons, from a metrological point of view, the use of the signal with frequency variations is better than a frequency to "something" conversion. From the measured frequency value, the IMU can directly estimate the acceleration magnitude. Additionally, by dead reckoning the velocity and position (or displacement) can be calculated. In other words, using initial conditions, after one integration of acceleration, the velocity is obtained; the position can be obtained after a second integration (Noureldin et al., 2013). Since position, velocity and acceleration depend of the frequency domain output from the accelerometer, fast and accurate frequency measurement is required. Usually, there are three accelerometers, one per each axis.

Some accelerometers that are commercially available are Vernier 3D-BTA (Vernier, 2019), Honeywell HG4930 (Honeywwell, 2019), Ellipse2 (SBG Systems, 2019) or Endevco 773 (MEGGITT Endevco, 2019). Some of their operating parameters are offered in Table 1. These sensors have a frequency domain output; for this reason, in order to use such sensors, proper frequency measurement is required. Besides of data in Table 1, the accelerometers based on MEMS are actively being researched and developed (Eling et al., 2015, Gao et al., 2017; Li et al., 2017, Li et al., 2019; Sekiya et al., 2016).

As it has been shown in previous works (Avalos-Gonzalez at al., 2018; Murrieta-Rico et al., 2014; Murrieta-Rico et al., 2015, Murrieta-Rico et al., 2016;), flying devices —with high maneuverability— experiment velocity variations in very short time. Some examples include space shuttles, rockets or spacecrafts. In such cases, the proper response of onboard control systems is dependent of the response

Figure 1. Elements of an INS

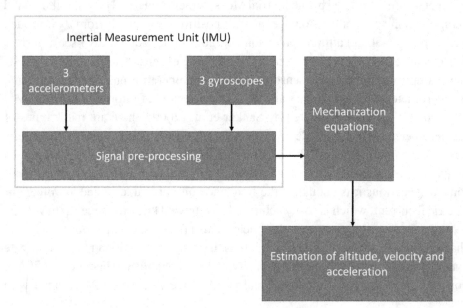

Table 1. Operating parameters of commercially available accelerometers

Model	Bandwidth [Hz]	Operating range
3D-BTA	100 Hz	+/-5g
Ellipse2	390	+/-16g
Endevco 773	2000	+/-200g
Omega ACC786A	14000	80 g

time of specific systems like an INS (Murrieta-Rico, et al., 2015). Particularly, the primary acceleration quantification depends of the measurement time, which is required to approximate the accelerometer frequency output. Therefore, even when signal processing methods could be used for reducing the error in INS parameters (Qi and Moore, 2002; Bruggemann et al., 2011; El-Sheimy et al., 2004; Parker and Finley et al., 2007; Pei et al. 2014, Zhao et al., 2011), the response time of the INS depends of the frequency measurement principle that is used.

Previous works of our research group have introduced the application of the principle of rational approximations (Avalos-Gonzalez at al., 2018; Murrieta-Rico et al., 2014; Murrieta-Rico et al., 2015, Murrieta-Rico et al., 2016;), for measurement of accelerometers frequency output. In these reports was shown that the accuracy in measurement is improved, and at the same time, the required measurement time is short enough to fulfill the requirements of an INS. However, all the previous work relating the rational approximations principle and the INS has focus only in one dimensional accelerometer. This fact is good enough for a basic model but lacks the realism of practical devices. In this work, the main novelty is the analysis of frequency measurement process for a three-axis accelerometer.

According to all stated before, the frequency estimation algorithm is one factor that could help to improve the response time and accuracy, from the data provided by the INS. In recent years, the principle of rational approximations has been proposed for measuring the output generated by the accelerometers inside the IMU. Moreover, there have been advances in theory related to the principle of rational approximations that could enhance its application to the IMU. This chapter presents the basic theory regarding the principle of rational approximations and its application to IMU. Additionally, implementation of recently published measurement theory is presented. Finally, different numerical scenarios are simulated and evaluated.

The Principle of Rational Approximations

In this section, we will explore the fundamentals of the principle of rational approximations and its application on a commercial accelerometer. For simplicity, let us consider a signal (S_x with a frequency to measure (f_x that has a value of 9 kHz. Such a value was selected because it is within an operating range shown in Table 1 (accelerometer Omega ACC786A).

When measuring with the principle of rational approximations, f_x is approximated, after the use of a reference signal S_0 whose frequency is known f_0. Before measuring, the silhouette of the signals must be adjusted, in such a way that $\tau \leq T_0/2$ for all the pulses in both S_x and S_0. The duration of pulses is denoted by τ such a value is also known as the pulse width and it must have the same duration in both input signals S_x and S_x.

After proper signal conditioning, both signals are compared using an AND-gate. This allows, in a very simple way, the search for coincidences of pulses. A coincidence means that during the time when a pulse is on high level in one signal (S_x or S_x, there is another pulse on high level in the other signal (S_x or S_x. This can be understood as two pulses overlapping, consequently we can say that the pulses have a "coincidence time" denoted as t_{0x}.

When the first coincidence occurs, the measurement process starts. For the principle of rational approximations, this implies that the coincidences, pulses in reference signal, and pulses in signal to measure are counted; they are denoted as n. P_n and Q_n respectively. It is important to note that in the first coincidence, $P_0 = Q_0 = n = 0$ and that the subindex n in $P_n Q_n$ indicates the coincidence number. Since $P_n Q_n$ values are paired by n we can say that in each coincidence a fraction is created, and it has the form P_n/Q_n. The signal coincidence process is illustrated in the Fig. 2.

The principle of rational approximations has its basic fundaments on mediant fractions. In this sense, a mediant fraction can be defined as the sums of the numerators and denominators of the two given fractions, namely P_n/Q_n and P_{n+1}/Q_{n+1}. Additionally, one of the properties exhibited by a mediant fraction is that

$$\frac{P_{n+1}}{Q_{n+1}} < \frac{P_{n+1} + P_n}{Q_{n+1} + Q_n} < \frac{P_n}{Q_n}. \tag{1}$$

The unknown frequency f_x [Hz] value can be approximated in each coincidence by

$$f_x = f_0 \frac{P_n}{Q_n}, \tag{2}$$

and the best approximation to f_x can be obtained in the mth mediant fraction.

$$f_x = f_0 \frac{\sum_m P_n}{\sum_m Q_n}. \tag{3}$$

For any measurement process, where the principle of rational approximations is used, the measurement time (M_t [Hz]) until the nth approximation is given by

$$M_t = Q_n T_0 \tag{4}$$

or

$$M_t = \frac{Q_n}{f_0} . \tag{5}$$

Considering f_r [Hz] as true frequency value of the signal, the relative error (β in measurement process is given by

$$\beta = 1 - \frac{P_n f_0}{Q_n f_r} . \tag{6}$$

From the Fig. 2, some interesting facts about the principle of rational approximations can be noted. The coincidences can in some cases last as maximum, the duration of pulse width, in this case we have a perfect coincidence $\tau = t_{0x}$; also, there is a case when the coincidences are "partial", and the coincidence time t_{0x} lasts less than the

Figure 2. Frequency measurement process using the principle of rational approximations: signal comparison of signal to measure
S_x with unknown frequency, f_x and reference signal; S_x with known frequency, f_x, when $f_0 = 10$ kHz and $f_x = 9$ kHz

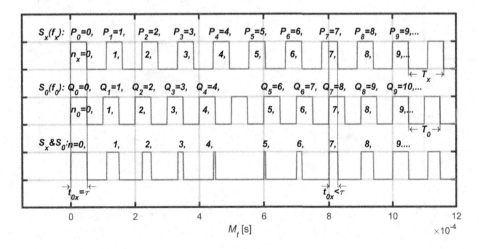

duration of a pulse width τ. In one hand, the existence of partial coincidences is what allows the existence of mediant fractions, but at the same time, since there are plenty of fractions, there is a high rate of apparition of mediant fractions; this means that approximations to f_x can be obtained almost since the firsts instants of measurement. On the other hand, the apparition of partial coincidences is reduced if the pulse width duration is shortened (Murrieta-Rico et al., 2016; Murrieta-Rico et al., 2018); consequently, the measurement uncertainty is reduced but the measurement time is increased.

Considering the numerical example in Fig. 2, we can compute that $5/6 < (5+6)/(6+7) < 6.7$ which is congruent with Eq. 1. Another important remark is that in the example shown in Fig. 2, both signals (S_x and S_0 are in phase condition (they start at the same time), this phenomenon leads to particularities like the existence of a perfect coincidence in $n=0$ and consequent apparition of fractions of the form $P_n/Q_n=1$ during the first instants of measurement. This behavior generates the higher uncertainty during the entire measurement process (Fig. 3a). This implies that approximations can be obtained when $P_n \neq Q_n$ which is the reason why in the previous numerical analysis, the fractions 5/6 and 6/7 were used. The value of β decreases in

Figure 3. Relative error in frequency measurement when $f_0=10kHz$ and $f_x=9kHz$: a) $M_t=0.1s$ and b) $M_t=10ms$

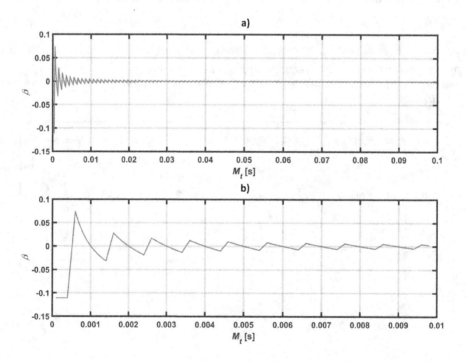

a time as short as 10 ms (Fig. 3b), to an absolute value 0.0023 which means 21 Hz if f_x=9 kHz. Also, it is expected to have a greater reduction of β is the measurement time increases, therefore for the practical task proposed in this work, it is important to have the best approximations in the shortest time.

An important property of the principle of rational approximations is the invariance of accuracy from jitter (Sergiyenko et al. 2011). In addition, this technique in difference to other known time frequency measurement methods (Johansson, 2005; Kalisz, 2003; Kirianaki et al., 2001), when the measurand value increases, the time required to measure decreases.

The versatility of the principle of rational approximations has been proved through several applications, mainly in sensors for the automotive industry (Sergiyenko et al., 2012), detection of specific chemical compounds (Murrieta-Rico et al., 2015), aerospace industry (Murrieta-Rico et al., 2015; Avalos-Gonzalez et al., 2018; Murrieta-Rico et al., 2014). Experimental implementation of this measurement principle requires the use of simple electronics such as logical gates, digital counters, digital memory circuits, and a data processing system. Experimental implementations include a FPGA based system (Balbuena et al., 2012), and logical circuits integrated with an Arduino development board (Avalos-Gonzalez et al., 2018).

In this section, the basics of the principle of rational approximations were reviewed and analyzed for a possible application in a real accelerometer. The next section will expand the unidimensional analysis of (Murrieta-Rico et al., 2014) to a three dimensions case.

Frequency Measurement of an Accelerometer Output

Since a moving object experiments variation on its position in the space, such variations can be quantified in the three axes of a reference frame (Fig. 4). Therefore, an accelerometer, which reacts in the three dimensions, allows to quantify the velocity and displacement vectors from direct acceleration measurement. As was stated above, the accelerometers have an electrical output that varies according to the acceleration experimented by the accelerometer.

The acceleration vector in the space can be represented as

$$\vec{a} = a_x \hat{i} + a_y \hat{j} + a_y \hat{k} \tag{6}$$

where a_x, a_y, and a_z represent the magnitude of acceleration in each axis. According to each manufacturer, there is a mathematical relationship between the acceleration experimented by the accelerometer, and the frequency generated by it. In general terms, we can say that the acceleration is a function of measured frequency: $a(f_x)$.

Figure 4. Considered scenario: An aircraft in displacement that experiences acceleration in any of its three axes

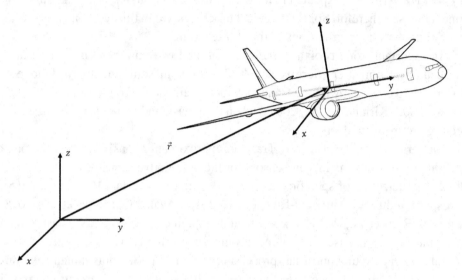

For a three-dimensional acceleration, the magnitudes associated to each axis are expressed as $a_{xx}(f_{xx})$, $a_{xy}(f_{xy})$, and $a_{xz}(f_{xz})$.

Considering an accelerometer with three frequency outputs (f_{xx}, f_{xy}, f_{xz}), the frequency in such signals can be approximated using the principle of rational approximations. The signal comparison process is illustrated in Fig. 5.

Like in Fig. 2, all the signals with frequency defined as f_{xx}, f_{xy}, f_{xz} are simultaneously compared with the reference signal with known frequency f_x. As a result, from Eq. 2, each frequency can be approximated as

$$f_{xx} = f_0 \frac{P_{nx}}{Q_{nx}}, \tag{7}$$

$$f_{xy} = f_0 \frac{P_{ny}}{Q_{ny}}, \tag{8}$$

$$f_{xz} = f_0 \frac{P_{nz}}{Q_{nz}}. \tag{9}$$

The Eqs. 7-9 show an interesting fact, since these three equations are dependent from the same f_0.

$$\frac{Q_{nx}}{P_{nx}f_{xx}} = \frac{Q_{ny}}{P_{ny}f_{xy}} = \frac{Q_{nz}}{P_{nz}f_{xz}} \tag{10}$$

This is a property previously reported in (Murrieta-Rico et al., 2019), where the simultaneous measurement of two signals was done. The relationship shown in Eq. 10 is relevant in the case, where after a first measurement, any of f_{xx}, f_{xy}, f_{xz} remains steady. In this case, only a stability analysis (Murrieta-Rico et al., 2015) is required rather than the approximation to any of such frequencies.

Another interesting parameter to consider is the relationship between the measurement time M_t and each coincidence time $t_{0xx}, t_{0xy}, t_{0xz}$. According to the Fig. 5, in the case of S_{xy} & S_0, there are perfect coincidences each four pulses (or cycles) of reference signal S_0 and each three pulses of S_{xy} according to Eq. 2, $f_{xy}=(3/4)(10$ MHz$)=7.5$ MHz. This result is difficult to obtain under real operational conditions, for this reason, in practice a suitable approach is the use mediant criterion.

Figure 5. Considered scenario: the signal coincidence process for an aircraft which can experience acceleration in any of its three axis when $f_0=10kHz$ and $f_{xx}=7kHz$, $f_{xy}=7.5kHz$, $f_{xz}=9.5kHz$ during a measurement time $M_t=12\times10^{-4}$, with $\tau=50\mu s$

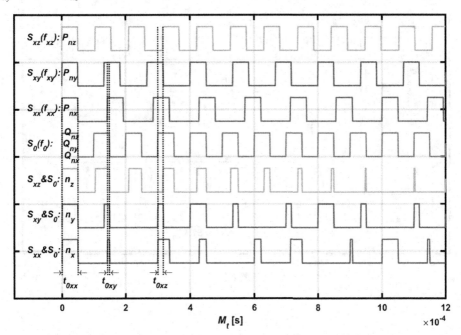

For the first four coincidences in S_{xy} & S_0, we can observe a similar pattern as in S_{xy} & S_0, this implies that $P_{ny}=P_{xy}$ during $n_y=n_x=1$, $n_y=n_x=2$ and $n_y=n_x=3$. However, the coincidence time of the pulses in S_{xy} & S_0, S_{xx} & S_0 is significatively different. In the case of $n_y=n_x=4$, n_y and n_x exist at different time, this implies that there is a different quantity of pulses when comparing P_{ny} and P_{nx} or Q_{nx} and Q_{ny} (Fig. 5). The reason of this is the difference of frequency values that we intend to approximate (f_{xy} and f_{xx}). An analysis of S_{xx} & S_0 shows that the best coincidence occurs when $P_{nx}/Q_{nx}=7/10$, which generates the value of $f_{xx}=7$ MHz. It is interesting to observe that if the first mediant fraction is calculated for S_{xx} & S_0, it has the same value of the previously calculated for S_{xy} & S_0. As it will be shown later, this is one of the reasons why the principle of rational approximations has the greatest error at the first instants of measurement process. In this measurement technique, the approximation to measurand is a continuous process, this is the reason why in Eq. 3 many (all the possible) mediant fractions are considered. This allows to properly approximate each case of f_{xy} and f_{xx}.

Figure 6. Considered scenario: the signal coincidence process for an aircraft which can experience acceleration in any of its three axis when $f_0=10$ kHz and $f_{xx}=7$ kHz, $f_{xy}=7.5$ kHz, $f_{xz}=9.5$ kHz during a measurement time $M_t=3.5\times10^{-3}$, with $\tau=50\mu s$

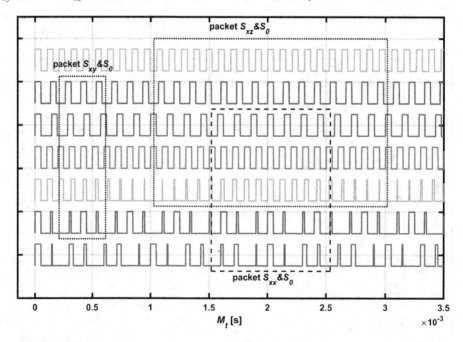

According to Fig. 5, in the case of S_{xz}, it is not possible to determine when occurs the second perfect coincidence. This means that the measurement time illustrated in Fig. 5 is not enough. Fig. 6 shows the signal coincidence process in a longer time than in Fig. 5. As it was presented by Murrieta-Rico, et al. (2016), the perfect coincidences appear at a constant rate, this means that if the reference and desired signal have constant frequencies, the perfect coincidences have a periodicity. This implies that also, the partial coincidences theoretically appear at the same rate. This generates regular packages of coincidences appearing at regular intervals, and the perfect coincidence can be considered as the center of the package. The presence of packages of coincidences is shown in Fig. 6, for each signal coincidence process. The packet corresponding to S_{xz} & S_0 has the longest duration from the three packages illustrated in Fig. 6. In principle, this can be attributed to the f_{xz}, f_0 values and how close are among them. This is not a determining factor as it shown in the packages of S_{xy} & S_0 and S_{xx} & S_0. Since S_{xy} & S_0 shows the measurement process of f_{xy}, and S_{xx} & S_0 the approximation of f_{xx}, it could be expected to have packages with more coincidences in S_{xy} & S_0 because it is approximating the value closer to f_0 than the process illustrated by S_{xx} & S_0. Fig. 6 shows that the packages corresponding to S_{xy} & S_0 have less coincidences than S_{xx} & S_0. Considering the whole signal comparison process, the number of coincidences increases if the measurand approaches to the standard value.

In the simulation presented in Figs. 5 and 6 is illustrated the case when f_{xx}=7 kHz, f_{xy}=7.5 kHz, f_{xz}=9.5 kHz. Considering Fig. 4, we can say that the aircraft is displacing to the front (x), it has a minimal displacement in y, and it has a significant speed variation on z. From Fig. 5, it is observed that for different frequencies, there is a behavior in the pulse coincidence train that generates a particular packet of coincidences. However, for all considered cases, there is a reduction in the relative error to almost a zero value, in a measuring time less than 10 ms.

Figure 7. Relative error calculation for the considered scenario when f_0=10 kHz and f_{xx}=7 kHz, f_{xy}=7.5 kHz, f_{xz}=9.5 kHz, with τ=50µs

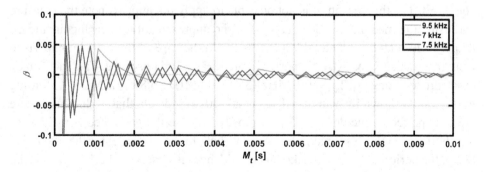

The accuracy of the analysis presented in Fig. 7 can be understood using the Eq. 6. In this case, the relative error β was calculated for each case presented in Fig. 7: $f_{xx}=7$ kHz, $f_{xy}=7.5$ kHz, $f_{xz}=9.5$ kHz. For all the three cases, the relative error is reduced to almost zero, in a time shorter than 10 ms. When the measurand has the values of: $f_{xx}=7$ kHz of $f_{xy}=7.5$ kHz, the rate in the decrement of β as a function of M_t is quite similar in both cases, this is a consequence of how similar are the packages of coincidences illustrated in Figs. 5 and 6. When observing $\beta(M_t)$ for $f_{xz}=9.5$ kHz it is evident that is very different from the relative error in $f_{xx}=7$ kHz, $f_{xy}=7.5$ kHz. As stated before, the amount of coincidences increases if the measurand value approaches to reference standard. This generates a "smother" decrement in $\beta(M_t)$ when $f_{xz}=9.5$ kHz than in the other cases. In general terms, the variations in how the error was reduced can be explained as consequence of the different kind of coincidences illustrated in Figs. 5 and 6.

In the measurement processes illustrated in Figs. 2, 5 and 6, the pulse width was adjusted to $\tau=T_0/2$ or $\tau=50\mu s$. This value was chosen according to the restriction that $\tau \leq T_0/2$ (Tyrsa & Zenya, 1983). For application in real systems, adjusting the pulse width at $\tau=T_0/2$ is not possible, this mainly because of physical factors related to the stability of frequency standards (Kundur, et al., 1994). For this reason, for implementation of the principle of rational approximations requires only to fulfill $\tau \leq T_0/2$, in addition if has been proved that if the duration of pulse width decreases, better approximations to measurand are obtained. In Fig. 8 is illustrated the signal coincidence process if the pulse width is $\tau=30\mu s$. The first observation is that there are less pulses of coincidence than in Fig. 5 (when $\tau=50\mu s$). Since the pulses now have a shorter duration, the probability for overlapping in time the pulses in desired signals (S_{xx}, S_{xy}, S_{xz}) and reference (S_0) has decreased. Consequently, the amount of pulses of coincidence and their maximum duration are reduced (Murrieta-Rico, et al. 2018).

An interesting consequence of shortening the pulse width duration is related to the application of mediant theory itself. The signal comparison from the z-axis (S_{xz} & S_0) shows a monotonous decrease in the coincidence time (t_{0xz}). Since for all the coincidences observed in S_{xz} & S_0, $P_{nz}=Q_{nz}$, in consequence the resulting fractions are equal to 1. For this reason, it is not possible to apply the approximate the measure and using the mediant fractions (Eq. 3) or the approximation in each coincidence (Eq. 2). This can be solved with a longer M_t. In the case of the signal comparison from the y-axis (S_{xy} & S_0), only perfect coincidences are observed. This fact can be easily proved using Eq. 2, $f_{xy}=(10\,MHz)(3/4)=(10\,MHz)(6/8)=7.5\,MHz$. According to the results shown by Murrieta-Rico et. al (2018) indicate that in this case, there are only perfect coincidences. This is mostly because the generated coincidences are shorter than the "optimal pulse width value". In the last case, for the z-axis (S_{xz} & S_0), the periodicity in the packet of coincidences is observed. Perfect and partial

Figure 8. Considered scenario: the signal coincidence process for an aircraft which can experience acceleration in any of its three axis when $f_0=10$ kHz and $f_{xx}=7$ kHz, $f_{xy}=7.5$ kHz, $f_{xz}=9.5$ kHz during a measurement time $M_t=12\times10^{-4}$, with $\tau=30\mu s$

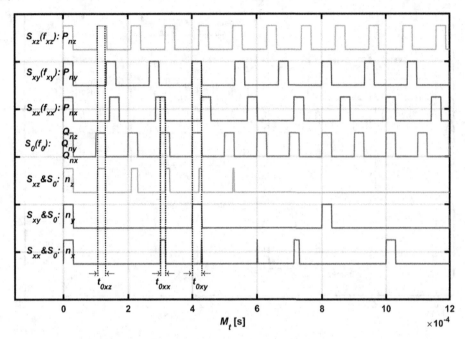

coincidences exist in the pulse coincidence train. The best approximation can be calculated when the second perfect coincidence exists, $f_{xy}=(10\,\text{MHz})(7/10)=7\,\text{MHz}$.

When the signal comparison process is observed during a longer time (Fig. 9), it can be observed how the packages of coincidences have less pulses than the ones shown in Fig. 6. This shows that the best approximations occur in the same time lapse (Figs. 6, 9), but there are less data to process if the pulse width is reduced. This is important for devices such as UAV, where there are low computational resources available for data processing. Additionally, from Fig. 9 it is observed that when the frequency of accelerometers that respond at the same time to different input stimuli is monitored, there are different behaviors for each measurement process.

This is a challenge for the implementation of this technique. Considering the scenario, where there are only perfect coincidences (as shown in Fig. 9, the coincidence pulse train corresponding to S_{xy} & S_0) could lead to a condition where the measurement could be obtained in a time as long as one second. These phenomena could lead to catastrophic consequences. For this reason, the pulse width cannot be reduced arbitrarily, such a time can be adjusted according to the maximum and minimal expected frequency value (Murrieta-Rico et. at, 2018).

Figure 9. Considered scenario: the signal coincidence process for an aircraft which can experience acceleration in any of its three axis when f_0=10 kHz and f_{xx}=7 kHz, f_{xy}=7.5 kHz, f_{xz}=9.5 kHz during a measurement time M_t=3.5×10⁻³, with τ=30µs

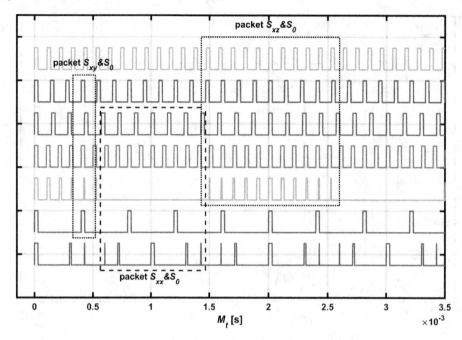

Since the number of coincidences decreased, the relative error in measurement decreased for the analyzed cases. The relationship between β and measurement time is illustrated in Fig. 10. From the results of these analysis, it can be concluded that if the frequency generated by the accelerometers varies between the maximum, and minimum value shown in Fig. 8, the measurand value can be approximated in a time shorter than 3.5 ms. As a consequence, from the analysis presented in this work, it was shown that the principle of rational approximations can be used inside the IMU.

ANALYSIS OF FREQUENCY SHIFT MEASUREMENT PROCESS

As stated before, accelerometers are sensors that generate a frequency which is proportional to acceleration. In practice, when an aircraft is in operation, the acceleration value experienced by the accelerometers is continuously changing. For this reason, it is necessary to analyze the process of measuring the signal with a dynamic frequency value. In order to study this scenario, consider the case when an airplane is landing.

Figure 10. Relative error calculation for the considered scenario when f_0=10 kHz and f_{xx}=7 kHz, f_{xy}=7.5 kHz, f_{xz}=9.5 kHz with τ=30μs

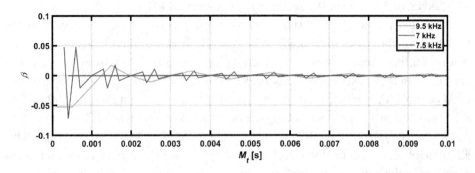

Figure 11. Frequency values in x-axis during aircraft landing (S_{xx} red track, S_0 blue track and S_{xx} & S_0 pink track): signal coincidence process a) with f_{xx}=7, 9.5 and 4 kHz, b) transition from 7 to 9.5 kHz in 20 ms, c) transition from 7 to 9.5 kHz in 2 ms

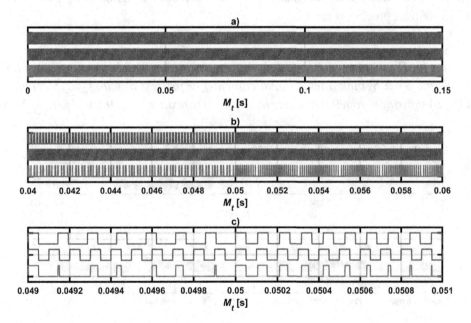

For the sake of simplicity, we can consider that the greatest deceleration is generated in the x-axis; therefore, the signal S_{xx} has different frequency values, which continuously appear in a short time. For this situation, three different frequency values are present in the signal S_{xx}: f_{xx}=7, 9.5 and 4 kHz. Each value corresponds to approach to the ground, first contact with the ground and reduced velocity after the first contact respectively. Figs. 11, 12 show the signal coincidence process during the measurement of f_{xx}. In simulations, f_{xx} starts at 7 kHz, at 0.05 seconds f_{xx} changes to 9.5 kHz, and finally at 0.1 seconds f_{xx} changes to 4 kHz.

After an analysis of the signal comparison process shown in Fig. 11a, the variations on the pulses of coincidence is slightly observed when $M_t > 0.1$ s. As stated before, for this simulation was considered that $f_{xx} = 7$ kHz when $0 < M_t < 50$ ms. The variations in the packets of coincidences can be observed in Fig. 11b, where before of $M_t = 50$ ms, the coincidences appear at a different rate than after $M_t = 50$ ms. As it is expected, if f_{xx} approaches to f_0, there are more coincidences. The last can be observed when $M_t > 50$ ms. In Fig. 11c a shorter M_t than Fig. 11b is shown. This allows to observe in detail the increment in the f_{xx} value. The variations in the amount of pulses in the coincidence packet is observed, additionally it can be observed the location of the kernel of the packet when there is a greater measurand value. The last observation now is presented in Fig. 11b when $M_t > 50$ ms. In this simulation, the pulse with (τ) has a duration of 50 µs, and for the analysis of the signal S_{xx}, $t_{0xx} = t_{ax}$; another consideration for this analysis is that when each variation of f_{xx} occurs, the measurement process is reset.

Figure 12. Frequency values in x-axis during aircraft landing (S_{xx} red track, S_0 blue track and S_{xx} & S_0 pink track): signal coincidence process a) with $f_{xx} = 7$, 9.5 and 4 kHz, b) transition from 9.5 to 4 kHz in 20 ms, c) transition from 9.5 to 4 kHz in 2 ms

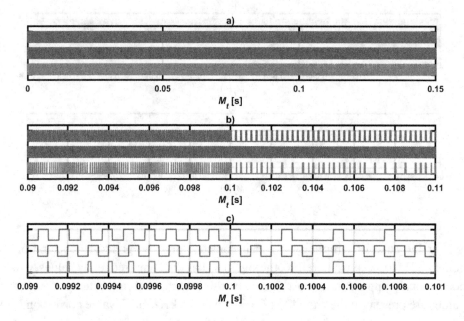

As stated by a previous a publication (Murrieta-Rico et al. 2016), the kernel of the packet of coincidences is located where $t_{0x}=\tau$. This allows to observe that when $M_t=50$ ms, and the pulses in both signals S_{xx}, S_0 coincide, there is a perfect coincidence and $t_{0x}=\tau$. This can be explained as a consequence of restarting the measurement process when the frequency shift occurred. Moreover, another important property can be observed. From the analysis of relative error shown in Figs. 3, 7, 10, the greatest value of β exists at the beginning of measurement time, this is a consequence of the fractions formed by the pulses in both input signals.

The reduction of f_{xx} is observed to occur after $M_t=0.1$ seconds (Fig. 12b), consequently the corresponding period (T_{xx}) increases. In other words, more spaced pulses are observed in S_{xx} (red tracks in Fig. 12). Another consequence of variations of the period in S_{xx} is observed if a shorter time window is observed. Particularly, in Fig. 12c the pulse train of coincidences shows that there is an important difference between the packets of coincidences before and after a sudden frequency shift. When f_{xx} changes from 9.5 to 4 kHz, there are less coincidences (Fig. 12c) than the previously discussed case (Fig. 11c). The relationship between the amount of coincidences and their duration determines the accuracy of measurement process (Murrieta-Rico et al. 2018), as a result, from observing the pulse train of coincidences in Fig. 12c, it can be expected to have better approximations than in Fig. 11c. This phenomenon is consequence of the difference in the magnitude of measurand and reference. In other words, if the value of f_{xx} decreases from the value of f_0, their probability to coincide decreases. Accordingly, when there is a coincidence it could have either a value near to pulse width or too short in comparison to τ. Another important remark can be done regarding the amount of coincidences in Fig. 12c. After the frequency shift there are basically two kind of coincidences, ones that last almost the duration of the pulse width and others that last a very short time. These last coincidences in most cases are not detected in practical applications, this is because these "short coincidences" do not last enough for being detected by the electronics, which is used in the instrumentation circuits. This phenomenon acts as a "natural filter", and the frequency estimation is improved.

As it as shown elsewhere (Murrieta Rico et al., 2016; Murrieta-Rico et al., 2018), information of the coincidence behavior can be obtained from an analysis of the coincidence time (t_{0x}) variations during measurement time (M_t). For the measurement process analyzed in Figs. 11 and 12, the analysis of coincidence time during frequency variations is shown in Fig. 13a. The detailed behavior of t_{0x} during M_t is shown after the first variation of f_{xx} (7 to 9.5 kHz) in Fig. 13b, and the second f_{xx} variation from 9.5 to 4 kHz is shown in Fig. 13c. The whole frequency measurement process shown in Fig. 13a illustrates how when $f_{xx}=7$ kHz and $f_{xx}=9.5$ kHz, there is a periodic behavior of t_{0x}. This obeys to the physical nature of the signal coincidence process and the apparition of the coincidence packets. Each packet appears at a periodical

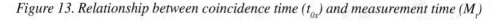

Figure 13. Relationship between coincidence time (t_{0x}) and measurement time (M_t)

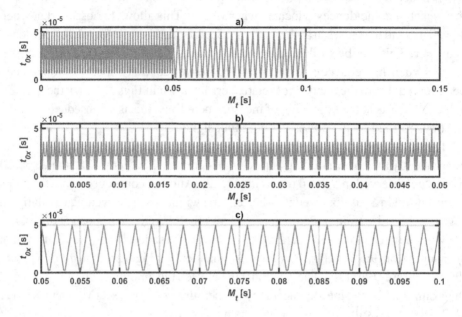

rate (Murrieta-Rico et al., 2016), and the kernel of the packet is located where the coincidence time has the closest duration to the pulse width from original signals. Nevertheless, in some cases there are plenty of pulses in the packet, at first sight this is the case when $f_{xx}=7$ kHz (Fig. 13b). Further understanding of this can be obtained if Fig. 11c is considered. While Fig. 11c shows a simulation of the physical process of signal comparison, Fig. 13b shows how long exist the coincidences in Fig. 11c, when $M_t<50$ ms. The peaks in Fig. 13b show how the pulses in the packet of coincidences do not have a monotonic increment or decrement, at the same time, this is also illustrated in Fig. 11c.

In the case when f_{xx} shifted from 7 to 9.5 kHz, the period of pulse packets apparition has increased. In first instance, it can be considered that the fraction where the best approximation can be expected is $f_{xx}/f_0=19/20$ (Murrieta-Rico et al., 2016), if $f_{xx}=9.5$ kHz and $f_0=10$ kHz. As a matter of fact, considering the same frequency reference value, in the previous case, when $f_{xx}=7$ kHz, $f_{xx}/f_0=7/10=14/20$. The increase in the duration of period for the packets of coincidences can be attributed $14/20<19/20$. As it is observed in Fig. 11c, when $M_t>50$ ms there are more coincidences after 50 ms (when $f_{xx}=9.5$ kHz), but this does not necessarily mean that better approximations are obtained.

Another important property is illustrated by the way in which t_{0x} increases in Fig. 13c. This indicates that the pulses of coincidence monotonically increase until they reach the value of τ, following the last, they also decrement monotonically until they reach the minimum detectable time.

In the last frequency variation, when f_{xx} changes from 9.5 to 4 kHz. There is only a continuous line (Fig. 13a when $M_t>0.1$ s). This can be understood as a consequence of the use of the "optimal pulse width value" (Murrieta-Rico et al., 2018). If there is continuous value of the coincidence time, and it approaches to the pulse width duration, then only "perfect coincidences" exist. The use of the optimal pulse width value allows to obtain the best approximations to the measurand in the second coincidence, but the time between coincidences increases (as it is shown in Fig. 12c when $M_t>0.1$ s).

The aim of the present work is to show how the principle of rational approximations could be used as a tool for improving the performance of an INS. Particularly, the improvement of frequency measurement from accelerometers is the objective of this work. In order to illustrate how all the theory presented until now serves to the purpose of this chapter, the simulation of the frequency estimation is shown in Fig. 14. In such a figure, a signal with three different frequency values for f_{xx} is shown. In all the three cases, the frequency is continuously measured and the approximation to the nominal value is shown. As it was shown in the analisis of Figs. 11-13, there are plenty more packet of coincidences when $f_{xx}=7$ kHz than when $f_{xx}=9.5$ kHz, but at the same time, in both cases is required almost the same time to converge to the best approximation to the measurand (Fig. 7). In the last case (Fig. 14 when $M_t>0.1$ s), the best approximation to measurand is obtained since the beginning of the measurement process.

The analysis shown in Fig. 14 illustrates how the principle of rational approximations is valuable tool, which allows to improve accuracy and resolution of the frequency measured from accelerometers inside IMU.

Figure 14. Frequency measurement process of f_{xx} in aircraft landing scenario

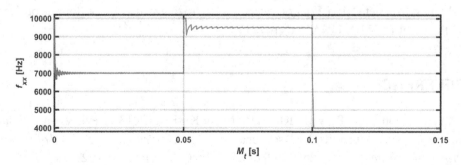

In this work it was shown how the principle of rational approximations can be used for improving measurements from accelerometer inside IMU. Consequently, the estimation of altitude and speed can be better approximated if the measurement time is shortened and the accuracy of the frequency measurement corresponding to the acceleration is improved. Nevertheless, the estimation of other frequency dependent parameters can be also improved after the use of the principle of rational approximations. One example is related to structural health monitoring in aircrafts, where sensors for structural damage based on a piezoelectric materials are used (Kessler et al., 2008, Yu et al., 2008; Ihn and Chang, 2008; Zhao et al., 2007).

CONCLUSION

In this work, an accelerometer with three frequency signals was studied. Particularly how its use could be improved using the principle of rational approximations. This study included the simulation of the measurement process and analysis of during aircraft landing. The results show that in less than 10 ms, the sensor outputs are properly approximated. Additionally, the physical and mathematical behavior of the signals during the coincidence process was analyzed. As a result, it was observed that if a signal has a steady frequency, after the first measurement, such a signal can be analyzed in terms of its stability, which allows to reduce the data processing in the INS. As future work, we expect to implement this technique in a real IMU and analyze in dept its functioning.

ACKNOWLEDGMENT

We thank the support of the Universidad Autonoma de Baja California (UABC) y Universidad Nacional Autonoma de Mexico (UNAM) for providing facilities for research and experimentation. Additionally, the authors would like to acknowledge the support of the PRODEP program for granting a postdoctoral scholarship for Fabian N. Murrieta Rico. Finally, we would like to thank the support for this project by the grant DGAPA-UNAM IN107817.

REFERENCES

Ahmad, N., Ghazilla, R. A. R., Khairi, N. M., & Kasi, V. (2013). Reviews on various inertial measurement unit (IMU) sensor applications. *International Journal of Signal Processing Systems*, *1*(2), 256–262. doi:10.12720/ijsps.1.2.256-262

Avalos-Gonzalez, D., Hernandez-Balbuena, D., Tyrsa, V., Kartashov, V., Kolendovska, M., Sheiko, S., . . . Murrieta-Rico, F. N. (2018). Application of Fast Frequency Shift Measurement Method for INS in Navigation of Drones. In *IECON 2018-44th Annual Conference of the IEEE Industrial Electronics Society* (pp. 3159-3164). IEEE. 10.1109/IECON.2018.8591377

Bruggemann, T. S., Greer, D. G., & Walker, R. A. (2011). GPS fault detection with IMU and aircraft dynamics. *IEEE Transactions on Aerospace and Electronic Systems, 47*(1), 305–316. doi:10.1109/TAES.2011.5705677

El-Sheimy, N., Nassar, S., & Noureldin, A. (2004). Wavelet de-noising for IMU alignment. *IEEE Aerospace and Electronic Systems Magazine, 19*(10), 32–39. doi:10.1109/MAES.2004.1365016

Eling, C., Klingbeil, L., & Kuhlmann, H. (2015). Real-time single-frequency GPS/MEMS-IMU attitude determination of lightweight UAVs. *Sensors (Basel), 15*(10), 26212–26235. doi:10.3390151026212 PMID:26501281

Endevco, M. (2019). *773 Triaxial Variable Capacitance Accelerometer*. March, 03, 2019, from MEGGITT Endevco Website: https://buy.endevco.com/773-accelerometer-1.html

Gao, Z., Ge, M., Shen, W., Zhang, H., & Niu, X. (2017). Ionospheric and receiver DCB-constrained multi-GNSS single-frequency PPP integrated with MEMS inertial measurements. *Journal of Geodesy, 91*(11), 1351–1366. doi:10.100700190-017-1029-7

Hernandez-Balbuena, D., Sergiyenko, O., Rosas-Méndez, P. L., Tyrsa, V., & Rivas-Lopez, M. (2012). Fast method for frequency measurement by rational approximations with application in mechatronics. In *Modern Metrology Concerns*. IntechOpen. doi:10.5772/23225

Honeywell. (2019). *HG4930 IMU*. March, 03, 2019, from Honeywell Website: https://aerospace.honeywell.com/en/products/navigation-and-sensors/hg4930

Huang, L., Yang, H., Gao, Y., Zhao, L., & Liang, J. (2013). Design and implementation of a micromechanical silicon resonant accelerometer. *Sensors (Basel), 13*(11), 15785–15804. doi:10.3390131115785 PMID:24256978

Ihn, J. B., & Chang, F. K. (2008). Pitch-catch active sensing methods in structural health monitoring for aircraft structures. *Structural Health Monitoring, 7*(1), 5–19. doi:10.1177/1475921707081979

Johansson, S. (2006). New frequency counting principle improves resolution. In *Proceedings of the 20th European Frequency and Time Forum* (pp. 139-146). IEEE.

Kalisz, J. (2003). Review of methods for time interval measurements with picosecond resolution. *Metrologia*, *41*(1), 17–32. doi:10.1088/0026-1394/41/1/004

Kessler, S. S., Jugenheimer, K. A., Size, A. B., & Dunn, C. T. (2008). *U.S. Patent No. 7,469,595*. Washington, DC: U.S. Patent and Trademark Office.

Kirianaki, N. V., Yurish, S. Y., & Shpak, N. O. (2001). Methods of dependent count for frequency measurements. *Measurement*, *29*(1), 31–50. doi:10.1016/S0263-2241(00)00026-9

Kundur, P., Balu, N. J., & Lauby, M. G. (1994). *Power system stability and control* (Vol. 7). New York: McGraw-Hill.

Li, T., Zhang, H., Gao, Z., Niu, X., & El-sheimy, N. (2019). Tight Fusion of a Monocular Camera, MEMS-IMU, and Single-Frequency Multi-GNSS RTK for Precise Navigation in GNSS-Challenged Environments. *Remote Sensing*, *11*(6), 610. doi:10.3390/rs11060610

Li, T., Zhang, H., Niu, X., & Gao, Z. (2017). Tightly-coupled integration of multi-GNSS single-frequency RTK and MEMS-IMU for enhanced positioning performance. *Sensors (Basel)*, *17*(11), 2462. doi:10.339017112462 PMID:29077070

Murrieta-Rico, F. N., Hernandez-Balbuena, D., Petranovskii, V., Nieto-Hipolito, J. I., Pestryakov, A., Sergiyenko, O., . . . Tyrsa, V. (2014). Acceleration measurement improvement by application of novel frequency measurement technique for FDS based INS. In *2014 IEEE 23rd International Symposium on Industrial Electronics (ISIE)* (pp. 1920-1925). IEEE. 10.1109/ISIE.2014.6864909

Murrieta-Rico, F. N., Hernandez-Balbuena, D., Rodriguez-Quiñonez, J. C., Petranovskii, V., Raymond-Herrera, O., Hipolito, J. I. N., . . . Melnyk, V. I. (2015). Instability measurement in time-frequency references used on autonomous navigation systems. In *2015 IEEE 24th International Symposium on Industrial Electronics (ISIE)* (pp. 956-961). IEEE. 10.1109/ISIE.2015.7281600

Murrieta-Rico, F. N., Mercorelli, P., Sergiyenko, O. Y., Petranovskii, V., Hernandez-Balbuena, D., & Tyrsa, V. (2015). Mathematical modelling of molecular adsorption in zeolite coated frequency domain sensors. *IFAC-PapersOnLine*, *48*(1), 41–46. doi:10.1016/j.ifacol.2015.05.060

Murrieta-Rico, F. N., Petranovskii, V., Sergiyenko, O. Y., Hernandez-Balbuena, D., & Lindner, L. (2017). A New Approach to Measurement of Frequency Shifts Using the Principle of Rational Approximations. *Metrology and Measurement Systems*, *24*(1), 45–56. doi:10.1515/mms-2017-0007

Murrieta-Rico, F. N., Sergiyenko, O. Y., Petranovskii, V., Hernandez-Balbuena, D., Lindner, L., Tyrsa, V., ... Karthashov, V. M. (2016). Pulse width influence in fast frequency measurements using rational approximations. *Measurement*, *86*, 67–78. doi:10.1016/j.measurement.2016.02.032

Murrieta-Rico, F. N., Sergiyenko, O. Y., Petranovskii, V., Hernandez-Balbuena, D., Lindner, L., Tyrsa, V., ... Nieto-Hipolito, J. I. (2018). Optimization of pulse width for frequency measurement by the method of rational approximations principle. *Measurement*, *125*, 463–470. doi:10.1016/j.measurement.2018.05.008

Noureldin, A., Karamat, T. B., & Georgy, J. (2012). *Fundamentals of inertial navigation, satellite-based positioning and their integration*. Springer Science & Business Media.

Parker, P. A., & Finley, T. D. (2007). Advancements in aircraft model force and attitude instrumentation by integrating statistical methods. *Journal of Aircraft*, *44*(2), 436–443. doi:10.2514/1.23060

Pei, F. J., Liu, X., & Zhu, L. (2014). In-Flight Alignment Using Filter for Strapdown INS on Aircraft. *The Scientific World Journal*. PMID:24511300

Qi, H., & Moore, J. B. (2002). Direct Kalman filtering approach for GPS/INS integration. *IEEE Transactions on Aerospace and Electronic Systems*, *38*(2), 687–693. doi:10.1109/TAES.2002.1008998

Sabatini, R., Cappello, F., Ramasamy, S., Gardi, A., & Clothier, R. (2015). An innovative navigation and guidance system for small unmanned aircraft using low-cost sensors. *Aircraft Engineering and Aerospace Technology: An International Journal*, *87*(6), 540–545. doi:10.1108/AEAT-06-2014-0081

Sankar, A. R., Das, S., & Lahiri, S. K. (2009). Cross-axis sensitivity reduction of a silicon MEMS piezoresistive accelerometer. *Microsystem Technologies*, *15*(4), 511–518. doi:10.100700542-008-0740-y

SBG Systems. (2019). *Ellipse2*. March, 03, 2019, from SBG Systems Website: https://www.sbg-systems.com/products/ellipse-2-series/

Schwarz, K. P. (1996). Aircraft position and attitude determination by GPS and INS. *International Archives of Photogrammetry and Remote Sensing*, *31*(B6), 67–73.

Sekiya, H., Kinomoto, T., & Miki, C. (2016). Determination method of bridge rotation angle response using MEMS IMU. *Sensors (Basel)*, *16*(11), 1882. doi:10.339016111882 PMID:27834871

Sergiyenko, O., Balbuena, D. H., Tyrsa, V., Mendez, P. L. A. R., Lopez, M. R., Hernandez, W., ... Gurko, A. (2011). Analysis of jitter influence in fast frequency measurements. *Measurement*, *44*(7), 1229–1242.

Sergiyenko, O. Y., Balbuena, D. H., Tyrsa, V. V., Mendez, P. L. R., Hernandez, W., Hipolito, J. I. N., ... Lopez, M. R. (2012). Automotive FDS resolution improvement by using the principle of rational approximation. *IEEE Sensors Journal*, *12*(5), 1112–1121. doi:10.1109/JSEN.2011.2166114

Tyrsa, V. E., & Zenya, A. D. (1983). Analysis of errors in frequency comparison by the pulse coincidence method. *Measurement Techniques*, *26*(7), 576–579.

Vernier. (2019). *3-Axis Accelerometer*. March, 03, 2019, from Vernier Website: https://www.vernier.com/products/sensors/accelerometers/3d-bta/

Yu, L., Santoni-Bottai, G., Xu, B., Liu, W., & Giurgiutiu, V. (2008). Piezoelectric wafer active sensors for in situ ultrasonic-guided wave SHM. *Fatigue & Fracture of Engineering Materials & Structures*, *31*(8), 611–628. doi:10.1111/j.1460-2695.2008.01256.x

Zhao, X., Gao, H., Zhang, G., Ayhan, B., Yan, F., Kwan, C., & Rose, J. L. (2007). Active health monitoring of an aircraft wing with embedded piezoelectric sensor/actuator network: I. Defect detection, localization and growth monitoring. *Smart Materials and Structures*, *16*(4), 1208–1217. doi:10.1088/0964-1726/16/4/032

Zhao, Y., Horemuz, M., & Sjöberg, L. E. (2011). Stochastic modelling and analysis of IMU sensor errors. *Archiwum Fotogrametrii, Kartografii i Teledetekcji*, 22.

ADDITIONAL READING

Barbour, N., & Schmidt, G. (2001). Inertial sensor technology trends. *IEEE Sensors Journal*, *1*(4), 332–339. doi:10.1109/7361.983473

Diaz, M. J. P., Mariano, C. E. A., & Beltran, A. A. Jr. (2014). Accelerometer-based Wave Motion Compensation on Ship Mounted Weaponry. *International Journal of Scientific Engineering and Technology*, *3*(5), 588–591.

Duncan, R. C., & Unnersen, A. S. (1964). Inertial guidance, navigation, and control systems. *Journal of Spacecraft and Rockets*, *1*(6), 577–587. doi:10.2514/3.27706

Farrell, J., & Barth, M. (1999). *The global positioning system and inertial navigation* (Vol. 61). New York: Mcgraw-Hill.

Goshen-Meskin, D., & Bar-Itzhack, I. Y. (1992). Unified approach to inertial navigation system error modeling. *Journal of Guidance, Control, and Dynamics*, *15*(3), 648–653. doi:10.2514/3.20887

Kayton, M., & Fried, W. R. (1997). *Avionics navigation systems*. John Wiley & Sons. doi:10.1002/9780470172704

Kourepenis, A., Petrovich, A., & Weinberg, M. (1991). Low cost quartz resonant accelerometer for aircraft inertial navigation. In *TRANSDUCERS'91: 1991 International Conference on Solid-State Sensors and Actuators. Digest of Technical Papers* (pp. 551-553). IEEE. 10.1109/SENSOR.1991.148935

KEY TERMS AND DEFINITIONS

Accelerometer: When this device is under the effect of an acceleration lower or greater than 1g, it generates an electrical signal proportional to the magnitude of such acceleration.

Dead Reckoning: This is the process for calculating the current position of an entity, after the use of a previously defined position. The variations in the current position can be estimated using the velocity of the moving entity.

IMU: Inertial measurement system.

Inertial Reference Frame: It refers to a coordinate frame in which Netwon's laws of motion are valid. For an inertial reference frame, it is considered that there is no acceleration or rotation.

INS: Inertial navigation system.

Chapter 3
Approaches to Development of Mechanical Design and Jumping Motion for a Wheeled Jumping Robot

Lyudmila Yurievna Vorochaeva
Southwest State University, Russia

Sergey Igorevich Savin
Innopolis University, Russia

Andrei Vasilievich Malchikov
Southwest State University, Russia

Andres Santiago Martinez Leon
Southwest State University, Russia

ABSTRACT

This chapter is dedicated to tackling the issues related to the design and locomotion control of a hybrid wheeled jumping monitoring platform. The studied robot consists of a body mounted on a wheeled platform and of a jump acceleration module. An approach to making design decisions regarding the structure of the investigated robot is proposed. To select the kinematic structure of the robot, classifications of possible variants of hybrid jumping platforms and accelerating modules are presented. Methods for controlling the function of the accelerating modules and the analysis of their work is carried out. Various implementations of jumping motion are discussed; these implementations are characterized by different combinations of relative links movements during various stages of motion. Each of the proposed jump motion types requires the development of a control system, which is also discussed in this chapter.

DOI: 10.4018/978-1-5225-9924-1.ch003

INTRODUCTION

The application of robots in performing tasks, traditionally imposed on people, results in lower costs and risks, and improves the quality of the accomplished work. The most typical example is the tasks of monitoring and survey of hard-to-reach areas. The use of robots for solving such problems can provide significant economic and social effects, allowing to automate a number of complex, labor-intensive and potentially dangerous activities, such as preparation and updating of maps and three-dimensional models of emergency situations, collection of data on the state of the environment in areas subjected to biological or radiation contamination, continuous monitoring of the environment, air and soil sampling, etc.

Therefore, one of the urgent tasks of modern robotics is the creation of small-sized monitoring devices with high mobility and the ability to move around complex terrain areas with debris and obstacles. Such devices include wheeled and tracked robots (Murphy et al., 2008; Kawatsuma, Fukushima, & Okada, 2012), walking robots (Yi et al., 2014; Calinon, Guenter, & Billard, 2007), crawling multi-link robots (Vorochaeva et al., 2017; Peters, Ahn, & Borkowski, 2002), flying (Tanaka et al., 2005; Vorochaeva et al., 2018) and jumping (Gilani & Ben-Tzvi, 2011; Tsukagoshi et al., 2005) systems. It should be noted that the dimensions of the obstacles that can be traversed are directly related to the design features and the dimensions of the robots (wheel radius, link lengths, ranges of relative angles, etc.). Flying and jumping systems have more opportunities to move on uneven terrain and can traverse a wider range of obstacles in size and shape. For flying systems equipped with propellers or wings, this range of obstacles is due to the thrust of the propellers or wings, while for jumping systems it depends on the takeoff speed and the inclination angle at which the takeoff occurs.

This study is focused on jumping robotic systems due to their advantages over flying devices. These advantages include better maneuverability and easier control when moving in confined spaces, as well as when moving between high-rise buildings, due to their lower exposure to turbulent air flows; jumping robots are also characterized by less noise and lower energy consumption.

The robot studied here is a combined system, including a body mounted on a wheeled platform with an acceleration module located inside, through which the jump is carried out. This chapter is devoted to general, basic issues related to jumping robots, and consists of the following parts: introduction of general terms and concepts of jumping motion, justification and selection of the combined wheel-jumping platform scheme, description of the design scheme of the device, description of the various variants of the jump implementation, development of the robot control system and the definition of permissible values ranges of the robot geometric parameters, providing the ability to carry out all previously proposed variants of the jump.

BACKGROUND: THE BASIC CONCEPTS OF A JUMPING LOCOMOTION

Jumping robots can be represented as consisting of a body 1 and an acceleration module 2 (figure 1 A). The principle of their movement is as follows: the relative accelerated movement of the acceleration module links leads to the situation when the object hits the surface and separates from it.

Consider the basic concepts of the jumping movement mode. Under the jump we understand the movement of the body with a temporary one-time planned loss of contact with the supporting surface, caused by repulsion from the support point with an initial velocity υ directed at an angle γ to the horizon. Characteristics of the jump are the following: height, length, angle of rotation of the device in flight. Under the height H and the length L of the jump we understand the distance traveled by the center of mass of the robot in the vertical and horizontal directions from the moment of takeoff from the surface to the end of the jump. The angle of rotation $\varphi_{,ax}$ of the device in flight is the greatest angle at which the robot body rotates relative to the angle at the time of the jump while taking off the surface (figure 1 B).

The characteristics of the jump are generally determined by the value of the robot's takeoff velocity υ from the surface and the direction of its vector relative to the horizontal line γ, as well as the velocity $\dot{\gamma}$ of change in the direction of this vector at the moment of takeoff:

$$\begin{cases} H = F(\upsilon,\gamma,\dot{\gamma}), \\ L = G(\upsilon,\gamma,\dot{\gamma}), \\ \varphi_{max} = R(\upsilon,\gamma,\dot{\gamma}). \end{cases} \tag{1}$$

Figure 1. A) Jumping robot scheme; B) jump characteristics

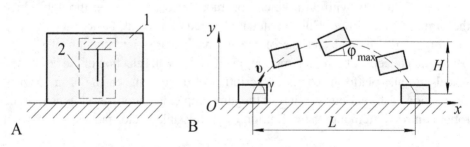

The robot jump can be divided into separate phases of movement as shown in figure 2. Each of these phases is characterized by certain movements of the robot links. The jump comes from the initial position in which the robot links do not make relative movements and contact with the surface. Positioning before takeoff is the change of relative positions of the robot links for making the jump with the required characteristics. Acceleration is characterized by accelerated movement of some links of the robot relative to others when the latter are on the surface. The takeoff corresponds to the position of the robot at the time of zeroing the normal reaction at the contact point of the robot with the supporting surface. The flight is a movement of the robot without contact with the supporting surface. The landing corresponds to the position of the robot at the time, when the robot begins to contact with the supporting surface. Positioning after landing is described by such relative movements of the robot links, which ensure the transfer of the robot to a position that allows it to maintain vertical stability after the jump. This position is the final for the given jump and the initial for the next one.

We assume that the jump starts at the moment of time t_n and ends at the moment of time t_k, and in the initial and final positions, the normal reaction between the robot links and the surface $N_n=N_k=N_0$, where $N_0=Mg$=const, M – the mass of the robot, g – the acceleration of gravity, and the ordinate of at least one point belonging to the robot, is zero $y_n=y_k=0$. Then each phase of the movement of the jumping robot can be described by three parameters (table 1, figure 3): phase time T_j, j=1-6, the value of the normal reaction N_i, $i\in[0,k]$, where $i=0$, when $t=t_n$, the smallest value of the coordinate point belonging to the robot, y_i.

Table 1 shows the following designations: t^* – the time of reaching the highest point of the jump, $\Delta t \rightarrow 0$ – the time of the robot's step, during which its motion is considered, a very small scale.

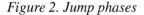

Figure 2. Jump phases

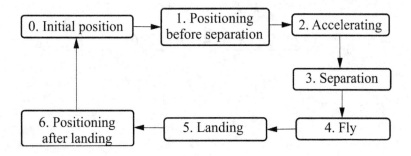

Table 1. Jump phases parameters

Phase number, j	Counter i	Phase time, T_j	Normal reaction, N_i	Ordinate, y_i
0	$i=0$	$T_0=t_n$	$N_i=N_0$	$y_i=0$
1	$i\in[1,k_1]$	$T_1\in(t_n,t_1]$	$N_i\approx N_0$	$y_i=0$
2	$i\in(k_1,k_2]$	$T_2\in(t_1,t_2]$	$N_i < N_{i-1}$	$y_i=0$
3	$i=k_3=k_2+1$	$T_3=t_3=t_2+\Delta t$	$N_i=0$	$y_i=0$
4	$i\in(k_3,k_4]$	$T_4\in(t_3,t_4]$	$N_i=0$	$y_i\begin{cases} > y_{i-1}, \text{ if } (y_i<H)\wedge(t_i\in(t_3,t^*)), \\ = y_{i-1}, \text{ if } (y_i=H)\wedge(t_i=t^*), \\ <y_{i-1}, \text{ if } (y_i<H)\wedge(t_i\in(t^*,t_4]). \end{cases}$
5	$i=k_5=k_4+1$	$T_5=t_5=t_4+\Delta t$	$N_i\approx N_0$	$y_i=0$
6	$i\in(k_5,k]$	$T_6\in(t_5,t_k]$	$N_i\approx N_0$	$y_i=0$

Figure 3. Jump phases diagrams

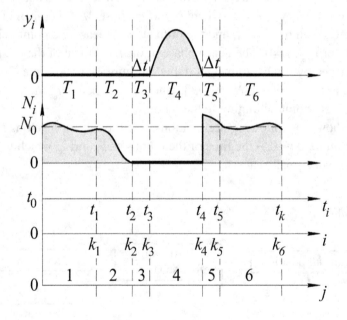

APPROACHES TO THE DESIGN OF THE JUMPING ROBOT

One of the urgent tasks of modern robotics is the creation of small-sized devices with high mobility and ability to move through difficult terrain with the presence of debris and obstacles. Such tasks can be successfully solved by jumping robots, additionally equipped with wheeled motors, that increases their maneuverability and speed by combining two modes of movement (Stoeter & Papanikolopoulos, 2006; Ackerman, 2012). Wheeled platforms are used to move over rugged terrain, and jumps are necessary to overcome obstacles that cannot be avoided in any other way. The fundamental task in the design of such combined monitoring platforms is a balanced and reasonable approach to the selection of their kinematic schemes.

Design Options of Wheeled Platforms

Due to the fact that the robot under investigation moves over rugged terrain, when implementing jumps, there is a high probability of acquiring an angular momentum, which will lead to rotation of the device when taking off the surface and in flight and to overturning when landing. Therefore, we will consider only the designs of wheeled platforms that can adequately function when changing the orientation of the robot in the jumps.

The main requirements for the wheeled platform are: small mass, ability to control all wheels, presence of damping properties, small number of drives (easy to operate), the robot body protection from hitting the surface, possibility of reverse movement. In general, all wheeled platforms can be divided into two types (figure 4).

Transformable platforms have several configurations, one of which allows the robot to move on wheels and the other (others) is used for the jump (figure 5 A). This is a four-wheeled platform (only two wheels are shown in the figure), two wheels of which are connected by a revolute joint to the body 1 of the device, and the other two − using the lever 4 (Vorochaeva, Malchikov, & Postol'niy, 2018). Separate drives are used to rotate the front and rear wheels, and an additional drive is required to rotate the lever. When rolling the platform on the surface, the lever is in a horizontal position, and at the moment of positioning before the jump, the robot

Figure 4. Wheeled platforms classification

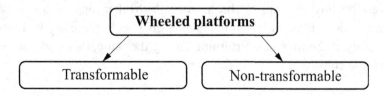

57

body together with the acceleration module 3 takes the position at the required angle to the horizon due to the convergence of the front and rear wheels when turning the lever. The disadvantages of such wheeled platform include the fact that in case device overturning of the jump will be made in the opposite direction compared to the previously implemented one.

The examples of non-transformable platforms are shown in figure 5 B-D. The first two of them have eight wheels 2 (only four wheels are shown in figure 5) (Jatsun, Volkova, & Vorochaev, 2013). The diameters of the wheels and the location of their axes in the body 1 are selected in such a way that the probability of hitting the body directly on the surface is very small. Each joint of wheels is driven by its own drive, so this wheeled platform requires four drives. The peculiarity of this platform type is that after landing the robot can interact with the surface with any two wheels belonging to one side of the body. This means that the next jump requires the acceleration module 3 turning relating to the body, as shown in figure 5 B, or the rotation of the body with the acceleration module relative to the surface with the help of additional support elements (levers) 4, as shown in figure 5 C. The second method of positioning will lead to the need of installing four supporting elements in the robot body, each of these elements will be equipped with its own drive. This will complicate the design of the robot and the drive control system, as well as increase the weight of the device. And when landing on the sides of the body AB and DE the next jump will be realized in the opposite direction to the previously implemented one.

The third construction is a four-wheeled platform (only two wheels are shown in figure 5 D) (Ackerman, 2012; Vorochaeva & Yatsun, 2015). The diameter of the wheels is such that the wheels "cover" the robot body in the height, so we obtain $2R>h$, where R − wheel radius, h − the body height, the number of drive wheels in this case is reduced to two. A feature of this construction is that after landing the robot can be on the surface only in two positions: on its basis, i.e. in the position in which it was before the jump (on the AE body side), or upside down (on the BD body side). Then the implementation of the next jump is possible in case of turning the acceleration module inside the robot body or by installing the body together with the acceleration module before the jump at the required angle to the horizon with the help of levers system or additional support elements, the same as for eight-wheeled platform.

The difference is that regardless of the landing option, the direction of the previous and the further jumps will be the same. With other constructions of wheeled platforms that do not "cover" the robot body with wheels, there is possibility to hit the robot body at the time of overturning during the jump, that's why we will not consider these constructions.

Figure 5. Wheeled platforms constructions: A) transformable platforms; B-D) non-transformable platforms

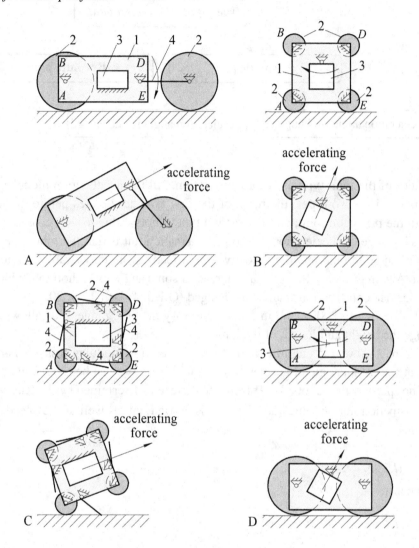

Design Options of Acceleration Modules

For making a reasonable choice of the acceleration module scheme of the robot, we will analyze their classifications according to a certain number of criteria: type of acceleration joint, possibility of changing the acceleration angle, the ability to rotate the acceleration module relative to the body, supporting surface with which contact at the acceleration moment is realized, contact type of acceleration module with the surface. Figure 6 shows the most general classification of acceleration

Figure 6. Classification of acceleration modules of the kinematic joint type

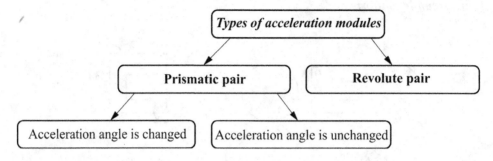

modules of the joint type used for acceleration. All the acceleration modules can be divided into two types: the robot of the first type accelerates before the jump due to the prismatic joint, and the second type robots – due to the revolute joint. The advantage of acceleration due to the prismatic joint is that the robot does not acquire an additional angular velocity, which leads to its undesirable rotation in flight. And in this case, the robot accelerates at some angle to the horizon, which in some device constructions can be unchanged (Gilani & Ben-Tzvi, 2011; Pradhan, 2009; Zhao et al., 2009), and in others may vary in some range, thus allowing to change the height and length of the jump.

Figure 7 shows the classification of the acceleration modules with a prismatic joint at a constant angle of acceleration of the device γ=const before the jump.

The speed of the robot takeoff from the surface is determined by the length l of the prismatic joint on which acceleration is carried out, as well as the speed $\dot{l}$ of its links:

$$\begin{cases} \upsilon = f(l, \dot{l},), \\ \gamma = \text{const}, \\ \dot{\gamma} = 0, \ \ddot{\gamma} = 0. \end{cases} \tag{2}$$

In such constructions the acceleration module is fixed in the robot body. This limits the use of such devices. Firstly, the implementation of a series of jumps is possible only if the robot always locates in the same position before the jump. This means that when landing, the robot must not roll over or be equipped with special mechanisms that allow it to take the initial position for the jump. Secondly, in such robots it is possible to control the height and length of the jump only by varying the acceleration force, but there is no possibility of adjusting the ratio between the height and length of the jump by changing the angle of acceleration of the device.

Figure 7. Classification of acceleration modules with a prismatic joint and a constant angle of acceleration

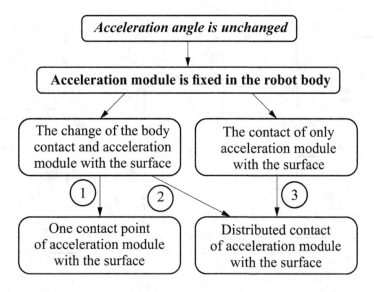

This significantly limits the application range of these devices from the point of view of the ability to overcome obstacles.

When considering the classification presented in figure 7, there are two options for jumping robots acceleration: with change of the body contact with the surface to the acceleration module contact and with the acceleration module contact with the surface. The first variant of acceleration is possible both with one point of contact of the acceleration module with the surface, in this case, the change of contact surfaces is necessary to prevent overturning of the device (figure 8 A), and in the distributed contact of the acceleration module with the surface (figure 8 B). Then in both cases, the acceleration module consists of two links of the prismatic joint 2, 3, but in the second one the link 3, interacting with the surface, is structurally made in the form of some contact platform. The second variant of acceleration is carried out only at the distributed contact of the acceleration module links with the surface (figure 8 C), the constructive scheme of the acceleration module remains the same, as in figure 8 B.

Let's consider the classification of acceleration modules with a prismatic joint, through which the jump is realized at different angles to the horizon (figure 9). Structurally, they can be installed in the robot body with the possibility of rotation relative to it, and without it. In the first case, the positioning before the jump occurs by the rotation of the acceleration module relative to the body (Salton, 2010; Volkova & Jatsun, 2013; Jatsun et al., 2014), and in the second one – due to the rotation of

Figure 8. Pictograms of acceleration modes of acceleration modules with a prismatic joint and a constant angle of acceleration

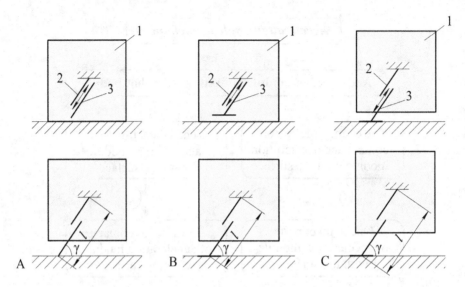

the body together with the acceleration module relative to the supporting surface (Ackerman, 2012; Vorochaeva & Yatsun, 2015; Volkova & Yatsun, 2013). Prismatic joint helps to realize the positioning of the acceleration module relative to the surface at an angle $\gamma \in [\gamma_{min}, \gamma_{max}]$, where γ_{min} and γ_{max} are the smallest and largest values of the rotation angle, and with the help of revolute joint takes place the acceleration with the further takeoff of the robot from the surface. The moment of takeoff can be characterized by the following formulas:

$$(\varphi_2 = \varphi_2^0) \wedge (\varphi_1 = \varphi_1^0) \wedge$$
$$y_{O1} = y_{K1} = 0 \tag{3}$$

For acceleration modules that rotate relative to the robot body, the acceleration methods correspond to those given in the previous classification (figure 7). The design schemes of such devices differ from those shown in figure 8 A-C by the introduction of an additional revolute joint between the body and one of the links of the acceleration module (figure 10 A), as well as by the complication of the acceleration module by introducing another link – foot (link 4) – with the possibility of turning relative to the second link of the prismatic joint (figure 10 B, C). In case when the acceleration module does not rotate relative to the body, the change in the acceleration angle can be achieved by rotating the acceleration module links

Figure 9. Classification of acceleration modules with prismatic joint and variable acceleration angle

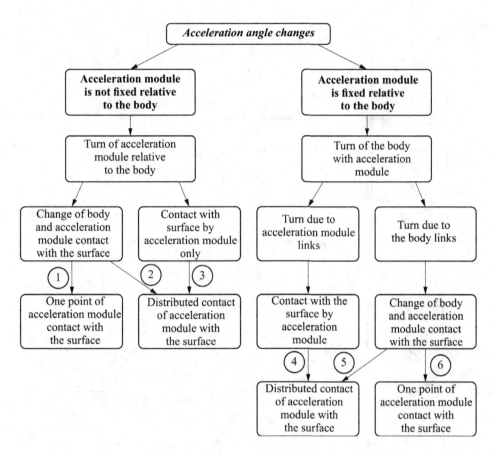

relative to the surface, as shown in figure 10 D, or by rotating the body with the acceleration module relative to the surface (figure 10 E, F).

The first of the considered acceleration options is possible for the design scheme of the acceleration module, in which the contact with the surface is carried out only by the links of the acceleration module, and this contact is distributed. In this case the jumping robot consists of a body 1, an acceleration module in the form of a prismatic joint 2, 3 and a foot 4, rotating relative to one of the prismatic joint links. The second variant of acceleration is possible when changing the contact surfaces of the body and of the acceleration module, the latter can have both a distributed and a point contact with the surface. In both cases the robot's body is equipped with special mechanisms 5, that help to its rotation relative to the surface, as shown in figure 10 E, F. The design of acceleration module at a point contact with the

Figure 10. Pictograms of acceleration modes of acceleration modules with prismatic joint and variable acceleration angle

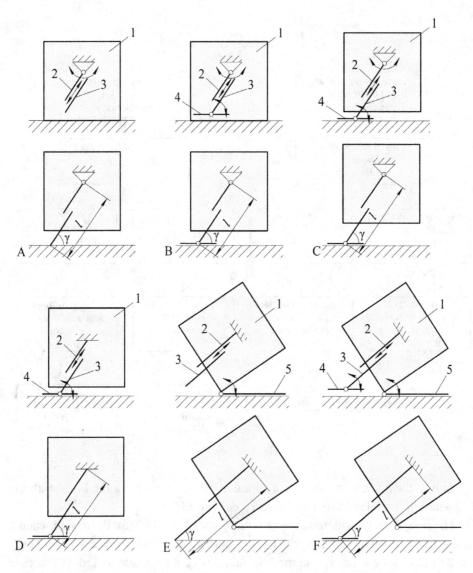

surface is the same as in figure 8 A, and at distributed contact it is supplemented with a rotary foot.

The last of the considered classifications of acceleration modules is shown in figure 11. Here are the acceleration modules with revolute joints. In this case the direction of the accelerating force depends on the relative angles between the links of the acceleration module and the corresponding angular velocities, which can be

Figure 11. Classification of acceleration modules with revolute joints

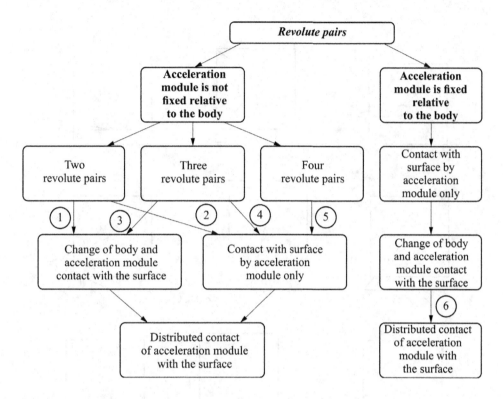

problematic to regulate. This means that jumping at exactly the right angle to the horizon is rather difficult.

As in the previous classification, the acceleration module can rotate relative to the body 1, and can be fixed. In the first case, the acceleration modules are divided by the number of revolute joints by which they are formed (2, 3 and 4 joints). For accelerating modules, consisting of two (Kovac, 2010; Poulakakis & Grizzle, 2009) and three (Vermeulen, 2004; Zeglin, 1991) revolute joints, acceleration is possible both at change of contact surfaces of the body and the module (figure 12 A, D), and at contact with a surface only by the acceleration module (figure 12 B, C, E), and in both cases this contact will be distributed. Structurally, the acceleration module with two revolute joints (figure 12 A, B) consists of a leg 2 and a foot 4, and with three revolute joints (figure 12 D, E) – of a leg formed by two links 2 and 3, and a foot 4. In the absence of the possibility of turning the acceleration module relative to the body, a scheme with two revolute joints is correct, where the robot leg consists of two links 2 and 3, and link 4 is a foot (figure 12 C).

Figure 12. Pictograms of acceleration modes of acceleration modules with revolute joints

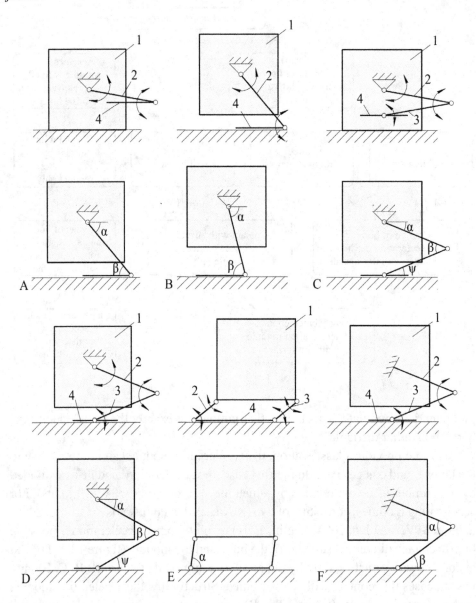

Then at the time of takeoff from the surface of the jumping robot with two revolute joints, the speed value υ and its direction γ are determined by a combination of relative angles α and β (and with three revolute joint – of angles α, β and Ψ). In addition, the implementation of the robot acceleration due to revolute joints leads to the fact that $\dot{\gamma} \neq 0$, $\ddot{\gamma} \neq 0$. Then the moment of takeoff is characterized by formulas

- for two revolute joints:

$$
\begin{cases}
\upsilon = f(\alpha, \dot{\alpha}, \beta, \dot{\beta}), \\
\gamma = g(\alpha, \dot{\alpha}, \beta, \dot{\beta}), \\
\dot{\gamma} \neq 0, \ \ddot{\gamma} \neq 0,
\end{cases} \tag{4}
$$

- for three revolute joints:

$$
\begin{cases}
\upsilon = f(\alpha, \dot{\alpha}, \beta, \dot{\beta}, \psi, \dot{\psi}), \\
\gamma = g(\alpha, \dot{\alpha}, \beta, \dot{\beta}, \psi, \dot{\psi}), \\
\dot{\gamma} \neq 0, \ \ddot{\gamma} \neq 0.
\end{cases} \tag{5}
$$

The acceleration module with four revolute joints is a pantograph, whose links 2 and 3 form a leg, and link 4 is a foot. Acceleration using such a module is carried out only with its distributed contact with the surface (figure 12 F). The relative angles between the links are equal α. The moment of takeoff of such robot is described by formulas:

$$
\begin{cases}
\upsilon = f(\alpha, \dot{\alpha}), \\
\gamma = g(\alpha, \dot{\alpha}), \\
\dot{\gamma} \neq 0, \ \ddot{\gamma} \neq 0.
\end{cases} \tag{6}
$$

The given review of acceleration modules and acceleration modes allows to make a reasonable choice of its design scheme, which will be used in the wheeled jumping robot.

Choice of the Kinematic Scheme for the Wheeled Jumping Robot

The design of the robot kinematic scheme includes the choice of wheeled platform and acceleration module scheme of the acceleration module and its acceleration mode.

First, let us focus on the choice of wheeled platform. Based on a comparison of the structural schemes of wheeled platforms, we can conclude that the scheme shown in figure 5 D is the most rational. This platform has less wheels and drives, that means less weight and a simpler control system. In addition, with such wheeled platform there are two possible options for positioning the robot before the jump: by rotating the acceleration module inside the body and by rotating the body together with the acceleration module relative to the surface, this allows us to consider a wider range of options for the jump. Therefore, in the designed robot we will use a wheeled platform with described design scheme.

Now let us move on to the choice of acceleration module scheme. To do this, we formulate some requirements:

- ability to rotate relative to the body in order to make the next jump after overturning the device,
- during acceleration there need to be a change of the contact surfaces of the body and the acceleration module to provide positioning before the jump at the location of the module inside the body,
- the contact of acceleration module links with the surface must be distributed to provide more exact positioning before the jump and to prevent unwanted rotation of the object in flight,
- the angle of acceleration must be a controlled value to provide overcoming of obstacles of various sizes and shapes.

On the basis of these requirements, as the acceleration module and the corresponding acceleration mode we choose the scheme shown in figure 10 B. Further the scheme of the wheeled jumping robot will have the form shown in figure 13 A, and its design is shown in figure 13 B.

The acceleration module is formed by three links: a foot 1 and links 2 and 3, forming a leg. Links 1 and 2, 3 and 4 (body) are a revolute joints, that are used for positioning robot, links 2 and 3 – a prismatic joint, due to which the acceleration of the device takes place. Wheels 5, 6 are installed in the body with the use of revolute and prismatic joints. Revolute joints provide the robot rolling as a wheeled

Figure 13. Scheme (A) and construction (B) of wheeled jumping robot

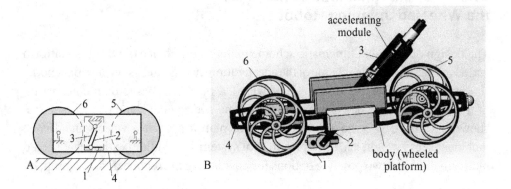

platform over flat terrain, and prismatic joints allow to regulate the clearance of the wheeled platform depending on the height of the obstacles to be overcome. In addition, due to these two joints, it is possible to control the trajectory of the robot and its orientation in flight.

A MATHEMATICAL MODEL OF THE WHEELED JUMPING ROBOT

On the basis of the studied robot construction we will move to the design scheme of the device and will present it in the form of the kinematic (figure 14) and power (figure 15) schemes.

We will consider the robot jump in such plane Oxy where the axis Ox is horizontal and the axis Oy is directed opposite to the action of the gravitational field. Let us assume that the links of the acceleration module $i=1\text{-}3$ are absolutely solid rods

Figure 14. Kinematic scheme of wheeled jumping robot

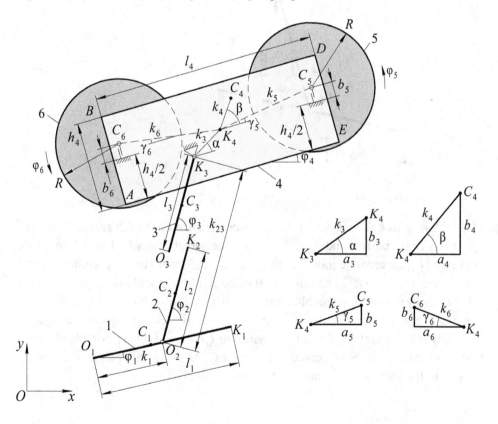

Figure 15. Force diagram of wheeled jumping robot

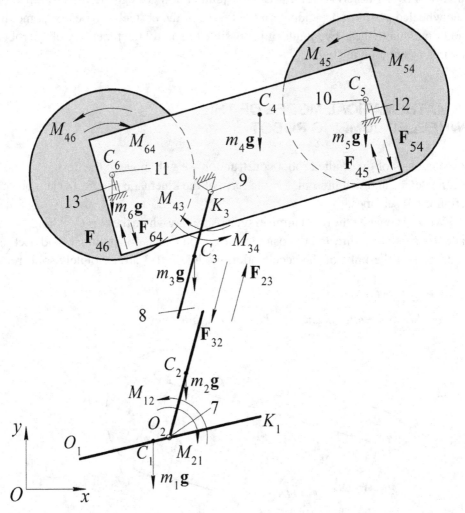

O_iK_i with lengths l_i and masses m_i, the mass centers mass of which are the centers of symmetry of the corresponding links and are located at points C_i. Links 1 and 2 are connected by the revolute joint 7 at the point O_2, the length of the prismatic joint 8 links 2 and 3 is equal k_{23}. The link 3 of the acceleration module is connected to the robot body by means of the revolute joint 9 at the point K_3, which is remote from the symmetry center of the body – of the point K_4 – at the distance k_3 – at the angle α. The body is a rectangle *ABDE* with dimensions $l_4 \times h_4$ and is also a completely solid body. Its center of mass is located at the point C_4, at the distance k_4, at the angle β relative to the point K_4, the mass of the body is equal to m_4.

At the points C_5 and C_6 of the robot body there are revolute joints 10 and 11, as well as prismatic joints 12 and 13, by means of which the wheels 5 and 6 of the radius R are installed. These points are the mass centers of the wheels m_i, $i=5,6$, and they are removed at distances k_5 and k_6 at angles γ_5 and γ_6 in opposite directions from the point K_4 of the body. The angles α, β, γ_5 and γ_6 are counted counterclockwise from the side AE of the robot body. To facilitate further descriptions of the coordinates of the points K_4, C_4, C_5, C_6 we will carry out the transition from the polar coordinate systems to Cartesian. The corresponding formulas have the form:

$$a_3 = k_3 \cos\alpha, \quad b_3 = k_3 \sin\alpha, \tag{7}$$

$$a_4 = k_4 \cos\beta, \quad b_4 = k_4 \sin\beta, \tag{8}$$

$$a_5 = k_5 \cos\gamma_5, \quad b_5 = k_5 \sin\gamma_5, \tag{9}$$

$$a_6 = k_6 \cos\gamma_6, \quad b_6 = k_6 \sin\gamma_6. \tag{10}$$

The position of the links $i=1$-4 is determined by the coordinates of their mass centers x_{Ci}, y_{Ci} and the angles φ_i of rotation relative to the axis Ox counterclockwise. For the body this angle is the angle between the side AE and the axis Ox. The position of the links $i=5,6$ is given by the coordinates of their mass centers x_{Ci}, y_{Ci} and the angles of rotation of the wheels φ_i. Due to the fact that links 2 and 3 form a prismatic joint, we assume that $\varphi_2=\varphi_3$.

As a base point for the construction of the kinematic chain of robot links, we will use the point O_1, its radius vector is equal to

$$\mathbf{r}_{O1} = [x_{O1}, \quad y_{O1}]^{\mathrm{T}}, \tag{11}$$

where x_{O1}, y_{O1} – the coordinates of the point projected on the coordinate axis.

Radius-vectors of the points O_2 and O_3 we will write in the following way (Buss, 2004; Featherstone, 2014):

$$\mathbf{r}_{O2} = \mathbf{r}_{O1} + T_1 \rho_{O1O2}^{(1)}, \tag{12}$$

$$\mathbf{r}_{O3} = \mathbf{r}_{O1} + T_1 \rho_{O1O2}^{(1)} + T_2 \rho_{O2O3}^{(2)}, \tag{13}$$

where T_i – rotation matrix, ρ – relative radius-vectors in the corresponding coordinate systems, the number of which is shown in the upper index in brackets:

$$T_i = \begin{bmatrix} \cos\varphi_i & -\sin\varphi_i \\ \sin\varphi_i & \cos\varphi_i \end{bmatrix}, \tag{14}$$

$$\rho_{O1O2}^{(1)} = [k_1, \quad 0]^T, \quad \rho_{O2O3}^{(2)} = [k_{23} - l_3, \quad 0]^T. \tag{15}$$

Radius-vectors of the points K_i, i=1-4, can be written in the following way:

$$\mathbf{r}_{K1} = \mathbf{r}_{O1} + T_1\rho_{O1K1}^{(1)}, \tag{16}$$

$$\mathbf{r}_{K2} = \mathbf{r}_{O1} + T_1\rho_{O1O2}^{(1)} + T_2\rho_{O2K2}^{(2)}, \tag{17}$$

$$\mathbf{r}_{K3} = \mathbf{r}_{O1} + T_1\rho_{O1O2}^{(1)} + T_2\rho_{O2K3}^{(2)}, \tag{18}$$

$$\mathbf{r}_{K4} = \mathbf{r}_{O1} + T_1\rho_{O1O2}^{(1)} + T_2\rho_{O2K3}^{(2)} + T_4\rho_{K3K4}^{(4)}, \tag{19}$$

where

$$\rho_{O1K1}^{(1)} = [l_1, \quad 0]^T, \quad \rho_{O2K2}^{(2)} = [l_2, \quad 0]^T, \tag{20}$$

$$\rho_{O2K3}^{(2)} = [k_{23}, \quad 0]^T, \quad \rho_{K3K4}^{(4)} = [a_3, \quad b_3]^T. \tag{21}$$

Radius-vectors of the mass centers of links are defined by the following formulas:

$$\mathbf{r}_{C1} = \mathbf{r}_{O1} + T_1\rho_{O1C1}^{(1)}, \tag{22}$$

$$\mathbf{r}_{C2} = \mathbf{r}_{O1} + T_1\rho_{O1O2}^{(1)} + T_2\rho_{O2C2}^{(2)}, \tag{23}$$

$$\mathbf{r}_{C3} = \mathbf{r}_{O1} + T_1\rho_{O1O2}^{(1)} + T_2\rho_{O2C3}^{(2)}, \tag{24}$$

$$\mathbf{r}_{Ci} = \mathbf{r}_{O1} + T_1\rho_{O1O2}^{(1)} + T_2\rho_{O2K3}^{(2)} + T_4\rho_{K3K4}^{(4)} + T_4\rho_{K4Ci}^{(4)}, \quad i\text{=4-6}, \tag{25}$$

where

$$\rho_{O1C1}^{(1)} = [l_1 / 2, \quad 0]^T, \quad \rho_{O2C2}^{(2)} = [l_2 / 2, \quad 0]^T, \tag{26}$$

$$\rho_{O2C3}^{(2)} = [k_{23} - l_3 / 2, \quad 0]^T, \quad \rho_{K4!4}^{(4)} = [a_4, \quad b_4]^T, \tag{27}$$

$$\rho_{K4!5}^{(4)} = [a_5, \quad b_5]^T, \quad \rho_{K4!6}^{(4)} = [-a_6, \quad b_6]^T. \tag{28}$$

The generalized coordinates of the jumping robot are position projections of the mass center of link 4 on the coordinate axes, the angles of inclination of links 1, 2 and 4 to the horizontal line, as well as the angles of rotation of the wheels (links 5 and 6) and the lengths of the prismatic joints. The vector of generalized coordinates describing the robot movement has the form

$$\mathbf{q} = [x_{C4}, \quad y_{C4}, \quad \varphi_1, \quad \varphi_2, \quad \varphi_4, \quad k_{23}, \quad \varphi_5, \quad \varphi_6, \quad b_5, \quad b_6]^T. \tag{29}$$

Let us consider the power scheme of the device (figure 15). The design of the robot provides the presence of seven drives. Four of them are installed in the revolute joints – points O_2, K_3, C_5, C_6 – and they generate joints of moments: M_{12} and M_{21} between the links 1 and 2, M_{34} and M_{43} between the links 3 and 4, M_{45} and M_{54} between links 4 and 5, M_{46} and M_{64} between the links 4 and 6. Moments between links 1 and 2, 3 and 4 provide positioning of the device by changing the relative positions of these links. Moments between links 4 and 5, 4 and 6 are necessary for rolling the wheeled platform on the surface, as well as to control the orientation of the robot during flight. The fifth drive is the drive of prismatic motion and forms a joint of forces F_{23} and F_{32} between links 2 and 3, providing acceleration of the device for taking off the surface. The sixth and seventh drives create forces F_{45} and F_{54}, F_{46} and F_{64}, allowing to change the clearance of the wheeled platform and control the trajectory of the robot in flight. In the mass centers of all links there are gravity forces $m_i g$.

The robot jump is considered to be taking place on a horizontal surface. When the robot interacts with the surface with its wheels, the contact is made at the points of the wheels K_5 and K_6 the radius-vectors of which are the following

$$\mathbf{r}_{Ki} = \mathbf{r}_{Ci} + \rho_{CiKi}, \quad \rho_{CiKi} = [0, -R]^T, \, i=5,6. \tag{30}$$

In the case when the foot is located on the surface, the interaction with the latter takes place at two extreme points O_1 and K_1 of link 1. The model of contact interaction with the surface is determined by the properties of the latter (absolutely solid, deformable, rough, smooth, etc.) and is not considered in this chapter.

Consider the general principles of developing a mathematical model of a jumping robot movement. The system of differential equations of motion of the device is written using the Lagrange equation of the second type (Buss, 2004; Featherstone, 2014)

$$\frac{d}{dt}\left(\frac{\partial T}{\partial \dot{q}_n}\right) - \frac{\partial T}{\partial q_n} = Q_n, \tag{31}$$

where T – kinetic energy of the system, q_n, n=1-10 – generalized coordinate, Q_n – generalized force in the coordinate q_n.

The kinetic energy of the system is determined by the formula

$$T = \sum_{i=1}^{6} T_i, \tag{32}$$

where i=1-6 – device links.

The kinetic energies of the links, each of which performs a flat motion:

$$T_i = m_i \frac{\dot{x}_{li}^2 + \dot{y}_{li}^2}{2} + \frac{J_i \dot{\varphi}_i^2}{2}, \tag{33}$$

where $J_{i=1\div3} = \frac{m_i l_i^2}{12}$, $J_4 = m_4\rho$, $J_{i=5,6} = \frac{m_i R_i^2}{2}$ – central moments of the links inertia, $\dot{x}_{li}$, $\dot{y}_{li}$ – the projection of the velocities of the links mass centers on the axis of absolute coordinate system, ρ – the distance between the mass center of link 4 and the most remote extreme point of the body.

The system of differential equations of motion of a wheeled jumping robot can be represented in matrix form:

$$\mathbf{A}\ddot{\mathbf{q}} + \mathbf{BC}\dot{\mathbf{q}} + \mathbf{D}\dot{\varphi}_2\dot{k}_{23} + \mathbf{H}\dot{\varphi}_4\dot{b}_5 + \mathbf{K}\dot{\varphi}_4\dot{b}_6 = \mathbf{F}, \tag{34}$$

where $\mathbf{A}$ – matrix of coefficients of the generalized accelerations, $\mathbf{B}$ – matrix of coefficients of the generalized velocities squared, $\mathbf{C}$ – diagonal matrix of the generalized velocities, $\mathbf{D}, \mathbf{H}, \mathbf{K}$ – the matrix of the coefficients at $\dot{\varphi}_2\dot{k}_{23}$, $\dot{\varphi}_4\dot{b}_5$, $\dot{\varphi}_4\dot{b}_6$, $\mathbf{F}$ – vector of generalized forces, $\mathbf{q}$ – vector of generalized coordinates, defined by formula (29).

THE JUMPING LOCOMOTION STAGES

We will consider the jump of the robot consisting of the following phases: positioning before the jump, acceleration, takeoff, flight, landing, positioning after landing, as shown in figure 2. Each of these phases can be formed by a different number of stages, and the phases themselves can be carried out sequentially or sequential-parallel. Let us consider various variants for the robot to jump, provided that at the time of its beginning the device body is on the surface, interacting with the last wheels, and the acceleration module is installed inside the body, the robot lands on the wheels. We will investigate the stage-by-stage variants of the jump, when each of the stages involves the imposition on the system of links due to its contact with the surface, as well as the lack of relative movements of the links. Figure 16 shows the possible variants for the implementation of the jump, according to which there are two basic variants: in the first case, all phases of the jump are carried out sequentially, and in the second one – positioning and acceleration are combined with each other. In sequential mode of the phases implementation, positioning can be achieved by specifying the desired position of the acceleration module both when it is inside the body and when the acceleration module is in contact with the support surface. The combination of the positioning and acceleration phases is possible when the acceleration module is on the surface and the stages are performed in sequential-parallel mode. Under the sequential implementation of the stages we understand the process when at this stage only one movement is carried out on the generalized coordinate. Sequential-parallel implementation of stages – a combination of stages at which the number of generalized coordinates exceeds one, with stages, during which the robot motion is described by one generalized coordinate. Figures 17 and 18 show the robot position pictograms during each of the considered stages of motion, and table 2 shows the generalized coordinates at each stage and the conditions for the completion of the stages.

Positioning and Acceleration of the Robot

Further we will consider each of the six proposed variants for positioning and acceleration, and the schemes of the stages of positioning and acceleration of the wheel are not shown for better clarity. Let the robot interact with the surface with the help of two wheels at the initial moment of the jump $y_{K5}=y_{K6}=0$, the acceleration module is inside the body, the prismatic joint has a minimum length $k_{23} = k_{23}^{\min}$. The coordinates of the mass center of the body $x_{C4} = x_{C4}^0$, $y_{C4} = y_{C4}^0$, $\varphi_4=0$. Links of the acceleration module are located at angles $\varphi_1=0$, $\varphi_2=\pi/2$. We will consider the stages of the jump, provided that the link of the robot interacting with the surface is stationary.

Figure 16. Variants of the jump implementation

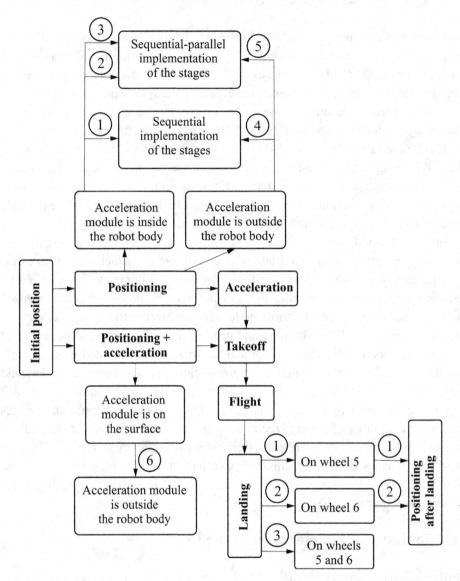

The first variant of positioning and acceleration is shown in figure 17 A. During the first stage of movement, the links of the acceleration module, rigidly connected to each other $\varphi_{21}=\varphi_2-\varphi_1=0$, $k_{23}=k_{23}^{\min}$, rotate relative to the body under the action of the moment M_{43} until the condition $\varphi_2=\varphi_2^0$ is fulfilled, where φ_2^0 – some angle of inclination of the links 2 and 3 to the horizon, under which the jump will be carried out. The robot contacts with the surface by the wheels $y_{K5}=y_{K6}=0$.

Table 2. Generalized coordinates and conditions for completion of the jump stages

Phase		Generalized coordinate	Conditions for phase or stage completion
Positioning and acceleration			
Variant 1	Positioning	φ_2	$\varphi_2 = \varphi_2^0$
		φ_1	$\varphi_1 = \varphi_1^0$
		k_{23}	$y_{O1} = y_{K1} = 0$
	Acceleration	k_{23}	$k_{23} = k_{23}^{\max}$
Variant 2	Positioning	φ_2, φ_1	$(\varphi_2 = \varphi_2^0) \wedge (\varphi_1 = \varphi_1^0)$
		k_{23}	$y_{O1} = y_{K1} = 0$
	Acceleration	k_{23}	$k_{23} = k_{23}^{\max}$
Variant 3	Positioning	$\varphi_2, \varphi_1, k_{23}$	$(\varphi_2 = \varphi_2^0) \wedge (\varphi_1 = \varphi_1^0) \wedge (y_{O1} = y_{K1} = 0)$
	Acceleration	k_{23}	$k_{23} = k_{23}^{\max}$
Variant 4	Positioning	k_{23}	$(y_{O1} = y_{K1} = 0) \wedge (k_{23} = k_{23}^{*})$
		φ_2	$\varphi_2 = \varphi_2^0$
		φ_4	$\varphi_4 = \varphi_4^0$
	Acceleration	k_{23}	$k_{23} = k_{23}^{\max}$
Variant 5	Positioning	k_{23}	$(y_{O1} = y_{K1} = 0) \wedge (k_{23} = k_{23}^{*})$
		φ_2, φ_4	$(\varphi_2 = \varphi_2^0) \wedge (\varphi_4 = \varphi_4^0)$
	Acceleration	k_{23}	$k_{23} = k_{23}^{\max}$
Variant 6	Positioning	k_{23}	$(y_{O1} = y_{K1} = 0) \wedge (k_{23} = k_{23}^{*})$
	Positioning + acceleration	$\varphi_2, \varphi_4, k_{23}$	$(\varphi_2 = \varphi_2^0) \wedge (\varphi_4 = \varphi_4^0) \wedge (k_{23} = k_{23}^{\max})$
Takeoff		υ_C, φ_C	$y_{O1} = y_{K1} > 0$
Flight		$x_{C4}, y_{C4}, \varphi_4, k_{23}, \varphi_5, \varphi_6, b_5, b_6$	$(y_{K5} = 0) \vee (y_{K6} = 0) \vee (y_{K5} = 0 \wedge y_{K6} = 0)$
Landing	Variant 1	φ_4	$(y_{K5} = 0) \wedge (y_{K6} \neq 0)$
	Variant 2	φ_4	$(y_{K5} \neq 0) \wedge (y_{K6} = 0)$
	Variant 3	–	$(y_{K5} = 0) \wedge (y_{K6} = 0)$
Positioning after landing	Variant 1	φ_4	$(y_{K5} = 0) \wedge (y_{K6} = 0)$
	Variant 2	φ_4	$(y_{K5} = 0) \wedge (y_{K6} = 0)$

Figure 17. Robot pictograms at the positioning and acceleration stages: A-F – variants 1-6

Figure 18. Robot pictograms at the time of takeoff (A), flight (B), landing (C-E) and positioning after landing (F, G) phases: C, F, - variant 1, D, G – variant 2, E – variant 3

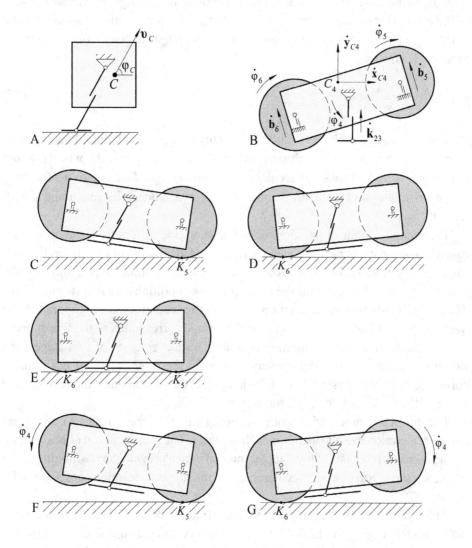

The second stage is as follows. Link 1 of the acceleration module rotates relative to link 2 under the action of the moment M_{21} until it becomes parallel to the surface, which corresponds to the condition $\varphi_1 = \varphi_1^0 = 0$ when jumping to the right and $\varphi_1 = \varphi_1^0 = \pi$ when jumping to the left. Links 2 and 3 of the acceleration module are fixed to each other and relative to the body: $k_{23} = k_{23}^{\min}$, $\varphi_2 = \varphi_2^0$. The robot interacts with the surface by the wheels $y_{K5} = y_{K6} = 0$.

In the third stage, links 1 and 2, stationary relative to each other $\varphi_1 = \varphi_1^0$, $\varphi_2 = \varphi_2^0$ move due to the force F_{32}, appearing in the prismatic joint, until link 1 is in contact with the surface: $y_{O1} = y_{K1} = 0$. I.e. at this stage there is a change of reference points of the robot. During the entire phase the robot is in contact with the surface by the wheels $y_{K5} = y_{K6} = 0$. At the time of completion of the stage, the length of the prismatic joint k_{23}^* is equal to

$$k_{23}^* = y_{C4}^0 / \sin \varphi_2^0 . \tag{35}$$

The fourth stage: link 1 is located on the surface $y_{O1} = y_{K1} = 0$, links 1 and 2, 3 and 4 are rigidly connected to each other: $\varphi_1 = \varphi_1^0$, $\varphi_2 = \varphi_2^0$, $\varphi_4 = 0$, the wheels do not rotate: $\dot{\varphi}_5 = \dot{\varphi}_6 = 0$. Links 3-6 accelerate relative to the surface due to the force F_{23} in the prismatic joint until the length of this joint reaches its maximum length $k_{23} = k_{23}^{\max}$.

The second variant of positioning and acceleration differs from the previously described one in its first stage that is a parallel implementation of the first and second stages of positioning, i.e. there is a simultaneous rotation of links 2 and 3 of the prismatic joint and link 1 until the corresponding conditions are met (figure 17 B). Then consistently two stages are implemented, the second stage corresponds to the previously described third one, and the third stage corresponds to the fourth one.

The third variant of positioning and acceleration consists of two stages, the first one represents a parallel implementation of the three stages of positioning, corresponding to the first variant of the jump, and the second one – the acceleration phase of the same variant of the jump (figure 17 C).

The following three variants of positioning and acceleration stages correspond to the acceleration module location on the surface. The fourth variant of positioning and acceleration consists of four stages, three of which provide the positioning of the robot, and the fourth one – its acceleration, all the stages are carried out sequentially (figure 17 D).

At the first stage of the jump the contact of the robot with the surface is carried out by wheels $y_{K5} = y_{K6} = 0$. Links 1 and 2, stationary connected to each other ($\varphi_1 = 0$, $\varphi_2 = \pi/2$), due to the force F_{32} are pushed out of the robot body until the contact with the surface is carried out by two points of link 1: $y_{O1} = y_{K1} = 0$, and the length of the prismatic joint exceeds the value (35) by some value k_{23}^{**}, sufficient for further positioning of the device:

$$k_{23}^{***} = y_{C4}^0 / \sin \varphi_2 + k_{23}^{**} . \tag{36}$$

During the second stage of the jump, link 1 of the robot is located on the surface $y_{O1}=y_{K1}=0$, and links 2 and 3 of prismatic joint, fixed relative to each other ($k_{23} = k_{23}^{***}$, $\varphi_{42}=\varphi_4-\varphi_2=\pi/2$), due to the moment M_{12}, rotate together with the body 4 until the link 2 reaches the required angle of inclination $\varphi_2 = \varphi_2^0$. During the third stage, link 1 ($y_{O1}=y_{K1}=0$) is located on the surface, links 1 and 2, 2 and 3 are fixed relative to each other ($\varphi_1=0$, $\varphi_2 = \varphi_2^0$, $k_{23} = k_{23}^{***}$). The device body rotates due to the moment M_{34} until its side AE is parallel to the surface, i.e. until the condition $\varphi_4 = \varphi_4^0 = 0$ is met. The fourth stage corresponds to the implementation of the robot acceleration and is as follows. Links 2, 3 and 4 accelerate relative to link 1 located on the surface ($y_{O1}=y_{K1}=0$) due to the force F_{23} of the prismatic joint to achieve its maximum length $k_{23} = k_{23}^{max}$. Links 1 and 2, 3 and 4 are interconnected: $\varphi_1=0$, $\varphi_2 = \varphi_2^0$, $\varphi_4 = \varphi_4^0$.

The fifth variant of positioning and acceleration (figure 17 E) differs from the fourth one in consisting of three stages, the first of which remains unchanged, the second one – represents the parallel implementation of the second and third stages of the fourth variant, and the third one – corresponds to acceleration.

The sixth variant of positioning and acceleration corresponds to the case when these two phases are combined (figure 17 F). The number of stages of movement is reduced to two. The first stage is the same as in the fourth variant of the jump, during which there is the pushing up of the acceleration module from the body, this stage corresponds to the positioning. The second stage – the combination of positioning and acceleration – includes three jumps subsequent for the fourth variant, i.e. there is a simultaneous positioning of the robot by turning the links of the prismatic joint and the body to achieve the required values of angles and acceleration of the robot by the prismatic joint elongation to its maximum length.

Other Phases of the Jump

The takeoff phase is an instantaneous transition from the acceleration phase to the flight phase (figure 18 A). This moment is characterized by the separation of the device link 1 from the surface ($y_{O1}=y_{K1}>0$) and the acquisition by the robot the initial flight speed υ_C applied to the center of mass of the device and directed at an angle φ_C to the horizon. The model of the robot hit on the surface arising from the takeoff is determined by the properties of the latter and is not considered in this chapter.

During the flight phase (figure 18 B) the robot moves in the direction separating from the surface, links 3 and 4 make a single body: φ_{42}=const, links 1 and 2, being also interconnected φ_{21}=const, are drawn into the robot body by the force F_{32} until the prismatic joint has its minimum length $k_{23} = k_{23}^{min}$. To control the device in flight, the following relative movements of the links can occur: rotation of the wheels with angular velocities $\dot{\varphi}_5$ and $\dot{\varphi}_6$ allows to change the orientation of the robot body, and the movement of the mass centers of the wheels relative to the mass center of the body due to the coordinates b_5 and b_6 provides a change in the flight trajectory. The flight phase ends when only one or both wheels of the robot start to contact the surface: $(y_{K5}=0) \vee (y_{K6}=0) \vee (y_{K5}=0 \wedge y_{K6}=0)$.

The landing phase has three different variants: landing on one of the wheels 5 or 6 (variants 1 and 2 – figure 18 C, D) or landing on two wheels at the same time (variant 3 – figure 18 E). At the time of the hit interaction with the surface all robot links are fixed relative to each other, i.e. the object is imposed with connections: φ_{21}=const, φ_{42}=const, $k_{23} = k_{23}^{min}$, $\dot{\varphi}_5 = \dot{\varphi}_6 = 0$. When landing on two wheels, the robot contacts the surface at two points K_5 and K_6, i.e., the condition for the completion of the landing stage is zeroing of the ordinates of the given points $(y_{K5}=0 \wedge y_{K6}=0)$. And at landing on one of the wheels, the coordinate of only one contact point K_5 or K_6 resets to zero. The landing phase is characterized by the acquisition of the angular velocity $\dot{\varphi}_4^*$ of rotation of the robot, determined by the hit theory, taking into account the properties of the support surface.

The positioning after landing phase is observed only if the robot lands on one of the wheels (landing variants 1 and 2 – figure 18 F, G). Then, during this phase, the robot turns, rolling the wheel on the surface, until the second wheel starts to contact the surface, i.e. the condition $(y_{K5}=0 \wedge y_{K6}=0)$ will not be fulfilled. When this phase is complete, the robot will be on the surface in the same position as the jump started, or it will be flipped by 180^0 relative to its initial position and ready for the next jump.

CONTROL SYSTEM

Further we will consider the structure of the control system of a wheeled jumping robot (figure 19). In this scheme there are the following main components that can be identified.

Operator's post – technical facilities for setting the strategy of the robot functioning, can be designed in the form of a remote or wired remote control. Additionally, it can be equipped with remote video monitoring installed on the robot body. The operator makes a decision to start a particular mode, to start the offline mode, etc.

Decision-making unit – the main core of the control system, implemented on the basis of a high-speed on-board microcomputer. The unit is designed for collecting and processing data from the local and global navigation system and from the operator's remote control and for set actions generation for electric drives and other equipment of the robot. In this case, the relative angles between the links of the jumping robot act as the set actions.

Figure 19. Structural diagram of a control system of a wheeled jumping robot

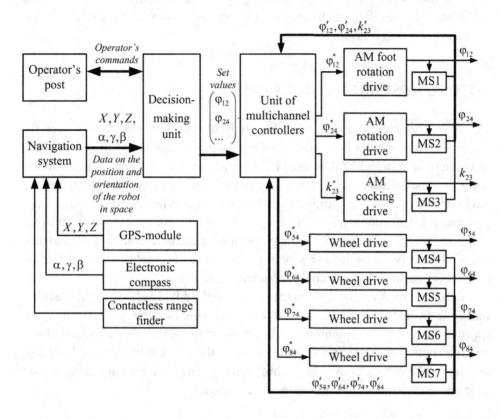

Unit of multichannel controllers is a set of subsystems that implement the set of actions by electric drives. It can include both motion sensors and force-torque circuits. Software solutions for controllers can be implemented both on the basis of the on-board computer described before, and in the form of separate independent modules.

Navigation system – a set of specialized sensors to determine the absolute angles of orientation of the body (α, γ, β) and the absolute coordinates of the robot (X, Y, Z). There can also be installed non-contact obstacle sensors based on ultrasonic or infrared sensor.

Measuring system (MS) – a contact switch that indicates the time of reaching a certain position by the working element of electric drive, as in the case of cocking actuator of the AM (acceleration module). More complex case is, for example, with turn drives of the AM or turn drives of a foot. Here can be used both a rotary encoder and current sensor required to determine the torque on the working body of the device. Incremental sensors – digital encoders are used to assess the rotation angle of the wheels. Data from measuring systems of the drives is received at a multichannel controller, which generates the required voltage for the electric drives, by means of which it drives the links.

FUTURE RESEARCH DIRECTIONS

The follow-up work includes a detailed study of each jumping motion stage for the robot. This includes development of algorithms for numeric simulation of the jumping motion, which will be used in the development of specialized simulation software. The software needs to allow varying the parameters of the robot and the environment (including the properties of the supporting surface and the aerodynamics parameters). This should allow the comparative analysis of the robot motion in the discussed here six variants of pre-jump positioning and acceleration. It should also allow to identify the optimal motion types according to a set of criteria (such as positioning accuracy, speed, energy efficiency, etc.).

A separate work can be done in the direction of studying of control of the robot during the flight stage of the jumping motion. This includes the control of the trajectories of individual points using the repositioning of the robot's center of mass, and the control of the robot's orientation using the rotation of the robot's wheels relative to the robot's body. This is important for implementing jumping motion over uneven terrain, and for jumping over obstacles.

CONCLUSION

This chapter focuses on the current topic – the main issues of design and movement of the wheeled jumping robot. It describes the basic principles of constructive and design schemes development of the device, determining the technique of the jump and the allocation of its phases and stages, as well as the design of the control system that provides the implementation of various variants for the jump. The chapter introduces the basic concepts of jumping motion, defines the most common and frequently used terms in the description of jumping, describes the characteristics of the jump, and presents its phases.

The chapter also considers an approach to the selection of a wheeled jumping robot scheme designed. This system is a combined platform consisting of a jump mechanism (acceleration module) and a wheeled module. For moving over the rugged terrain, the platform uses a wheeled module, and for overcoming obstacles – acceleration module, which allows to increase its maneuverability, travel capability and movement speed. To select the kinematic scheme of the robot there were developed classifications of possible variants of wheeled platforms and acceleration modules, as well as ways to implement their acceleration on the basis of the requirements imposed on them, their analysis was carried out, advantages and disadvantages were revealed.

For the robot under study the constructive and design schemes are developed, the generalized coordinates which describe the device movement are determined. There is also a detailed analysis of various jump implementations, due to its scheme, and varying in combining the relative movements of the parts during certain stages of the movement. To describe the movements of jumping robot more precisely the following phases of the jump are distinguished: positioning, acceleration, takeoff, flight, landing, positioning after landing. Each of these phases includes from one to several stages, and both phases and stages of the jump can be described by one or more generalized coordinates.

For the implementation of each of the proposed variants of the jump, the system of the device automatic control is designed, the setting and control actions, controlled values are determined, the type of controller and algorithms for its functioning providing the required laws of the links movement are selected.

ACKNOWLEDGMENT

The study is supported by RFBR project N° 18-31-00075.

REFERENCES

Ackerman, E. (2012). Boston dynamics sand flea robot demonstrates astonishing jumping skills. *IEEE Spectrum Robotics Blog, 2*(1).

Buss, S. R. (2004). Introduction to inverse kinematics with jacobian transpose, pseudoinverse and damped least squares methods. *IEEE Journal of Robotics and Automation, 17*(1-19), 16.

Calinon, S., Guenter, F., & Billard, A. (2007). On learning, representing, and generalizing a task in a humanoid robot. *IEEE Transactions on Systems, Man, and Cybernetics. Part B, Cybernetics, 37*(2), 286–298. doi:10.1109/TSMCB.2006.886952 PMID:17416157

Featherstone, R. (2014). *Rigid body dynamics algorithms*. Springer.

Gilani, O., & Ben-Tzvi, P. (2011, January). Bioinspired Jumping Mobility Concepts for Rough Terrain Mobile Robots. In *ASME 2011 International Mechanical Engineering Congress and Exposition* (pp. 207-214). American Society of Mechanical Engineers. 10.1115/IMECE2011-64050

Jatsun, S., Loktionova, O., Volkova, L., & Yatsun, A. (2014). Investigation into the influence of the foot attachment point in the body on the four-link robot jump characteristics. In *Advances on Theory and Practice of Robots and Manipulators* (pp. 159–166). Cham: Springer. doi:10.1007/978-3-319-07058-2_18

Jatsun, S. F., Volkova, L. Yu., & Vorochaev, A. V. (2013). Investigation of acceleration modes of the four-link jumping device. *Izvestia Volgograd State Technical University, 24*(127), 86–92.

Kawatsuma, S., Fukushima, M., & Okada, T. (2012). Emergency response by robots to Fukushima-Daiichi accident: Summary and lessons learned. *IndustrialRobot: AnInternational Journal, 39*(5), 428–435.

Kovac, M. (2010). *Bioinspired jumping locomotion for miniature robotics*. Academic Press.

Murphy, R. R., Tadokoro, S., Nardi, D., Jacoff, A., Fiorini, P., Choset, H., & Erkmen, A. M. (2008). Search and rescue robotics. In *Springer handbook of robotics* (pp. 1151–1173). Berlin: Springer. doi:10.1007/978-3-540-30301-5_51

Peters, J. F., Ahn, T. C., & Borkowski, M. (2002, October). Obstacle classification by a line-crawling robot: A rough neurocomputing approach. In *International Conference on Rough Sets and Current Trends in Computing* (pp. 594-601). Springer. 10.1007/3-540-45813-1_79

Poulakakis, I., & Grizzle, J. W. (2009, May). Modeling and control of the monopedal robot thumper. In *Robotics and Automation, 2009. ICRA'09. IEEE International Conference on* (pp. 3327-3334). IEEE. 10.1109/ROBOT.2009.5152708

Pradhan, S. S. (2009). *Design and Implementation of Energy Pumping Mechanism and Stabilizing Control on a One-Legged Hopping Robot*. Department of Mechanical EngineeringIndian Institute of Technology, Bombay.

Salton, J. R. (2010). *Urban Hopper // SPIE Defense*. Orlando, FL: Security and Sensing.

Stoeter, S. A., & Papanikolopoulos, N. (2006). Kinematic motion model for jumping scout robots. *IEEE Transactions on Robotics, 22*(2), 397–402. doi:10.1109/TRO.2006.862483

Tanaka, H., Hoshino, K., Matsumoto, K., & Shimoyama, I. (2005, August). Flight dynamics of a butterfly-type ornithopter. In *Intelligent Robots and Systems, 2005. (IROS 2005). 2005 IEEE/RSJ International Conference on* (pp. 2706-2711). IEEE. 10.1109/IROS.2005.1544999

Tsukagoshi, H., Sasaki, M., Kitagawa, A., & Tanaka, T. (2005, April). Design of a higher jumping rescue robot with the optimized pneumatic drive. In *Robotics and Automation, 2005. ICRA 2005. Proceedings of the 2005 IEEE International Conference on* (pp. 1276-1283). IEEE. 10.1109/ROBOT.2005.1570291

Vermeulen, J. (2004). *Trajectory generation for planar hopping and walking robots: An objective parameter and angular momentum approach. Vrije Universiteit Brussel.*

Volkova, L. Y., & Jatsun, S. F. (2013). Studying of regularities of movement of the jumping robot at various positions of a point of fixing of a foot. *Nelineinaya Dinamika, 9*(2), 327–342. doi:10.20537/nd1302009

Volkova, L. Y., & Yatsun, S. F. (2013). Simulation of motion of a multilink jumping robot and investigation of its characteristics. *Journal of Computer and Systems Sciences International, 52*(4), 637–649. doi:10.1134/S1064230713030155

Vorochaeva, L. Y., Efimov, S. V., Loktionova, O. G., & Yatsun, S. F. (2018). Motion Study of the Ornithopter with Periodic Wing Oscillations. *Journal of Computer and Systems Sciences International, 57*(4), 672–687. doi:10.1134/S1064230718040147

Vorochaeva, L. Y., Malchikov, A. V., & Postol'niy, A. A. (2018). Approaches to designing wheeled jumping robot. *Extreme robotics and conversion tendencies*, 308-316.

Vorochaeva, L. Y., Panovko, G. Y., Savin, S. I., & Yatsun, A. S. (2017). Movement Simulation of a Five-Link Crawling Robot with Controlled Friction Forces. *Journal of Machinery Manufacture and Reliability, 46*(6), 527–535. doi:10.3103/S1052618817060152

Vorochaeva, L. Y., & Yatsun, S. F. (2015). Mathematical simulation of the controlled motion of the five-link wheeled jumping robot. *Journal of Computer and Systems Sciences International, 54*(4), 567–592. doi:10.1134/S1064230715030168

Yi, S. J., McGill, S., Vadakedathu, L., He, Q., Ha, I., Rouleau, M., . . . Lee, D. D. (2014, September). Modular low-cost humanoid platform for disaster response. In *Intelligent Robots and Systems (IROS 2014), 2014 IEEE/RSJ International Conference on* (pp. 965-972). IEEE. 10.1109/IROS.2014.6942676

Zeglin, G. J. (1991). *Uniroo--a one legged dynamic hopping robot* (Doctoral dissertation). Massachusetts Institute of Technology.

Zhao, J., Yang, R., Xi, N., Gao, B., Fan, X., Mutka, M. W., & Xiao, L. (2009, October). Development of a miniature self-stabilization jumping robot. In *Intelligent Robots and Systems, 2009. IROS 2009. IEEE/RSJ International Conference on* (pp. 2217-2222). IEEE. 10.1109/IROS.2009.5353949

ADDITIONAL READING

Burdick, J., & Fiorini, P. (2003). Minimalist jumping robots for celestial exploration. *The International Journal of Robotics Research*, *22*(7-8), 653–674. doi:10.1177/02783649030227013

Dubowsky, S., Kesner, S., Plante, J. S., & Boston, P. (2008). Hopping mobility concept for search and rescue robots. *Industrial Robot: An International Journal*, *35*(3), 238–245. doi:10.1108/01439910810868561

Lee, L. W., Lo, Y. H., Lyu, Y. F., Chen, R. P., & Hsu, C. C. (2018). *U.S. Patent Application No. 15/291,327.*

Niiyama, R., Nagakubo, A., & Kuniyoshi, Y. (2007, April). Mowgli: A bipedal jumping and landing robot with an artificial musculoskeletal system. In *Robotics and Automation, 2007 IEEE International Conference on* (pp. 2546-2551). IEEE. 10.1109/ROBOT.2007.363848

Sayyad, A., Seth, B., & Seshu, P. (2007). Single-legged hopping robotics research—A review. *Robotica*, *25*(5), 587–613. doi:10.1017/S0263574707003487

Stoeter, S., Rybski, P. E., Gini, M. L., & Papanikolopoulos, N. (2002, September). *Autonomous stair-hopping with Scout robots* (pp. 721–726). IROS.

Tammepõld, R., Fiorini, P., & Kruusmaa, M. (2011). *Attitude Control of Small Hopping Robots for Planetary Exploration: theory and Simulations* (pp. 1–8). Netherlands: ESA/ESTEC.

Weiss, P. (2001). Hop… hop… hopbots!: Designers of small, mobile robots take cues from grasshoppers and frogs. *Science News*, *159*(6), 88–91. doi:10.2307/3981566

KEY TERMS AND DEFINITIONS

Acceleration: Accelerated movement of some parts of the robot relative to others when the latter are on the surface.

Angle of Rotation in Flight: Greatest angle at which the robot body rotates relative to the angle at the time of the jump while taking off the surface.

Flight: Movement of the robot without contact with the supporting surface, during which relative movements of the links are possible in order to make a landing in one way or another.

Height of the Jump: Distance traveled by the center of mass of the robot in the vertical direction from the moment of takeoff from the surface to the end of the jump.

Jump: Movement of the body with a temporary one-time planned loss of contact with the supporting surface, caused by repulsion from the support point with an initial velocity directed at an angle to the horizon.

Landing: Position of the robot at the time of transition from the flight phase to the positioning phase after landing, when the robot begins to contact with the supporting surface.

Length of the Jump: Distance traveled by the center of mass of the robot in the horizontal direction from the moment of takeoff from the surface to the end of the jump.

Positioning After Landing: Relative movements of the robot links, which ensure the transfer of the robot to a position that allows it to maintain vertical stability after the jump and which appears final for the given jump and the initial for the next one.

Positioning Before Takeoff: Change of relative positions of the robot links for making the jump with the required characteristics.

Takeoff: Position of the robot at the time of transition from the acceleration phase to the flight phase, i.e. at the time of zeroing the normal reaction at the contact point of the robot with the supporting surface.

Chapter 4
Design of a Garbage Collection Robot

Tawanda Mushiri

(iD) https://orcid.org/0000-0003-2562-2028
University of Zimbabwe, Zimbabwe

Emmison Gocheki
University of Zimbabwe, Zimbabwe

ABSTRACT

The Arduino is programmed to control the robot navigation. The Garbage Collection Robot is designed to collect solid waste at public places (schools, workplaces, and parks) and residential areas. The design of the robot is such that when it starts, it maneuvers as per programmed route. The Garbage Collector can sense by means of capacitive proximity sensors if the obstacle is living (for example, a human being) or non-living (for example, vehicle) and then gives appropriate warning signals like flashing light, hoot, or voice commands. The robot is equipped with vision capabilities in order for it to detect colors, namely green, red, yellow, blue, and black for organics, plastic, metal, paper, and glass, respectively. When the GCR sees a particular color code on garbage container, it picks up the bin, carries it in its carriage, then offloads it at a desired station to wait for recycling or final dumping.

INTRODUCTION

Garbage (solid waste) management poses a huge challenge both in industry and in urban areas for most countries over the world (Machale, 2015) as shown in Figure 1. Though today there exist several technologies of garbage collection, disposal and

DOI: 10.4018/978-1-5225-9924-1.ch004

Figure 1. The world infested with garbage. (Courtesy of WHO)

recycling, a more efficient method is required in order to maintain a green, safe, and healthy environment. The effects of the drastic growth in population density of residential area and the demand for metropolis environmental protection generate a challenging backbone for waste management (Jamelske, 2005). There are still several significant problems being encountered in collecting, transporting and processing (sorting and recycling) residential and industrial waste as most of the work is entirely manual, requiring much effort and time, especially in developing countries (Loughery, 2018). The uneasiness of waste management procedures is undoubtedly of primary interest to municipal local agencies (Hanshar, 2013). This article seeks to implement robotics and automation to come up with an efficient, safe and productive process of garbage collection, disposal and/or sorting for recycling. This project would lead to a low-cost waste management technique as the robot would be set to work repeatedly without much cost, besides the routine maintenance costs.

BACKGROUND

A robot can be comprehended as a programmable, self-guided device constituting electronic, electrical, and/or mechanical components(Rakesh, A, & Ajay, 2013). Essentially, a robot is a machine that can function as a substitute for some living agent. Robots are mostly desirable for particular work functions, unlike humankind, they can never get tired, they can perform, Dirty, Dangerous, Dull and Dear (4D's) tasks

(Salmador, Cid, & Novelle, 1989). uncomfortable or dangerous physical conditions, they can do operate well in confined or airless conditions, they never get bored when doing repetitive work, and they can never get themselves distracted from an ongoing task at hand. A robot is a powerful and reliable machine and can perform well in areas of hot temperature where humankind can get tired and sick after working for considerably long hours(Rakesh, A, & Ajay, 2015). A typical vehicle that is used in garbage collection is shown in Figure 2. This technique of garbage collection is entirely manual as humans have to pick up waste and load it in the payload.

The current technique of collection, transportation, sorting and/or disposal of industrial solid waste is mainly manual, taking a lot of efforts, time consuming, unsafe and posing hazards to humankind and environment as shown in figure 3.

The garbage collection robot should be able to:

1. Detect garbage tins for collection.
2. Pick up the tins successfully, pack and hold them in its payload.
3. Transport the tins and offload them at a dumping site or recycling station.
4. High speed stability.
5. Low speed maneuverability.

LITERATURE REVIEW

In this section, the designers will take a review on the findings and works of a considerable number of researchers on garbage collection management. This chapter will also look onto the garbage types, scattering and collection methodologies, and will explain which methodologies lead to sustainable environment. The designer

Figure 2. Present day garbage truck (Courtesy of the Environmental Management Agency)

Figure 3. Current garbage collection technique (Courtesy of Harare Town Council)

shall take a look into the current situation and challenges encountered by both international and regional urban councils pertaining garbage collection. A thorough discussion shall also be conducted pertaining the world-wide view on waste management and health related issues. The policies, regulations and laws that govern Solid waste management will be explained too. The discussion extends to look on the performance analysis and benchmarking parameters in the assessment of the current garbage collection technologies used by urban councils internationally and locally. The environmental aspects and impacts related to waste management will be discussed also. A thorough look will also be taken into the collection systems in both developing and developed countries.

GARBAGE

Garbage is commonly perceived as any refuse other than effluents from industry. It constitutes mainly of decomposable and organic waste from preparation, storage, handling and serving of food (Engineering, 2013). Garbage can also be imprecisely defined as non-soluble, semi-solid or solid material derived from agricultural waste, sewage sludge, demolition refuse, industrial waste, municipal refuse, and mining residues (Jamelske, 2005).

Garbage Types

Waste can be classified into various categories including organic waste, paper, metallic, plastics, hazardous waste and non-hazardous waste (Salmador et al., 1989).

This is a very important consideration for planning, especially when it comes to the formulation of a waste recycling strategy and also in designing landfills (UN Habitat, 2013). For instance, if the solid waste in a landfill contains more of organic content, it might be necessary to make means for capturing the released gas which can then be used as a fuel. Nevertheless, when the general objective of waste management is deposition of waste into landfills, then knowledge of the composition of solid waste might seem not be of great importance(Loughery, 2018).

How Garbage is Scattered

The scatter or distribution and quantity of garbage generated largely depends on the population size of the area (Zimbabwe National Statistics Agency, 2016). Taking Zimbabwe for example, with the municipal councils arranged according to the size of population as shown in Fig 4. The largest being Harare, it is expected to generate the highest amount of garbage(UNEP, 2016).

Figure 4. Population of 16 Zimbabwean urban areas (Zimbabwe National Statistics Agency, 2016)

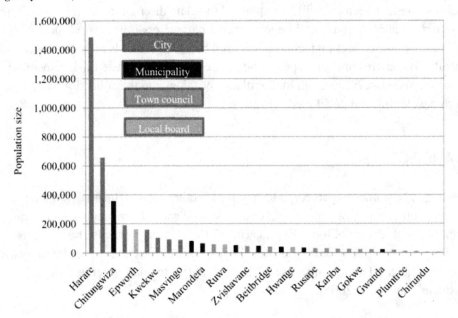

Garbage Composition

Generally, it is unsurprisingly expected to note that urban areas with many industries are likely to release vast amounts of metallic waste. For example, the EMA 2013 statistics reveal that the waste composition in Bulawayo (second largest Zimbabwe city) counts to 50% plastic waste, 40% food and biomass waste, 10% paper waste, and 5% metallic. Considering Chinhoyi, a municipality in Zimbabwe has waste composition of 45% biomass and food waste, 23% plastic waste, 12.5% of metallic waste and 7% of paper waste(Zimbabwe National Statistics Agency, 2016). The metal composition in Bulawayo was low due to non-functionality of the industries for almost a decade as a result of declination in economic status. Figure 5 shows the solid waste generation rate per each city in Zimbabwe.

GARBAGE COLLECTION SYSTEMS

Systems in Developing Countries

Receptacles: A receptacle can be defined as a container for storing waste temporarily. A suitable receptacle should be durable, have a lid and not easily get damaged by pets and/or rodents(Torres-garcía, Rodea-aragón, Longoria-gandara, Sánchez-garcía, & González-jiménez, 2015). In some developing countries like Zimbabwe,

Figure 5. Solid waste rates of generation for urban areas of Zimbabwean in the year 2014

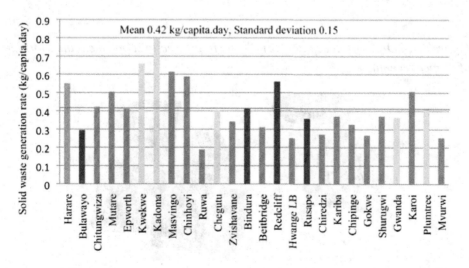

municipality councils take the responsibility to ensure that proper receptacles are available at each property, such as a residential area or industrial site.

For instance, City Council of Bulawayo used to do supply of metal bins to residents. Even so, due to high cost of bins, residents tend to use temporary storage facilities such as plastic bags. Approximately 48% of the Bulawayo residents have been using the plastic bags for waste collection and temporary storage(UNEP, 2016).

An estimation of 44% was found to be the coverage of metallic bins. The author suggested that the charge of the bins should lie on a nominal fee in order that members in the community could afford to get them. This was found to be the same scenario with Gweru, the town of Midlands in Zimbabwe. It was suggested that the municipal authority should ensure that the waste bins are available to community members for sustainability in waste management. Provision of receptacles to residents by municipal authorities is of great importance as this would cater for the financially decent who cannot afford to buy the receptacles (Özturan, Bozanta, Basarir-Ozel, Akar, & Coşkun, 2015).

In Chinhoyi, polythene bags, metal bins, propylene sacks have for long been used as receptacles. The collected waste was then disposed into pits in the backyard. In Chinhoyi, about 22% of the households were using bins they have been provided by the urban council, 26% using sacks, 19% using the plastic bags, 25% depositing waste into pits, and 8% making use of other alternatives of receptacles, such as plastic buckets and boxes. Serious scattering of waste in Harare, Bulawayo, Chinhoyi and Gweru provide sufficient evidence that the garbage collection techniques used by the city council are not effective enough to maintain a garbage free environment. In a case study by Harare City Council (2013), the coverage of suitable receptacles in high density suburbs of Sakubva in Mutare, Zimbabwe was about 64% (Engineering, 2013). Coverage of the receptacles is defined by the equation:

Figure 6. a.) Ordinary receptacle b.) Plastic bags used to supplement the shortage of proper receptacles

(a)   **(b)**

$$Coverage\ of\ receptacles = \frac{p}{q} \times 100\%$$

where p = Total count of occupied establishments in the area of service and q = Total count of occupied establishments using proper receptacles

The residents were using improper receptacles, for example sacks and card boxes which constitute about 26%. The Harare City Council also did point out that it takes seven days to fill-up the right receptacle. This is usually the time delay between any two successive collections. cardboard boxes and sacks fill up quickly resulting in waste overspills and this attracted flies, mosquitoes, and rodents and thus speeding up the spread of certain diseases. Those residents using sacks dump the waste illegally in order to create space for some other waste and also to prevent the infestation by flies.

Garbage Trucks and Problem Definition

These technologies are mainly used in developing countries. The truck is loaded manually by local residents or workers of the municipal authority. This method poses enormous health hazards to people (refer to Figure 3).

Receptacles

Developed countries also use receptacles but advanced in the sense that they encourage waste sorting in the early stage of waste disposal by assigning color distinctions to receptacles (for example, green receptacles collect bio-degradable garbage while yellow ones collect non-biodegradable matter (Hayawi, 2016).

Front Loader

The Front Loader garbage collection vehicle constitutes automated forks mounted on the front which the operator carefully controls and aligns them with sleeves on the garbage container using a gearstick or some set of gearlevers. This is an advanced methodology and it eliminates the risk of health threatening to humans.

GARBAGE COLLECTION EFFICIENCY

Garbage (solid waste) collection efficiency is the ratio of the quantity of garbage collected to the garbage generated. This is an important indicator when assessing solid

Figure 7. Early stage waste sorting

Figure 8. Typical front loader in the first world

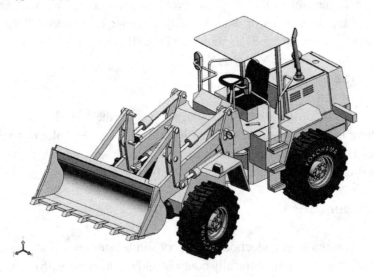

waste (garbage) management system (Apoorva, Prabhu, Shetty, & Souza, 2017). Such an indicator answers the question regarding whether municipal councils do collect all the garbage which is generated or not. The role of municipal authorities is to make sure that all the waste generated gets collected in the right time for the purpose of energy recovery, recycling, composting, and disposal in order to maintain a green, health and safe environment. Focusing on health reasons, all the garbage should get collected, meaning the collection efficiency must be 100%. The mean garbage collection efficiency for the 17 chosen cities in the world is 89% with a standard deviation of 17 as depicted in Fig 9 (I, Much, Faithful, You, & Palumbo, 2017).

$$\textit{Garbage collection efficiency} = \frac{y}{x} \times 100\ \%$$

where x = Total garbage generated (tones/month), and y = Total amount of garbage collected by city council (tones/month)

Developing Countries

In urban areas of developing countries, approximately 30% to 60% of the solid waste remain uncollected with less than 50% of the population being served (UNEP, 2016). In most low-income (developing countries), approximately 80 to 90 percent of the budget for waste management is taken up by garbage collection only. This is essentially because developing countries usually face poor financial backup mechanisms and the available funding are not sufficient to embark on investment in other systems of waste management. In developing cities Asia, lack of machinery resulted in the reduced garbage collection efficiency, that is 70% of all household wastes gets collected. In developing countries, Municipal Local Bodies (MLBs) spend 60 to 70

Figure 9. Garbage collection efficiency for 17 selected world cities(UNEP, 2016)

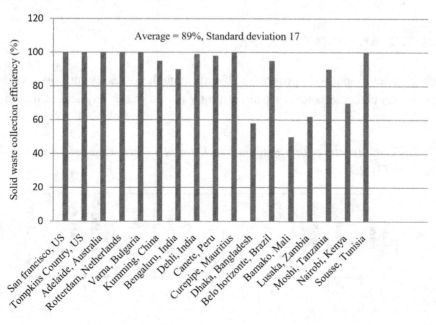

percent of money on garbage collection and not more than 5% is spent on treating and disposing waste, owing to poor financial mechanisms (Chandrasekara, Rathnapriya, Handwriting, & Recognition, 2012). Countries in Africa face big challenges in garbage collection in urban areas. For instance, Blight (1996) mentioned the issue of inadequate machinery and the utilization of old and inappropriate equipment as attributed to inefficient waste management in low-income developing countries. Case studies by Mumbai *et al.* (2015) gave an estimate of the coverage of garbage management services as approximating 58% in cities of Kenya due to lack of proper equipment and old-aged machinery (Salmador et al., 1989).

Developed Countries

In high-income (developed) countries, garbage collection takes up less than 10% of the waste management budget, which allows huge funding to be assigned to waste treatment technologies. Direct community involvement in these developed countries reduces the cost of garbage collection and facilitates recycling and recovery of waste (UNEP, 2016).

Tariff structures: The system financial sustainability is an important measurement when carrying out an evaluation of the performance of garbage collection system. When running a business, it is of great importance to have the knowledge of the costs and also to prepare the budget before embarking on the business (Saravana, Sasi, Ragavan, & Balakrishnan, 2016). Table 1 shows the structure of the tariff of the few chosen world cities.

ROBOTIC ANATOMY SYSTEMS

A robot is a multi -purpose, programmable machine. It possesses anthropomorphic (human-like) characteristics, that interestingly resemble the physical structure of

Table 1. Tariff structure for waste management of 11 chosen world cities (US$) (Courtesy of WHO)

City	Adelaide, Australia	Belo Horizonte, Brazil	Rotterdam, Netherlands	San Francisco, USA	Tompkins, USA	Varna, Bulgaria	Delhi, India	Dhaka, India	Lusaka, Zambia	Moshi, Tanzania	Nairobi, Kenya	Average
Monthly Fee (US$)	8	6	33	22	15	8	0.8	1	18.80	1	0.20	8

a human(A.R. Akkar & Najim A-Amir, 2016). The robots can also offer response to sensory signals similarly as humans do. Human-like characteristics such as the mechanical arms can be used for various industrial tasks. Sensory devices allow the robot to communicate and also interact with other machinery and also to take simple decisions. The commercial and the technological advantages of robotic systems are:

- Robots can substitute humans in hazardous and/or uncomfortable working environments.
- A robot can perform its work consistently and repeatedly which is of course difficult for humans to go through over long period of continuous working.
- Robots can also be reprogrammable. When the current task is accomplished, a robot can then be reprogrammed and get it equipped with necessary tooling to carry out a different task.
- Robots can also be interfaced with the computer systems and/or other robotic systems. In the present day, wire-less control can be used to control robots. This has greatly enhanced the efficiency and productivity of automated systems in industry (Liqing & Qingjiu, 2012).

Robotic Anatomy

Robot anatomy looks into the study of various joints, links and also other aspects concerning the physical construction of the manipulator.

Manipulator: The manipulator consists of a series of joints and links. The manipulator is the arm mechanism, created by a combinatorial sequence of joint and linkage (Nurlansa, Istiqomah, Astu, & Pawitra, 2014). Usually, it is used to describe the entire robot itself, excluding the power supply system and the controller. Each of the joints is connected to two linkages, namely input and output links.

End effector: An end effector is also known as the hand and is attached to the end of the robot arm or wrist. End effectors include grippers, vacuum cups, spray guns, welding tools and electro-magnetic pick-ups, their performance being vital to precision and repeatability (Weiland, Boekhoff, & Staloch-schultz, 2017).

Joints: Robot joints are described as either rotational or translational. Rotational joints have a rotary action along the joint axis and are also referred to as revolute. Translational joints have a linear or sliding motion along the joint axis and are also known as prismatic. Joint provides controlled relative movement between the input link and output link.

Actuators: Actuators are also known as drives. These devices convert electrical, pneumatic and hydraulic energy into motion of the robot. Today, actuators are very fast, accurate AC servo-drives, while the robot base rotates using a harmonic drive or, less commonly, ring gear.

Kinematics: It is the actual arrangement of joints/axes and rigid links in the robot, as well as being the study of motion in robotics. Common robot kinematics, or configurations, include Cartesian, Articulated, Parallel and SCARA.

Work envelope: This is the total volume of space that the end effector of the manipulator can reach and is also known as workspace and work volume (American, America, & American, 2017). The size and shape of the work envelope is determined by the robot kinematics and the number of DOF; it should be large enough to accommodate all the points the end effector needs to reach.

Degrees of freedom: This is the number of independent movements the end effector can make along the axes of its coordinate system. For example, movement along the X Y Z coordinates only constitutes 3 DOF, whilst adding rotation around the Z axis equals 4 DOF. This term is often confused with degrees of mobility (see below). The robot's complexity can be classified according to the total number of degrees-of-freedom they possess (Watanasophon & Ouitrakul, 2017).

Degrees of mobility: While DOF are often incorrectly determined by simply counting the number of independent joints on the robot, this is more accurately expressed as degrees of mobility (DOM). Thus, an industrial robot has a maximum of 6 DOF, but might actually have, say, 9 DOM.

Axis/axes: An axis is a line across which a rotating body turns. Two axes are required to reach any point in a straight plane, while three axes (X Y Z) are needed to reach any point in space(Schultz & Miner, 2016). Three further axes (roll, pitch and yaw) are needed to control the orientation of the end of the robot arm or wrist.

DESIGN CONSIDERATIONS

Lagrange Equations

Lagrange equations can be simply defined as a set of equations of motion of a mechanical system which relate the kinetic energy of the system to its N generalized coordinates q_i such that $i = 1, 2, 3$. The degrees of freedom N is fully described by this set of generalized coordinates (Dautenhahn, 2017). The coordinates are unrestricted, and are independent, meaning they have no geometrical or kinematical relationship. Lagrange's equations can be stated as:

$$\mathbf{Q_J} = \frac{\mathbf{d}}{\mathbf{dt}} \left(\frac{\partial \mathbf{T}}{\partial \dot{\mathbf{q}}_j} \right) - \frac{\partial \mathbf{T}}{\partial \mathbf{q}_j} + \frac{\partial \mathbf{D}}{\partial \dot{\mathbf{q}}_j} + \frac{\partial \mathbf{V}}{\partial \mathbf{q}_j}$$

where $\dot{q}_J$ = generalized velocities; T= system kinetic energy; V= system potential energy; Qj= generalized force in the j-th equation; and D= Rayleigh dissipation function.

If the system constitutes of *l* bodies. Then Q_j are defined by:

$$Q_j = \sum_l F_l \cdot \frac{\partial r_l}{\partial q_j} + \sum_l M_l \cdot \frac{\partial \acute{E}_l}{\partial \dot{q}_j}$$

where F_l = external forces, M_l = external moments on the lth body And r_l = the position vector drawn to the point of application of F, ω_l = the angular velocity of the lth body about the axis of application of the momentum of concern.

The results of the equations give the equations of motion of the system. The Lagrange equations shall be used during the designing of motion aspects of the manipulator

Grashoff's Law: Considering a four-bar-linkage. Let the smallest link be denoted by S, and the longest link denoted by L and the remaining two links represented by M_1 and M_2.

If L+S < M_1 + M_2

…then depending on whether S is in connection with the ground from one end, two ends, or from no end, then the mechanism is described as under the type illustrated in Figure 10.

If L+S > M_1+M_2

…then the mechanism is of a rocker-rocker type.

Figure 10. Interpretation of crank and rocker mechanisms

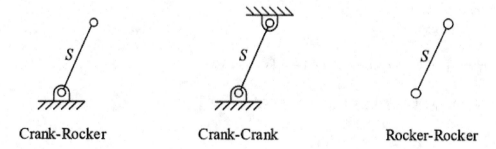

Crank-Rocker   Crank-Crank   Rocker-Rocker

KINEMATICS ANALYSIS AND MODELING OF THE ROBOTIC ARM

Kinematics defines the relationship between manipulator link, velocities, and accelerations and positions. In the kinematics pertaining manipulator position, we have two distinct problems to solve, that is forward kinematics, and the inverse kinematics (Sharukh, 2015).

Forward Kinematics: The forward kinematics refers on how to determine the position of the end-effector relative to the base of the arm for given joint angles. The solution can be found by determining the transformation matrices of the robotic arm from one to next link according to Denavit-Hartenberg (D-H) coordinate system.

Each of the joints per frame has 1 degree of freedom, which can be represented by just a single number, and that is the rotational angle when dealing with a revolute joint, that is, $(\theta_0, \theta_1, ..., \theta_n, \theta_{n-1})$. The cumulative effect of all the joint variables can be calculated by taking consideration starting from link 0 (the base) to link n (Hanshar, 2013).

In order to keep the initial manipulator position configuration as home position, in which all angles at the joints are zero, all the joint variables (θi, got adjusted +90° or −90° as illustrated in Table 2 of the Denavit-Hartenberg parameters (Craig, 2018). Considering the aforesaid adjustment and working based on the convention of Craig (Sajjad, Talpur, & Shaikh, 2012), the homogenous transformations that relate the end-effector (gripper) position and its orientation with the global system of coordinates is defined by:

$$_6^0T = {}_1^0T\,{}_2^1T\,{}_3^2T\,{}_4^3T\,{}_5^4T\,{}_6^5T = \begin{bmatrix} r11 & r12 & r13 & px \\ r21 & r22 & r23 & py \\ r31 & r32 & r33 & pz \\ 0 & 0 & 0 & 1 \end{bmatrix} \tag{1}$$

Note: *The joint variables in the equation above were determined with respect to home position.*

Where:

$$r_{11} = c_6[c_{45}(s_1s_3 + c_1s_2c_3) - s_{45}c_1c_2] - s_6[s_1c_3 - c_1s_2s_3]$$

$$r_{21} = s_6[c_1c_3 + s_1s_2s_3] - c_6[c_{45}(c_1s_3 - s_1s_2c_3) + s_{45}s_1c_2]$$

$$r_{31} = c_6(s_{45}s_2 + c_{45}c_2c_3) + c_2s_3s_6$$

Figure 11. Robot kinematic model

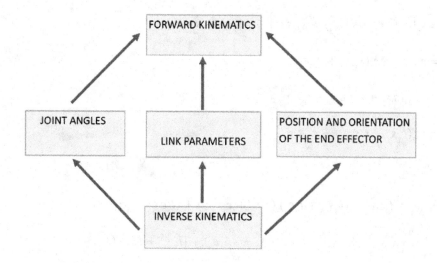

Figure 12. D-H parameters interpretation

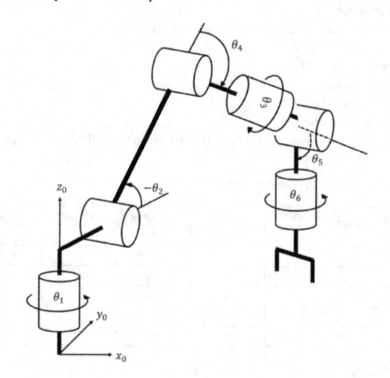

$$r_{12} = -s_6[c_{45}(s_1s_3 + c_1s_2c_3) - s_{45}c_1c_2] - c_6[c_3s_1 - c_1s_2s_3]$$

$$r_{22} = s_6[c_{45}(c_1s_3 - s_1s_2c_3) + s_{45}s_1c_2] + c_6[c_1c_3 + s_1s_2s_3]$$

$$r_{32} = c_2s_3c_6 - s_6(s_{45}s_2 + c_{45}c_2c_3)$$

$$r_{13} = -s_{45}(s_1s_3 + c_1s_2c_3) - c_{45}c_1c_2$$

$$r_{23} = s_{45}(c_1s_3 - s_1s_2c_3) - c_{45}s_1c_2$$

$$r_{33} = c_{45}s_2 - s_{45}c_2c_3$$

$$p_x = l_2c_1 + l_3c_1c_2 + (l_5s_{45} + l_4s_4)(s_1s_3 + c_1s_2c_3) + (l_5c_{45} + l_4c_4)(c_1c_2)$$

$$p_y = l_2s_1 + l_3s_1c_2 - (l_5s_{45} + l_4s_4)(c_1s_3 - s_1s_2c_3) + (l_5c_{45} + l_4c_4)(s_1c_2)$$

$$p_z = l_5(s_{45}c_2c_3 - c_{45}s_2) + l_4(c_2c_3s_4 - c_4s_2) - l_3s_2 + l_1$$

Note on notation:

r_{ij} represents any element of the rotation matrix,

p_{ij} represents any element of position vector,

c_i represents $cos(\theta_i)$,

s_i represents $sin(\theta_i)$,

c_{ij} represents $cos(\theta_i + \theta_j)$, and

s_{ij} represents $sin(\theta_i + \theta_j)$.

Note on reference: All the relevant link transformations are illustrated in Appendix.

Table 2. Denavit-Hartenberg parameters

Joint	Link	a_{i-1} min	α_{i-1} degree	d_i mm	degree
0-1	1	0	0	l_1	θ_1
1-2	2	l_2	-90	0	θ_2-90
2-3	3	0	-90	l_3	θ_3
3-4	4	0	-90	0	θ_4-90
4-5	5	l_4	0	30	θ_5+90
5-6	6	0	-90	$-l_5$	θ_6

Where:

a_i: The distance between z_i and z_{i+1} measured in the z_i direction

α_i: Twist angle measured between z_i and z_{i+1}

d_i: Offset distance measured from x_i to x_{i+1} taken along z_i

θ_i: Angle measured between x_i and x_{i+1} taken about z_i

Inverse Kinematics (Iterative): The method of Iterative Inverse Kinematics, IIK consists of deriving 2 simultaneous non-linear equations that describe, in mathematical nature, the geometrical configuration of the robotic arm. Usually the first and the second joint angles of the manipulator, that is θ_1 and $\theta2$ are used to derive such equations(Shepherd et al., 2017). For any method of inverse kinematics method, both the position and the orientation of the end-effector should be known. When the two simultaneous equations are found, the remaining problem would be to solve the non-linear system of equations simultaneously. The entire solution is given by the roots of that system of equations. A bisectional method is used to do such calculations (Alex et al., 2017). Once the joint angles θ_1 and θ_2 are obtained, joint angle number 4 is determined through the application of Law of cosines. The other remaining joint angles can be computed using the kinematics equations.

Work envelope (workspace of the manipulator): The workspace of the manipulator can be defined as a field of all points that can be reached by the end-effector. Knowledge of the workspace is very important as it helps to determine the limit of the manipulator when operating. An error in calculation of joint angles is likely to occur when a position outside the workspace is given when performing inverse kinematics of the system. To determine the boundaries of the workspace of the manipulator, we shall use the Abdel-Malik method. The method comprises of two steps. The first step is the calculation of the Jacobian then secondly; we have to determine whether the surfaces are boundaries or not(American et al., 2017). The identification of a permissible direction of motion whilst crossing the surfaces will help achieve the second step.

The manipulator arms are designed to move in certain orientation controlled by joint types. The three common motion types are roll, pith and yawl as shown in Figure 13.

Figure 13. Joint orientation (Shepherd et al., 2017)

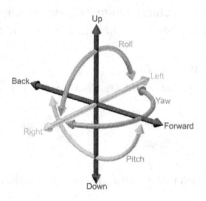

Singular parametric surfaces: Singular configurations are also simply known as singularities and these are defined by that loss of the number of degrees of freedom within the system (The NXT Generation, 2017). The idea of singular configurations is a critical phenomenon because these configurations may represent not only the boundary locus of the work envelope, but also regions for which the end-effector experiences motion difficulties. Upon reaching these singular surfaces, the movement of the end-effector becomes limited(Polytechnic, 2015). For instance, if the manipulator tip is placed at the work envelope boundary, it cannot move further. Analysis of the Jacobian matrix will help to find such singularities.

EXISTING TECHNOLOGIES

There are several garbage collection technologies in existence today. In this section, we are going to discuss on a few of most interest. All of them are robotic type systems.

Magnetic handed vehicle: The Magnetic Hand type is a kind of a pick and place robot (Rakesh et al., 2015). For waste management technologies, this robot is fitted to a vehicle with a payload. The resulting system is the termed the magnetic hand mobile pick and place robotic system. This kind of robot can be simply represented by essentially three basic subsystems namely:"

Layout of the robot
1. Moveable base
2. Rotational manipulator
3. Magnetic gripper

Autonomous robot: The autonomous garbage collection robot consists of the manipulator and a payload. The garbage is collected into the payload and transported to the desired destination. The mechanism has the advantage that it has a wide and unlimited reach radius as the vehicle can move independently.

This garbage collection robot constitutes four main features namely:

1. Drive train to move the robot.
2. A mechanism to lift garbage tins (marked as balls) off of the floor and into the payload.
3. The payload into which the garbage tins are placed.
4. A camera for object identification.

These features are controlled by the Arduino microcontroller (Rakesh et al., 2015).

Figure 14. The schematic of the magnetic handed vehicle

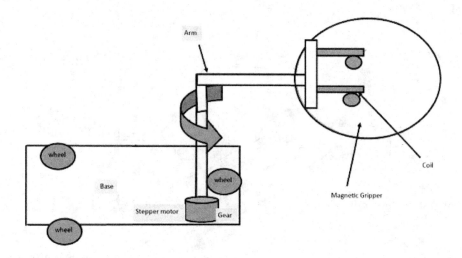

THE GARBAGE CONTAINER (BIN)

A garbage container is a receptacle for temporary storage waste, and is basically made of metal or tempered plastic. Some common names are garbage can, dust bin and trash can. The words basket, rubbish, and bin are more dominant in British English, whilst can and trash are more common in the American English (Alex et al., 2017). In the year 1875, the first personal bins got introduced in the United Kingdom to bring about a controlled system of garbage collection.

Nowadays, company's such as Simpro are advancing drastically in garbage bin design to ensure effective means of solid waste management. One of the kind of bins made by this company is the 240-liter Wheeler bin depicted in Figure 16.

Recommended applications of the Wheelie bin include:

- Municipal waste management
- Household refuse and recycling.
- Schools, and colleges and university campuses
- Hospitals
- Rest homes
- Commercial setups and offices
- Hospitality industry, restaurants and hotels
- Industrial waste
- Materials handling in the commercial realm
- Virtually any purpose that requires a standard wheeled mobile garbage container,

Figure 15. a.) The autonomous robot b.) the autonomous flow chart

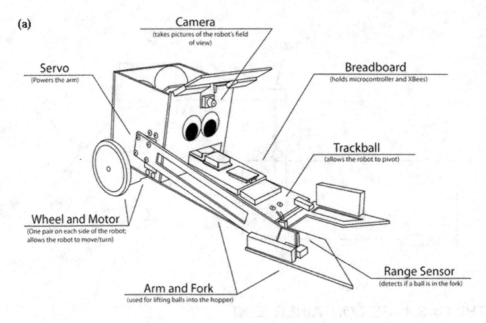

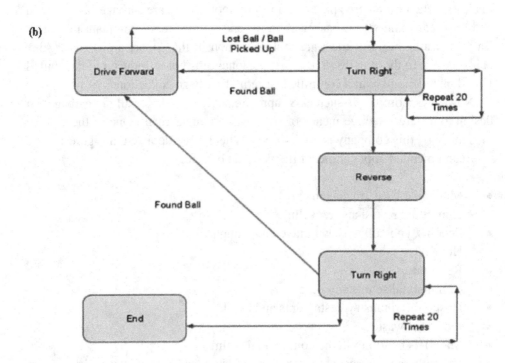

Figure 16. Simpro wheeler bin

Classic 240 litre wheelie bin, ideal for recycling and kerbside waste collection

Compliant with EN840 and AS4123 quality requirements

Incredibly versatile and popular product, huge range of accessories available

Large, thick handle for excellent control and user comfort

Double angle rail for greater safety when emptying

Reinforcing ribs also prevent bins from jamming together

Built-in RFID chip nests for identification and weighing systems

Injection-moulded from specially designed HDPE plastic - fully recyclable and resistant to decay, frost, heat and chemicals

Proprietary UV stabilisation technology ensures excellent ageing characteristics

Suitable for all DIN lifting equipment, designed to withstand high levels of mechanical stress

Externally positioned wheels provide excellent stability and prevent accumulation of dirt

Rounded internal floor to assist emptying

Corrosion-resistant steel axle and quiet-running solid rubber tyres

Requires just 0.36m2 of floor space

The 240-litre Simpro's Wheelie bin is supplied with a wide range of features, options and accessories.

- Standard colors: Black, Grey, Dark Green, Nature Green
- Non-standard colors: Red, Blue, Lime Green, Yellow
- Other colors are available upon consumer request, whilst minimum order quantity also applies.
- Custom markings and imprints
- Optional venting of the lid to allows air circulation, thereby reducing weight of the contents if it is organic.
- Optional Gravity Latch, that is it automatically secures the lid when standing vertically
- Color coding of bins: Standard colors of rubbish and recycling bins

Figure 17. Bin color codes (Courtesy of Simpro)

To simplify matters in order for the public to carry out an effective recycling, councils around the world, *including The Packaging Forum, who are the in*itiators of public site recycling schemes, and also key partners of the industry have come to an agreement to use standard color codes (shown in Figure 17) to indicate which bin is meant for rubbish, which is designated for paper recycling, food waste recycling et cetera. The use of standard color codes for recycling purposes at home, at work and at public places, will significantly minimize confusion to the public, and will increase recycling and thereby reducing contamination.

Figure 18. Simplified color coding

Each passing day, the once standard color coding is getting more simplified as depicted by figure today. For instance, other than putting plastic, glass and paper in the same bin, new technologies have made these materials to be placed into separate bins, and such a practice will make garbage sorting and recycling more effective and relatively less costly.

POWER SOURCE

The two common types of batteries used in powering automobiles are the dry and metal-acid batteries. An example of a dry type is a Lithium Ion Battery, and that of a metal-acid type is a Lead acid Battery. The advantages of the Lithium Ion (Li Ion) battery over the Lead Acid type are in Table 3.

TECHNICAL SPECIFICATIONS

The specifications and technical data of the bin and the robot were discussed and are described in Figure 19 and Table 4.

BASIC PROCEDURE IN ROBOT DESIGN

a)   *Segmentation:* This deals with calculating the number of segments the robot should consist of.
b)   *Anatomy:* This deals with the mechanical structure of the robot.
c)   *End effector:* This focuses on the end effector and its characteristics
d)   *Softw*are: The designers will use a countable number of software including
   ◦   *Soli*dWorks
   ◦   ANSYS
   ◦   RoboAnalyzer
   ◦   MATLAB
   ◦   Arduino
e)   *Controller:* Decision is made on which type of controller and actuator to use.
f)   *Programming Language:* Decision is made on what language to use in programming the machine.
g)   *Stability Analysis:* This deals with the analysis on dynamic stability of the robot.

Table 3. Power source types

Characteristic	Lithium Ion Battery	Lead Acid Battery
Usable Capacity	About 95% usable during a robot maneuver	About 45% of the capacity usable
Weight	Light	Heavy
Charging	▪ Can accommodate fast charging to 100% ▪ Can be charged to a percentage lower than 100% without damaging the battery.	▪ Need some absorption phase for storing a final 20% ▪ Always need charge to 100% otherwise the battery gets damaged.
Efficiency	More efficient even under high temperature conditions	Low efficiency in harsh conditions

Figure 19. Technical specifications for the bin under study

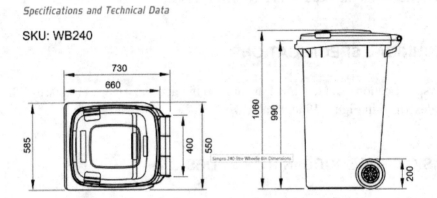

Specifications and Technical Data

SKU: WB240

Table 4. 240 litre wheelie bin specifications

Model	WB240
Volume	240 litres
Maximum load weight	96 kg
Length	660 mm
Width	585 mm
Height	1060 mm
Weight	13 kg
Handling specifications	1×φ25 mm plastic grab handle, EN 840-compliant reinforced lifting comb
Materials	High-density polyethylene (HDPE), cadmium-free
Colours	Dark green body with red lid (options: any combination of black, nature green, dark green, grey, red, lime green, light blue, yellow, or custom colours)
Wheels	2× outset φ200 mm HDPE plastic wheels, non-marking solid rubber tyres, steel axle
Applicable standards and certifications	EN 840, AS4123, ISO 9001, ISO 14001, Conformité Européene (CE)

EFFICIENCY OF GARBAGE COLLECTION

The main obje*ctive of the intend*ed design is to collect garbage safely and efficiently. More attention will be paid on designing a robot with surpassing mobility features in varying terrain so as to ensure efficient garbage collection. By equipping the robot with the best hand to pick garbage containers, the functionality of the device will be brought to a significantly high level of convenience and hence effectiveness (Chandrasekara et al., 2012).

In this work, a countable number of garbage collection robots which already did hit the market will be considered analytically. Through a close critic of the possible solutions, the best solution to the problem at hand shall be obtained and further developed. The careful critic will be done with the aid of the Binary Dominance Matrix (whereby each design factor amongst functionality, cost, complexity, et cetera is compared against each other factor).

DESIGN DATA

The data presented in the table 5 and 6 gives a description of the technical specifications of the garbage collection robot. The technical specifications are categorized into the Skeleton (the framework) and the Interior (the power and control).

THE PICK AND PLACE ROBOT

This kind of garbage collection *robot w*as chosen and is made to *use eithe*r compressed air or electrical energy to power the grippers to open or close. The garbage collector is fitted with its own payload. The main disadvantage with this mechanism is that the payload is not divided into compartments for proper garbage sorting, hence the machine collects one type of garbage at a time.

Table 5. Design specifications of the skeleton

SPECIFICATION	DESCRIPTION
Length	2000mm
Width	1200 mm
Height	1600 mm
Weight	700kg
Wheel system	Rubber track with metallic wheels
Speed	15 km/hr.

Table 6. Design specifications of the interior

MODULE	SPECIFICATION
Power supply	Battery 24V DC charging from PVC
Power consumption by microcontroller	5V DC
Computer software	Windows 10 Enterprise
Display	Liquid Crystal Display (LCD)
Microcontroller	PIC18F4550
Programming language	C language
Communication (Bluetooth and microcontroller)	RS232
Communication (PC and Camera)	Ad-hoc
Communication (PC and Robot)	Bluetooth
Camera	IP wireless
Wheel control	DC motor

Figure 20. The pick and place robot

A mechanical gripper is essentially a kind of actuator gripping solution that constitutes fingers or tooling jaws that grasp a piece. Grippers are able to pick up, hold, release and place objects during execution of an action. The most common types of mechanical grippers are 2-jaw angular and 2-jaw parallel grippers. Parallel grippers are so called as they open and close whilst parallel to the target object while angular grippers can move their gripping jaws wider than the parallel jaws and they require more space.

MERITS
- Can be customized into a variety of designs to suit any application.
- Can work well with non-magnetic or porous materials.
- Can perform well even under harsh environments.

- ◦ The robot has a simple mechanism, thus easy to maintain.
- ◦ Compressed air is a cheap fuel type.

DEMERITS

- ◦ There is need for tool changing for non-adaptive grippers so as to suit different products.
- ◦ With pneumatic grippers, it is either fully open or fully close. Thus the gripping force should be cautiously limited to avoid marking/damaging the product whilst at the same time doing the required job.
- ◦ Takes up considerably more space.

Development of Chosen Solution

The Garbage Collection Robot constitutes many technologies in order to achieve its intended purpose of collecting garbage and then sorting it at a dump site. The complete task is automated. The machine comprises of tools varying from mechanical to electronic, and such technological advancement like port programming. In addition, equipment including IR sensors, DC motors, motor drive system, et cetera will make it possible to come up with an automated garbage collection robot. This chapter will explain the various technologies to be used in the design and development of the machine.

Garbage Container Selection

The Wheelie Bin in Figure is of capacity 240 liters and was selected to be the garbage container for this project. This kind of bin is manufactured by Simpro. The bin is mobile, and is used in schools, households and business campuses around the globe for garbage collection and recycling. The Wheelie bin is made by injection-molding from HDPE plastic which resists heat, decay, frost, and chemicals. It is mounted with a special type corrosion-resistant high speed steel axle and solid (tubeless) rubber tires to ensure a considerably long service life even in the extremely harsh environments. The net weight m b of the bin is 13kg.

Capacity (Volume): Volume of space V of the bin is given by the formula:

$V=L \times W \times H$, $V=0.575 \times 0.730 \times 1.060$, $V=0.444935$(say $0.445m^3$)

Figure 21. Bin dimensions

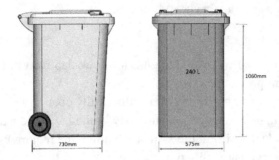

GARBAGE DESIGN MASS

Before embarking in the det*ailed de*velopment of the robot itself, let us take [a] look into the garbage properties and its tin. Residential waste essentially comprises of metals, glass, plastics and food waste. The robot will be designed to manipulate the largest possible mass of the garbage. These materials are classified according to their density as in Table 7.

The mass design will base on the material with the highest product of compaction factor, f and specific density, ρ'. From the table above, food material has the highest $\rho' \times f = 1.921$ hence it was taken *as* the design material and was used to determine the design mass.

We are going to determine the design mass of the garbage using the relation.

$$\text{Density } \rho = \frac{Mass\left(m_g\right)}{Volume\left(v\right)},$$

Table 7. Characteristics of different types of garbage

Material (Waste)	Design case	Density (kg/m³)	Specific density, ρ' (Relative to water)	Compaction Factor, f	
Metal	Metal	7750 - 8940	8.940	0.20	1.780
Glass	Float glass	2000 - 8000	8.000	0.12	0.960
Plastic	PVC	1000 - 1467	1.467	1.11	1.628
Organics	Food material	96 - 194	0.195	9.85	1.921

$$m_g = \rho v,$$

$$m_g = 194 \times 0.445,$$

$$m_g = 86.33 \text{kg (say 87kg)}$$

The maximum load weight of the 240-liter Wheeler bin is 96kg. Therefore, the design garbage mass of 87kg is safe since it is well below the maximum load capacity of the bin. The gross mass Mgross of the load is the sum of the net mass of bin and the mass of the garbage:

$$M_{gross} = m_b + m_g,$$

$$M_{gross} = 13 + 87,$$

$$M_{gross} = 100 \text{kg}.$$

MATERIAL SELECTION

The robotic links should be designed so that they relatively have a greater strength to mass ratio so as to achieve the minimal weight of the manipulator and at the same time producing arms that can withstand the weight of the garbage tin. From Table 7, the Aluminum Alloy 300-H16 will be used owing to its relatively high yield strength though with slightly higher density than the least dense metal. The base of the manipulator should have a considerable weight so as to bring stable support to the manipulator so it will be made of Cast Alloy steel due to its relatively moderately high tensile strength and density. Moreover, cast iron is preferable as it can be easily and cheaply machined.

Table 8. Material properties

Material	Shear Modulus (G Pa)	Elastic modulus (G Pa)	Tensile strength (M Pa)	Yield strength (M Pa)	Density (kg/m³)
ASTM A36 Steel	79	200	400	250	7850
Aluminum 300-H16	25	69	180	670	2730
Aluminum 6061 Alloy	26	69	124	551	2700
Cast Alloy Steel	78	190	448	241	7300
Steel AISI 304	75	190	517	206	8000

Figure 22. Arm configuration calculations

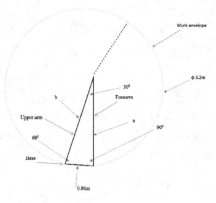

Determining Lengths of Arms from Simplified Arm Configuration

The arm will be designed in such a way that at the outmost loci of the end effector, the picking and placing of the garbage container is still done perfectly. This is made possible by first determining the lengths of the manipulator links. The upper arm will be rigid at 30 degrees from the horizontal and the forearm at a maximum of 50 degrees from the horizontal at full stretch.

$$\text{Tan}60° = \frac{a}{0.5}, = 0.86 \tan 60,$$

a = 1.4896 (**say 1.5m**)

and

$$\text{Cos}60° = \frac{0.86}{b},$$

$$b = \frac{0.86}{Cos60°},$$

= 1.72m (**say 2m**)

Verdict: Therefore, the design lengths are **1.5m for the forearm** and **2m for the upper arm**.

Determining the Weights for the Links

Weight for each arm,

$$W = \rho g V,$$

where ρ = density of the material = 2730kg/m³; g = Acceleration due to gravity = 9.81m/s²; V = Volume of the matᵉrial

The length of link l_1 = 1.5m, l_2 = 2m Thickness of links t_1 = 0.02m. Thus

$$W_1 = \rho \times g \times \pi r^2 l = 2730 \times 9.81 \times [\pi \times (0.03^2 - (0.03 - 0.02)^2) \times 1.5] = \mathbf{100.96N},$$

$$W_2 = 2730 \times 9.81 \times [\pi \times (0.04^2 - (0.04 - 0.01)^2) \times 2], = \mathbf{117.79N}$$

Note that, W_1, W_2 are weights of the arms (links) without the end effector and the load.

In the diagram of forces shown in Figure 25, MM is the mass of the servomotor to be installed in the elbow, and ME + L is the combined mass of end effector ME and load ML.

$$\Sigma F_y = \left[M_{(E+L)} \times 9.81 + W_1 + M_m \right] - Wy = 0$$
$$= \left[(5 + 100) \times 9.81 + 100.96 + 1 \times 9.81 - Wy = 0 \right]$$

$$W_y = 1140.82N$$

$$\Sigma Fy = \left[M_{(E+L)} \times 9.81 + W_1 + M_m + W_2 \right] - E_y = 0$$
$$= \left[(5 + 100) \times 9.81 + 100.96 + 1 \times 9.81 + 117.79 \right] - E_y = 0$$

$$E_y = 1258.61N$$

$$\Sigma M_W = -\frac{W_1 \times l_1}{2} - (E + L)(l_1) + M_W$$

$$\Sigma M_W = -\frac{100.96 \times 1.5}{2} - (5 + 100)(1.5) + M_W,$$

$$M_W = 233.22Nm$$

Figure 23. Arm configuration

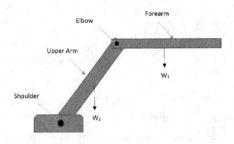

Figure 24. Arm dimensions

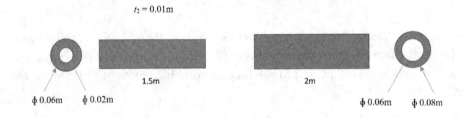

Figure 25. Diagram of forces

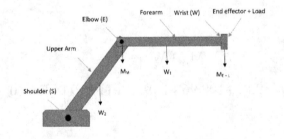

Figure 26. Diagram for moments calculation

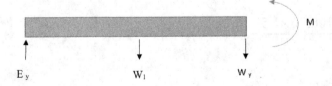

$$\sum M_E = -W_1\left(\frac{l_1}{2}+l_2\right)-(5+100)\times 9.81(l_1+l_2)-W_1\left(\frac{l_2}{2}\right)+M_E,$$

$M_g = 378.75\textbf{Nm}$

$M_S = (M_E + L)(l1 + l2),$

$M_S = (5 + 100)(1.5 + 2),$

$M_S = 262.52\textbf{Nm},$

$M_{CR} = (W_2)\times 0.05 = 117.79\times 0.05,$

$M_{CR} = 5.8895\textbf{Nm},$

$M_E = M_{E+L}\times l_1,$

$M_E = (5 + 100)\times 1.5$

$M_E = 157.5\textbf{Nm}$

POWER SUPPLY SELECTION

The source of power to the motors, manipulator and the control system is power pack of 3, 12.8V, 90Ah Lithium Iron Phosphate (LFP/ LiFePO4) rechargeable battery. The nominal supply voltage of the LFP cell amounts to 3,2V (lead-acid type: 2V/ cell). Hence, a 12,8V LFP type battery constitutes 4 cells in series connection.

The batteries will be in a series connection so as to provide a greater Voltage output. The battery has the following specifications (refer to Appendix).

If the battery suddenly and unexpectedly runs flat, power can be supplied from an external source by means of an umbilical cable.

SELECTION OF THE ACTUATOR (SERVOMOTOR)

The manipulator links are designed to be driven by servomotors. A careful selection of appropriate servomotor will give us the proper actuation of the robotic arms. The two joints we have in motion are the Shoulder and Elbow and the torques associated with these two joints are M_S=262.5**Nm** and M_g=378.75**Nm** respectively.

Table 9. Battery specifications

Technology	Specification
Length	293mm
Width	139mm
Height	235mm
Weight	12kg
Nominal Voltage	12.8 V
Capacity	90Ah
Chemistry	Lithium Iron
Cycle Life 50% DOD	5000 Cycles
Charge Voltage (recommended)	< 14.5 V
Operating Temperature	Discharge: -20°C to 50°C, Charge: 5°C to 50°C

Figure 27. LiFePO$_4$ battery open circuit voltage curve

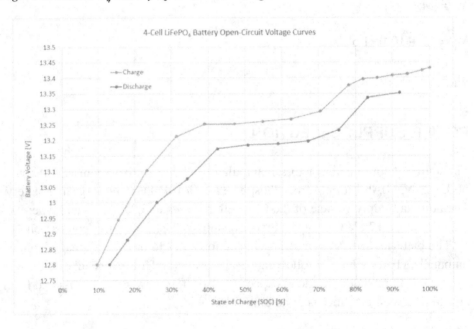

The Siemens Servomotors Catalogue, in the Appendix was used to select the motor for the robot manipulator. The suitable asynchronous motor selected was of the type **1PH7 184-B** which has a rated torque of **390Nm** (which is pretty well above M_S and M_E) and its rated voltage (of 271 V) and rated amperage (of 51 A) are appropriate for a relatively small industrial type robotic manipulator. The motor has a rated rotational speed of **400 rpm** and the rated power output of **51 kW**.

Robot Structures

In this section we will carry out the design of shafts and keys for coupling the manipulator arms. We will design the shafts and keys in such a way that the arms are properly linked to the actuators. The arms are fitted with taper lock bushing to ensure a convenient mate with shafts for torque transmission.

The shoulder and elbow joints use the same kind and size of actuator hence we will use the same size of keys and shafts. This follows that we will need the same kind of fixtures to keep in stock, thus making maintenance easier. Considering a torque rating of 390Nm for the motor and an *ulti*mate yield strength of 685MPa for the shaft mate*rial,* we can determine the shaft diameter.

Shafts

$$T = \frac{\pi}{16} \times \tau \times d^3$$

where T = Torque, τ = Shear Stress, d = Diameter of shaft

$$390 = \frac{\pi}{16} \times 685 \times 10^6 \times d^3, \text{d} = 0.0143, = 14.3 \text{ mm}$$

Verdict: The standard shaft diameter is 15mm

Keys

Design of the Key Under Shear Stress

For the shaft diameter obtained, the corresponding key thickness is 6mm, from the Appendix

$$T = L \times w \times \tau_k \times \frac{d}{2},$$

where L = Length of key, W = Width of key = 6mm, τ_k = Shear stress of key, d = Diameter of shaft,

$$390 = L \times 0.006 \times 685 \times 10^6 \times \frac{0.015}{2}, \text{L} = 0.0127\text{m}, = 12.7 \text{ mm}$$

Design of Key Under Crushing

$$T = L \times \frac{t}{2} \times c_k \times \frac{d}{2},$$

where c_k = Yield stress / stress factor,

$$390 = L \times \frac{0.006}{2} \times \frac{685 \times 10^6}{2} \times \frac{0.015}{2}, \text{L} = 0.0506 \text{ m}, = \textbf{50.6 mm}$$

Verdict: Standard size of key is length 55mm, and width 6mm.

We will use square keys to ensure protection of device against overloading. They also allow for easy assembling and disassembling during maintenance and repairs. In addition, square keys help align shafts perfectly.

DESIGN OF END EFFECTOR

It is the end effector that interacts directly with the object. The end effector is also known as the hand of the robot. There are different kinds of end effectors namely the grippers, grinders, sprayers, welders and vacuum. In this project, we choose a gripper of the active type. An active type gripper controls the gripping force hence preventing the garbage tin from getting damaged. A gripper of a pick and place robot is simple to design and manufacture. A gripper can be powered by pneumatics and hence minimizing the running cost

MATERIAL SELECTION

Material selection should be done carefully as this will result in reduction of the cost of the robot. Materials selection is done so carefully that the minimum design requirement is not compromised. In this project we will use a combination of iron, mild steel and reinforced plastic. The gripper and upper body can be made of fiber reinforced plastic (FRP). FRP provides adequate strength for holding and lifting the object, and also significantly reduces the weight of the robot. The selection of material usually varies on the load to be lifted. However, in case we only want to reduce the cost and ensure ease of manufacturing, regardless of the resulting weight, we can safely use iron.

DESIGN OF THE CHASSIS AND THE WHEEL SYSTEM

The Chassis

The chassis (base) of the garbage collection robot will be made of *High Tensile Steel (HTS)*. This kind of steel has a relatively high tensile strength of about 517MPa. The (HTS) will be the most appropriate as the chassis will be under high stresses by supporting large masses of batteries, manipulator, motors and the load.

The Wheel System

The wheels will be out-surfaced with Goodyear Heavy Duty Off Road tire to ensure stable cruising even on a rough, uneven terrain. To each wheel, there will be attached encoders to help in the sending of feedback (about rolling speed) to the controller. The controller will be controlling the speed of the robot depending on the information sent from the proximity sensors.

The robot wheel system will constitute four wheels, two in the front and two in the rear. The wheels are linked to an independent suspension system to ensure stable movement on uneven terrain. Each of the wheels will be controlled by its own motor for rolling and braking. The front wheel system will be linked to a swerve drive system. The swerve mechanism will provide the desired turning capability for the robot. The two wheels effectively work as a single wheel as they are driven by the same motor and roll at the same speed. The pretty separation and independence of the wheels was to ensure stability of the vehicle.

Figure 28. The wheel system

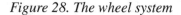

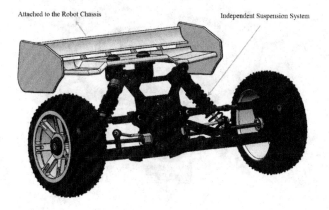

DESIGN OF SHOULDER JOINT (ORIENTATION CONTROL)

In this section, a detailed design of the Shoulder joint will be done.

Drive Configuration

An illustration is better explained by the Figure 29.

Motion Study

The motion executed at the base of the manipulator is either yawl or pitch. The two motions will be studied below.

Figure 29. Configuration of joint motion control mechanism

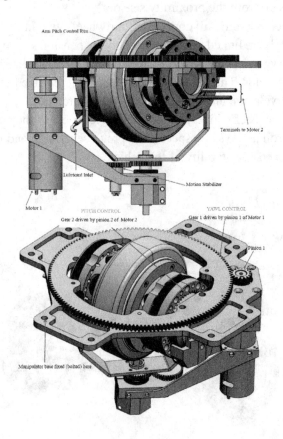

Figure 30. a.) Yawl at the shoulder b.) pitch at the shoulder

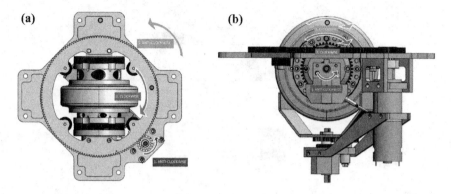

Yaw: Sequence of Motion Execution

Objective: To execute left (anti-clockwise) yawl.
1. Pinion 1 driven anti-clockwise by Motor
2. Gear 1 driven clockwise by Pinion 1
3. Base Plate driven anticlockwise.

Pitch: Sequence of Motion Execution

Objective: Lifting arm from the shoulder.
1. Pinion 2 driven anti-clockwise by Motor 2
2. Gear 2 driven clockwise by Pinion 2
3. Rim driven clockwise by Gear 2 (Note: Gear 2 is fixed to the Rim)

DENAVIT-HARTENBERG (D-H) PARAMETERS

The D-H parameters (Figure 31) can be used to define the assignments to joint and all the parameters that define the robot. These are four parameters related to a certain convention used for connecting reference frames in design to all the links in a spatial chain of a kinematic field, or a manipulator.

Table 10 depicts the related 6 joints parameters the robotic arm has and will be used to determine the position and the orientation of the body which is crucial in finding the composition of coordinate transformation in between consecutive frames.

Figure 31. Denavit-Hartenberg parameters

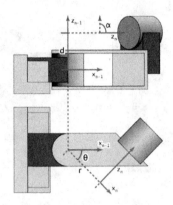

Table 10. 6 degrees of freedom manipulator joint limits

Joint	Lower Limit	Upper Limit
1	-160	160
2	-120	95
3	-160	160
4	-119	119
5	-119	119
6	-180	180

Figure 32. Global and the moving coordinate links of the manipulator arm in home position

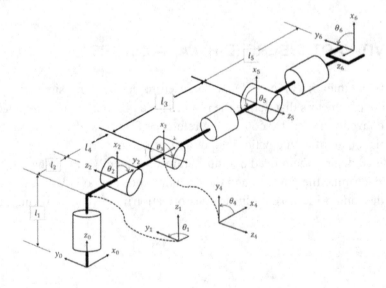

Figure 33.

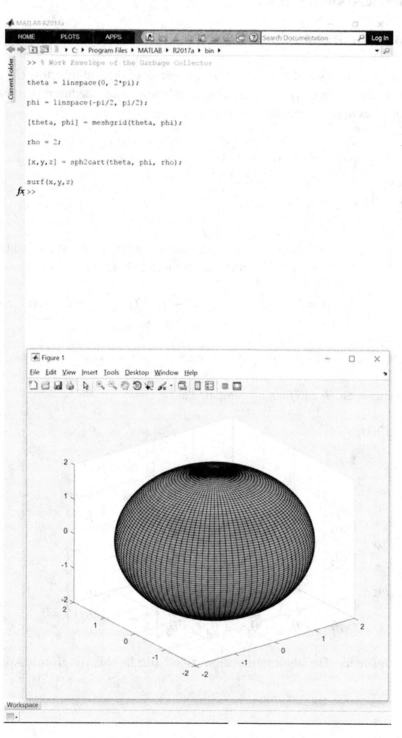

MANIPULATOR DESIGN

Computing the Jacobian MATRIX

By taking a close look at the global position vector of the end-effector ($G\theta$) for the 6-DOF manipulator extracted from Equation below, it was deduced that the joint angle $\theta6$ d es not have an effect on such position:

$$G\theta = \begin{bmatrix} p_x \\ p_y \\ p_z \end{bmatrix} = \begin{bmatrix} f_1(,_1,_2,_3,_4,_5) \\ f_2(,_1,_2,_3,_4,_5) \\ f_3(,_2,_3,_4,_5) \end{bmatrix}$$

where f_1, f_2, f_3 are Cartesian coordinate position functions of all the joint angles θ_1, θ_2, θ_3, θ_4, and θ_5 as indicated in the expressions below:

$$f_1(\theta_1, \theta_2, \theta_3, \theta_4, \theta_5) = l_2 c_1 + l_3 c_1 c_2 + (l_5 s_{45} + l_4 s_4)(s_1 s_3 + c_1 s_2 c_3) + (l_5 c_{45} + l_4 c_4)(c_1 c_2)$$

$$f_3(\theta_2, \theta_3, \theta_4, \theta_5) = l_5(s_{45} c_2 c_3 - c_{45} s_2) + l_4(c_2 c_3 s_4 - c_4 s_2) - l_3 s_2 + l_1$$

The Jacobian, J_θ can then be computed as:

$$[D] = [J_\theta][D_\theta]$$

where:

$$D = \begin{bmatrix} dpx \\ dpy \\ dpz \end{bmatrix}, \begin{bmatrix} \dfrac{\partial f_1}{\partial \theta_1} & \dfrac{\partial f_1}{\partial \theta_2} & \dfrac{\partial f_1}{\partial \theta_3} & \dfrac{\partial f_1}{\partial \theta_4} & \dfrac{\partial f_1}{\partial \theta_5} \\ \dfrac{\partial f_2}{\partial \theta_1} & \dfrac{\partial f_2}{\partial \theta_2} & \dfrac{\partial f_2}{\partial \theta_3} & \dfrac{\partial f_2}{\partial \theta_4} & \dfrac{\partial f_2}{\partial \theta_5} \\ \dfrac{\partial f_3}{\partial \theta_1} & \dfrac{\partial f_3}{\partial \theta_2} & \dfrac{\partial f_3}{\partial \theta_3} & \dfrac{\partial f_3}{\partial \theta_4} & \dfrac{\partial f_3}{\partial \theta_5} \end{bmatrix}, D, = \begin{bmatrix} d\theta_1 \\ d\theta_2 \\ d\theta_3 \\ d\theta_4 \\ d\theta_5 \end{bmatrix}$$

Note on reference: The whole formulation of the J_θ (the Jacobian) is given in Appendix.

Work Envelope

Since when crossing singular parametric surfaces, may result in motion difficulties, then the center of the manipulator should be avoided. Finally, the work envelop of the robot manipulator was described by 3 parametric surfaces, constrained by the joint values below:

$f^{(1)}(\theta_1, \theta_2); -160 \leq \theta_1 \leq 160$ and $-120 \leq \theta_2 \leq 95$

$f^{(2)}(\theta_2, \theta_4); -100 \leq \theta_2 \leq 95$ and $0 \leq \theta_4 \leq 41$

$f^{(3)}(\theta_2, \theta_4); -100 \leq \theta_2 \leq 95$ an$_d$ $-41 \leq \theta_4 \leq 0$

where: $f^{(1)}$, $f^{(2)}$ and $f^{(3)}$ are the global position vectors of the end-effector and constrained by some sets of singularities listed respectively below:

$s^{(1)} = [\theta_4 = 0, \theta_5 = 0]$

$s^{(2)} = [\theta_1 = 160, \theta_3 = 90, \theta_5 = 0]$

$s^{(3)} = [\theta_1 = -160, \theta_3 = 90, \theta_5 = 0]$

The work envelope was generated in Mat Lab and is shown in Figure below. The work envelope of the robot is described by the volume space of a sphere of radius 3.5m. This follows that the wrist can reach any point within a radius of 3.5m

Verdict: The work envelope of the Garbage Collector is a sphere of radius **3.5m**

MANIPULATOR CONFIGURATION

The diagram below illustrates the arm **config**uration of the Garbage Collection Robot.

Number of Degrees of Freedom

The Number of Degrees of Freedom of the robotic arm was calculated from the Gruebler equation

$N_{DOF} = 3(n-1) - 2j_p - j_h$

Figure 34. The work envelope of the 6-DOF robotic manipulator: a) isometric; and (b) cross sectional views

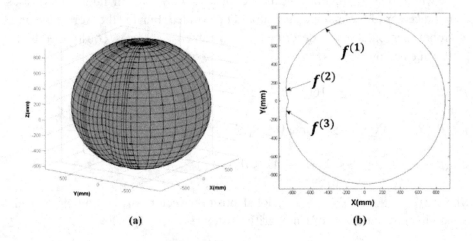

(a)   (b)

where N_{DOF} = Number of Degrees of Freedom, n=number of links, j_p=primary joints (sliding or pinned), j_h=higher-order joints (gear or cam). Therefore

$$N_{DOF} = 3(5 - 1) - 2(3) - 0 = 6$$

The device has 3 concurrent and prismatic (rotary) joints, together with 3 axes in its wrist, adding to 6 Degrees of freedom (**6 DOF**). The end effector (hand) can easily be manipulated to take any orientation in the entire work envelope (workspace) which is almost spherical as depicted in Figure below. The ability to reach over the garbage bin and twist and/ or tilt the hand, makes it ideal for picking, empting and placing the garbage container back to its designated position. The end effector (hand) can easily be manipulated to take any orientation in the entire work envelope (workspace) which is spherical. The ability to reach over the garbage bin and twist and/ or tilt the hand, brings it ideal for picking, empting and placing the garbage container back to its designated position.

THE BIRTH OF THE SCAVENGER-THE GARBAGE COLLECTION ROBOT

The Designer of the Scavenger used metal, plastic and rubber to construct the skeleton of the robot. He then installed electronic components, and developed a program to aid the execution of operation.

Figure 35. Early stages of prototyping

PRINCIPLE OF OPERATION

The principle of operation of the garbage collector is depicted in a flow chat in Figure below. The Robot uses color detecting sensors to see the garbage bins to pick up.

SIMULATION OF THE MANIPULATOR IN MATLAB

Simulation allows a detailed deep study of the structure, function and characteristics of a system. As the complex the system under investigation increases. The role of simulation becomes more important as the complexity of the system to be observed increases. The appropriate simulation tools can enhance design, operation, and development of the robotic systems.

Figure 36. The flow chart of operation of the garbage collector

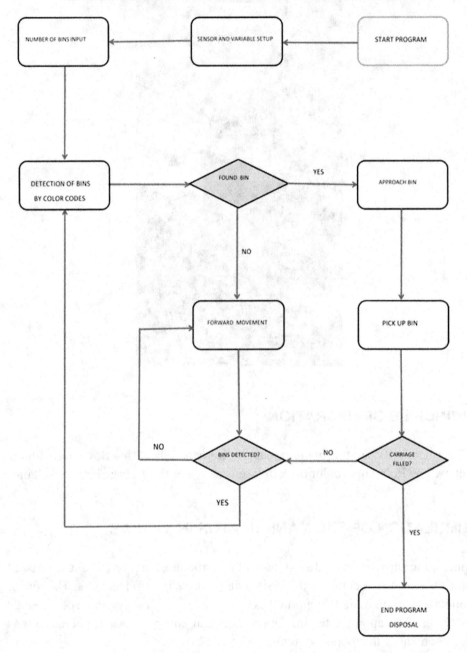

FLOW CHART OF OPERATION FOR THE SCAVENGER—THE GARBAGE COLLECTION ROBOT

Code Generation

The code for the simulation of the arm was generated in the Integrated Development Environment (IDE) namely Code Blocks using C++ Language. The code was generated as shown in Figure 37.

Kinematics

The forward and inverse kinematics of the manipulator was done in **Mat Lab** software and the results shown in Figure 38 were obtained.

NODAL ANALYSIS

Nodal analysis of the manipulator was carried out in Mat Lab (see Figure 39).

SYSTEM PERFORMANCE

Resolution

Resolution in robotics is defined as a feature that is determined from the robot control unit and this feature mainly depends on the position feedback sensing. Programming resolution can be defined as the smallest possible position change in the robot programs. It is also known as the Basic Resolution Unit (BRU). In case of IRB2000 ABB robot (class of the Garbage Collection Robot), it is about **0,125 mm** on the linear axis.

Control resolution can be defined as the smallest position increment that can be sensed by the feedback device. Assuming there is some optical encoder directly attached to the rotating shaft and do emit 1000 pulses in 1 revolution made by the shaft. Such an encoder will be evidenced to emit 1 pulse per each 0.36° of the angular displacement made by the shaft. Therefore, the current axis of rotation has a control resolution of unit **0.36°**. It can be confidently said all angular changes less than this 0,36° will be undetectable.

Verdict: In order to achieve the best performance by optimizing the system resolution, the designer opted to make the programming resolution equal to the control resolution.

Figure 37. Code generation in code blocks IDE

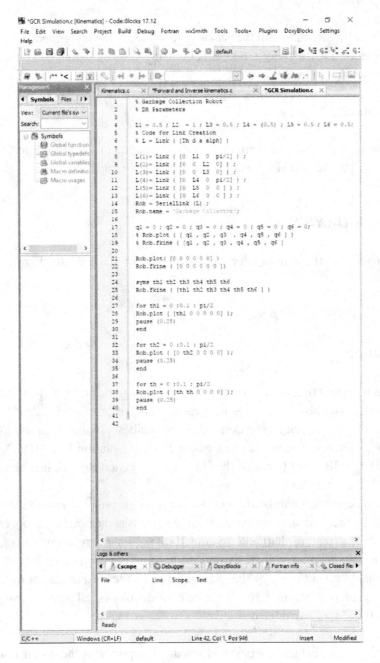

Design of a Garbage Collection Robot

Figure 38. Plot of the garbage collector

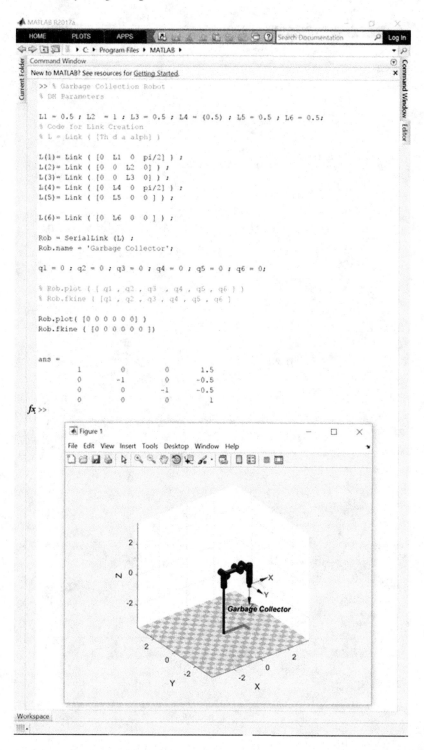

Figure 39. Nodal analysis of the manipulator

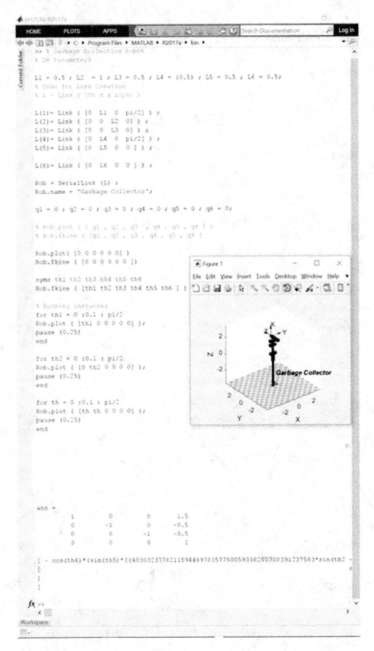

Accuracy

Accuracy is demonstrated by the ability of the robot to place its wrist at an intended target within the work space. Accuracy can be precisely defined by its relation to spatial resolution (Figures 40 and 41).

System inaccuracy of the Garbage Collection Robot is taken to be:

$$Inaccuracy = \frac{1}{2}BRU\left(Basic\ Resolution\ Unit\right) = \frac{1}{2} \times 0.125 = 0.0625,$$

$$Accuracy\ of\ the\ Robot = \frac{BRU + Mechanical\ Accuracy}{2} = 1 - 0.0625 = 0.9375$$
$$= 94\%$$

DETAILED DESIGN OF COMPONENTS

Robot Joints

The joint plays a pivotal role in robot design. It permits relative motion between each pair of links or arms of robot arms. The joint provides governed relative motion between the two links (that is input and output). There are various types of joints including linear joints, rotational joints, orthogonal joints, revolving joints and twisting joints. Out of these joints, the rotational type of joints was found to be the easiest to manufacture and best suits the demand. Therefore, hinged joints will

Figure 40. Accuracy depicted in two-dimensional frame

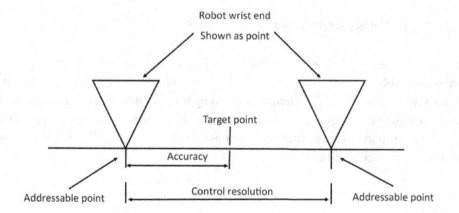

Figure 41. Accuracy related to spatial resolution with mechanical inaccuracies represented as statistical distribution

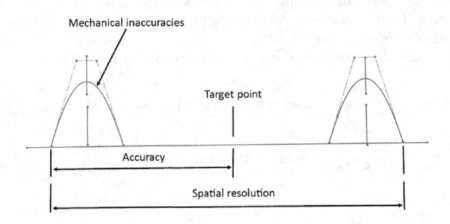

be used. Hinged joints are of relatively low cost and at the same time can satisfy what we require. However, various joint type will be used depending on the weight of load to be lifted.

Design of Body

The body positions the object in work envelope of the robot. Through the implementation of value engineering, the body can be designed in such a way that its weight will be of minimum value through the use of less material. We need to optimize both material and manufacturing cost. To achieve this, the arm should be made up of various components which are then assembled together, hence saving material and significantly reducing the cost.

Design of Wrist Assembly

The wrist assembly is used to orient the object within the work envelop. At the wrist assembly, an end effector is attached and this does the gripping job. The wrist assembly is a 3 DOF system, described by roll, pitch and yaw. To take care of some design complexities, the end effector should be directly attached to the arm by means of hinged joints. Such design considerations can significantly impact in minimizing the cost of the robot.

DESIGN OF CONTROL SYSTEM

In this section, the design of the system is analyzed. The system will be made to constitute several modules including DC motors, motor controller, IR sensors, remote controller unit, radio transmitter and radio receiver. The main part, which connects all these modules, are the micro controller ICs. It acts as the brain of the system, performing all the processing between the inputs and the outputs.

The Microcontroller

The Arduino microcontroller will be used to establish control of the garbage Collection Robot (GCR). The advantages of Arduino microcontroller over its counterparts is its relatively cheap price in the market, easiness in reprogramming. Arduino software will be used to feed the machine control code. Signals sent from color code detectors and proximity sensors will be used in controlling the device using the program below.

NAVIGATION CONTROL CODE

Sensors

Sensors will be used to relate measured variables to some electric signals in the digital form in order that the obtained information is then fed into the microcontroller for processing. The angles and position of the arms will be controlled by the proprioceptor sensor. Since the robot should be autonomous, and goes for a mission on its own, it should be equipped with proximity sensors to avoid collision with other traffic.

An example of the arduino code for capacitive proximity sensor can be seen in Figure 42. The circuit of the capacitive proximity sensor can be seen in Figure 43.

Setup of Components

- 10 M Ohm resistor was installed between the pins 2 and 4. Sensor wire was connected to pin 2. The wire was then connected to an aluminum foil.
- Pin 9 was connected to LED and then to GND. (The brightness of the LED was controlled by the pulse-width modulation (PWM).) Same setup was done for pin 10.

Figure 42. Proximity sensor test code

```
Garbage_Colection_Robot_Proximity_Sensor.txt.ino.ino | Arduino 1.8.5 (Windows Store 1.8.10.0)     —     □     ×
File Edit Sketch Tools Help

Garbage_Colection_Robot_Proximity_Sensor.txt.ino.ino §

CapacitiveSensor   cs_4_2 = CapacitiveSensor(4,2);
// 10M resistor between pins 4 & 2, pin 2 is sensor pin
// add a foil if desired

int receivePin = 2;
int sendPin = 4;
int ledPin1 = 10;
int ledPin2 = 9;

void setup()
{
   cs_4_2.set_CS_AutocaL_Millis(0xFFFFFFFF);
   // above: turn off autocalibrate on channel 1 - just as an example
   Serial.begin(9600);
   pinMode(ledPin1, OUTPUT);
   pinMode(ledPin2, OUTPUT);
}

void loop()
{
   long start = millis();
   long total =  cs_4_2.capacitiveSensor(30);

   Serial.print(millis() - start);
   // check on performance in milliseconds

   Serial.print("\t");
   // tab character for debug window spacing

   Serial.print(total);
   // print sensor output

   Serial.println();

   // parameters to set level for PWM output
   float level1;
   float level2;
   float threshold1 = 0;
   float threshold2 = 100;
   float max = 3000;

   if (total < threshold1) {
      level1 = 0;
   }
   else {
      level1= map(total,threshold1,max,0,1023);
   }

   if (total < threshold2) {
      level2 = 0;
   }
   else {
      level2= map(total,threshold2,max,0,1023);
   }

   analogWrite(ledPin1,level1);
   analogWrite(ledPin2,level2);

   delay(1);   // arbitrary delay to limit data to serial port
}
Done Saving.
```

Figure 43. Circuit of the capacitive proximity sensor

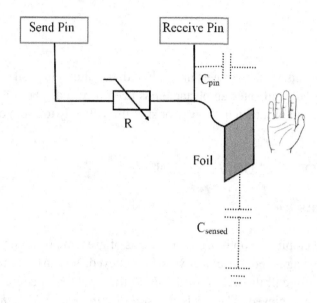

Procedure and Discussion of Results

- A resistor of a high value of 10M Ohm was used between the send pin and the receive pin.
- The Capacitive Sensing Library was used to measure the time elapse for an RC circuit to get back to state 1 after it has been forced to state 2. The time is a several multiples of RC (Time Constant). Therefore, with fixed resistance R, the time measured is the Capacitance Measurement C.
- The Capacitance of a Human body was estimated to lie between the values 100 and 400 p F. If the resistance is 10 M Ohms, and C is 100 pF, it then follows that RC is 1 m s.
- Resistor R effects sensitivity, experiment was performed with values in the range 50K Ohms – 50M Ohms. It was observed that larger values of R values yield larger values of the sensor.
- Receive pin was set to be the sensor pin and aluminum foil was attached to achieve higher sensitivity of the Capacitance.
- The threshold 1, threshold 2, and maximum values were determined by experimentation. In the developed code above, LED 1 was set to show response when the hand was considerably close. LED 2 could show response to anything.

- That helps when there is need to manually reset Arduino when the unit gets bizarre

Conclusion

When the hand gets close to or gets in touch with the aluminum foil sensor, a LED light turns on. This indicates an obstacle has been detected. The robot receives a warning signal and will then turn away or stop immediately to avoid collision.

CONTROL OF THE DRIVE TRAIN

Design Goals

The idea of developing an optimized driveline is of great importance though seems to be somewhat vague. For the objective to be achieved, certain design goals should be set. In order to call the design of the drive train optimal, there are essentially two goals to be achieved. The goals are *stability at relatively high speeds* and *maneuverability at relatively low speeds*.

High Speed Stability

In the field of robotics, a speed above 3 meters per second can be called as high speed.t then follows that, the design should be such that the drivetrain and the chassis will be capable of maintaining that high speed while moving in a circular path of lane of 1meter with a radius of 3 meters.

Low Speed Maneuverability

To achieve maneuverability at low speeds, the driveline should be capable be capable of executing specific operations, and at standstill or considerably low speeds. The driveline should be capable to achieve zero radius turning with respect to a point inside the chassis.

   With the vision to achieve the above mentioned goals, respective program codes will be developed using Arduino Software.

Swerve Drive Electrical Assembly

The electrical assembly of the robot swerve drive wheel system constitutes:

- Arduino mega 2560 microprocessor and board
- Standard 12 V battery,
- Bump sensor
- Speed controller
- A 300° potentiometer.

A potentiometer was used to check how far left wise or right wise the wheels turn.

- **Encoder** (see Figure 44)
- **Potentiometer** (see Figure 45)
- **Potentiometer and Encoder Testing** (see Figure 46)

Conclusion

As was expected, the driveline of the robot was capable of achieving a turning of *zero radius* about the rear swerve wheel system. When the wheel system turned to the condition of zero radius, the robot could seamlessly rotate in either clockwise or anti-clockwise direction through application of a throttle in forward or reverse direction. The experiment shows the goal of zero radius turning was achieved by the robot being able to turn about a certain point inside the chassis.

STRESS ANALYSIS (VON MISES)

A detailed stress analysis (von Mises) for critical components was carried out in SolidWorks. The v*on Mises stress* is a computed value used to check if some given material is liable to yield or fracture after application of some external load. A safe design has a yield stress higher greater than the von Misses stress. The stress analysis was done for the following components:

1. Upper arm
2. Shoulder blade
3. Wrist
4. Motor shaft.

Upper Arm

Static Nodal Stress Analysis

Figure 47 illustrates the static nodal stress analysis of an upper arm.

Figure 44. Arduino code for encoder

```
Encoder_for_GBR | Arduino 1.8.5 (Windows Store 1.8.10.0)          —  □  ×
File Edit Sketch Tools Help

Encoder_for_GBR §

// Connect Encoder to Pins encoder0PinA, encoder0PinB, and +5V.

int val;
int potPin = 2;
int encoder0PinA = 3;
int encoder0PinB = 4;
int encoder0Pos = 0;
int encoder0PinALast = LOW;
int n = LOW;
long lastChange = 0;
long timeNow = 0;
int lastPos = 0;
int rpm;

void setup() {
  pinMode (encoder0PinA, INPUT);
  pinMode (encoder0PinB, INPUT);
  Serial.begin (9600);
  Serial.print("test!");
}


void loop() {
  n = digitalRead(encoder0PinA);
  if ((encoder0PinALast == LOW) && (n == HIGH)) {
    if (digitalRead(encoder0PinB) == LOW) {
      encoder0Pos--;
    } else {
      encoder0Pos++;
    }
  }
  timeNow = millis();
  long timedif = timeNow-lastChange;
  if(timedif > 1000){
  lastChange = timeNow;
  rpm = (60*(encoder0Pos-lastPos))/(360.0)
  Serial.println(rpm);
  lastPos = encoder0Pos;
  }
  encoder0PinALast = n;
}

Done Saving.
```

Figure 45. Arduino code for potentiometer

```
Potentiometer_for_Garbage_Collection_Robot_Swerve_Drive | Arduino 1.8.5 (Windows Store 1.8.10.0)    —    □    ×
File Edit Sketch Tools Help

Potentiometer_for_Garbage_Collection_Robot_Swerve_Drive §

#include <Servo.h>

int potPin = 2;  // select the input pin for the potentiometer // select the pin for the LED
int value;
float angle;
Servo servo;
int goal = 122;
int error;
int command;

void setup() {
  servo.attach(10);
  Serial.begin (9600);
  Serial.print("test!");
  //goal = Serial.read();
}

void loop() {

  if(Serial.available()>0){
      goal = Serial.parseInt();
    }
  value = analogRead(potPin);
  angle = map(value, 40, 970, 0, 360);
  servo.write(110);
  error = goal - angle;
/*
  if() > 5){
  if((goal < 180) && (angle > 180) {
    if (360 - angle) + goal < (goal - angle) {
      servo.write(104);
    }
    else{
      servo.write(84)
    }
  }
  }*/

  if(abs(error) > 180){
    error = error - 360;
  }
  servo.write(94-error);

  /*
  if( error > 5 && error < 180) || (error > -360 && error < -180) ){
    servo.write(94);
  }else if( error < -5 && error > -180 || (error > 180 && error < 360) ){
    servo.write(104);
  }else{
    servo.write(94);
  }
  */

  //command = error*.3 + 90;
  //servo.write(command);
  Serial.print(error);
  Serial.print(" and ");
  Serial.println(angle);
}
```

Figure 46. Test code for potentiometer and encoder

```
Potentiometer_and_Encoder_Testing | Arduino 1.8.5 (Windows Store 1.8.10.0)
File Edit Sketch Tools Help

Potentiometer_and_Encoder_Testing §

#include <Encoder.h>
#include <Servo.h>
// select the input pin for the potentiometer // select the pin for the LED
int potPin = 2;
int value;
float angle;
Servo angleMotor;
int goalAngle = 100;
int goalSpeed = 10;
int angleError;
int speedError;
int command;
Servo speedMotor;
int speedM;
int speedPower = 0;
int encoder0PinA = 3;
int encoder0PinB = 2;
volatile int encoder0Pos = 0;
volatile int encoder0PinALast = LOW;
int n = LOW;
long lastChange = 0;
long timeNow = 0;
int lastPos = 0;
Encoder WheelReader(3, 2);

void setup() {
    angleMotor.attach(10);
    speedMotor.attach(9);
    Serial.begin(9600);
    Serial.print("test!");
    pinMode (encoder0PinA, INPUT);
    pinMode (encoder0PinB, INPUT);
    attachInterrupt(0, changey, CHANGE);
}

void changey(){
    n = digitalRead(encoder0PinA);
    if ((encoder0PinALast == LOW) && (n == HIGH)) {
        if (digitalRead(encoder0PinB) == LOW) {
            encoder0Pos--;
        } else {
            encoder0Pos++;
        }
    }
    encoder0PinALast = n;
}

void loop() {
    //speedMotor.write(110);
    if(Serial.available()>0){
        //goalAngle = Serial.parseInt();
        goalSpeed = Serial.parseInt();
    }
    value = analogRead(potPin);
    angle = map(value, 40, 970, 0, 360);
    angleError = goalAngle - angle;
    if (abs(angleError) > 180){
        angleError = angleError - 360;
    }
    //angleMotor.write(94-angleError);

    //Get Speed
    timeNow = millis();
    long timedif = timeNow-lastChange;
    //Serial.println(encoder0Pos);
    if (timedif > 100){
        Serial.print(timedif);
        Serial.print("    ");
        speedM = encoder0Pos-lastPos;
        Serial.println(speedM);

        speedError = goalSpeed - speedM;
        speedPower -= (speedError/5);
        speedMotor.write(speedPower);
        Serial.print("    ");
        Serial.println(speedPower);
        lastChange = timeNow;
        lastPos = encoder0Pos;
    }
}
```

Figure 47. Static nodal stress analysis

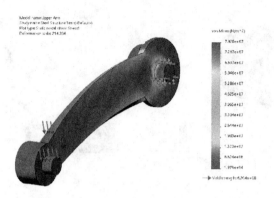

Results Interpretation

The highest von Mises stress is evidenced to occur around the region where the Upper Arm is attached to the Shoulder Blade and this maximum von Mises stress is $7.928 \times 10^7 N/m^2$ which is well below the yield stress for alloy steel of 6.204×10^8 N/m^2. Therefore, the design is safe.

Static Displacement and Strain Analysis

Figure 48 illustrates the static displacement and strain analysis of an upper arm.

Figure 48. Static a.) displacement b.) strain analysis

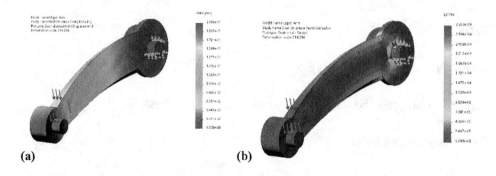

(a)   (b)

Figure 49. Static a.) nodal stress b.) displacement analysis

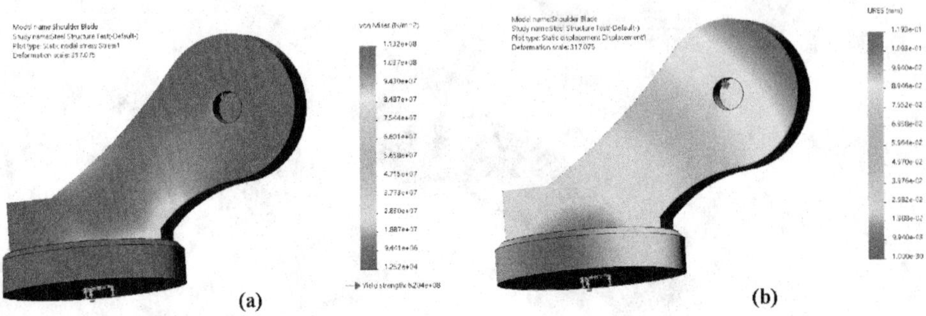

(a)   (b)

Shoulder Blade

Static Nodal Stress Analysis

Figure 49 illustrates the static nodal stress analysis of a shoulder blade.

Results Interpretation

The highest von Mises stress spread evenly away the shoulder and this maximum von Mises stress is 1.132×10^8 N/m^2 which is well below the yield stress for alloy steel of 6.204×10^8 N/m^2. Therefore, the design is safe.

Static Strain Analysis

Figure 50 illustrates the static strain analysis of a shoulder blade.

Wrist

Static Nodal Stress Analysis

Figure 51 illustrates the static nodal stress analysis of a wtist.

Results Interpretation

The highest von Mises stress occurs at the hinge where the wrist links the forearm and this maximum von Mises stress is 3.775×10^4 N/m^2 which is well below the yield stress for ductile iron of 5.515×10^8 N/m^2. Therefore, the design is safe.

Figure 50. Static strain analysis

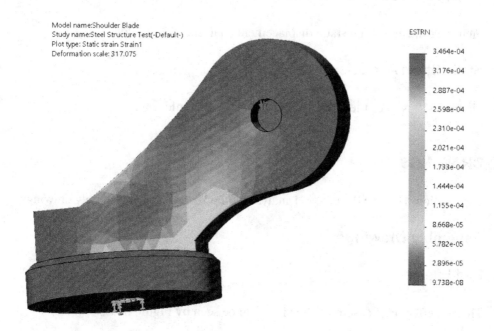

Figure 51. Static nodal stress analysis

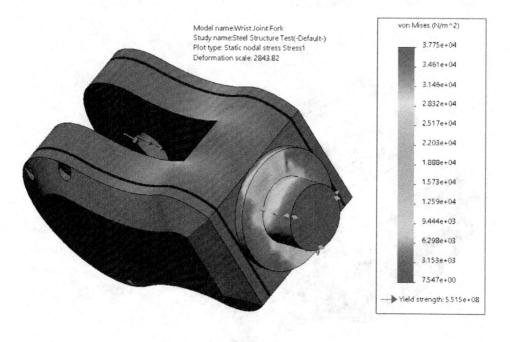

Static Displacement Analysis

Figure 52 illustrates the static displacement analysis of a shoulder blade.

Static Strain Analysis

Figure 53 illustrates the static strain analysis of a shoulder blade.

DRAWINGS

This section will give the detailed analysis of the Working and Assembly drawings.

Assembly Drawings

End Effector

The end effector of assembly drawings can be seen by Figure 54.

Figure 52. Static displacement analysis

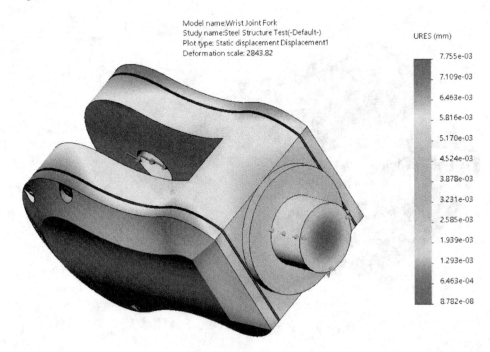

Figure 53. Static strain analysis

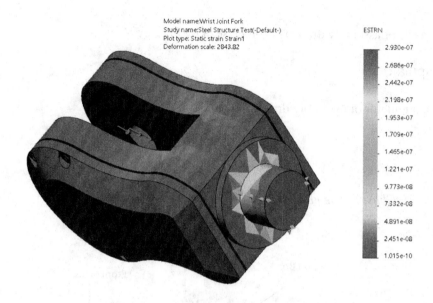

Figure 54. End effector

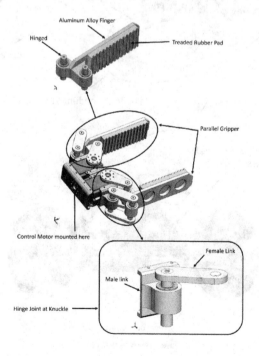

Chassis

The chasis of assembly drawings can be seen by Figure 55.

Manipulator

The manipulator of assembly drawings can be seen by Figure 56.

Figure 55. Chassis assembly

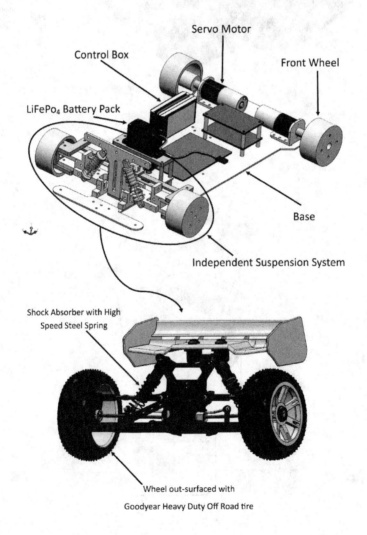

Figure 56. Manipulator system

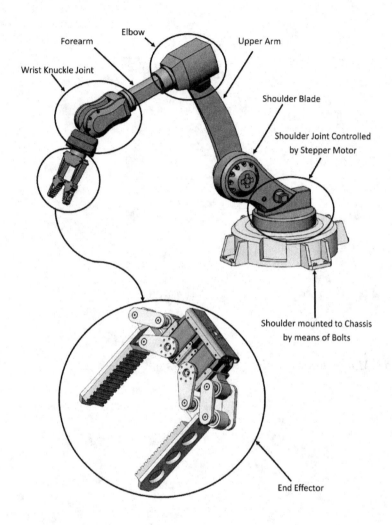

Working Drawings

End Effector

The end effector of working drawings can be seen by Figure 57.

Manipulator

The manipulator of working drawings can be seen by Figure 58.

Figure 57. Assembly drawing

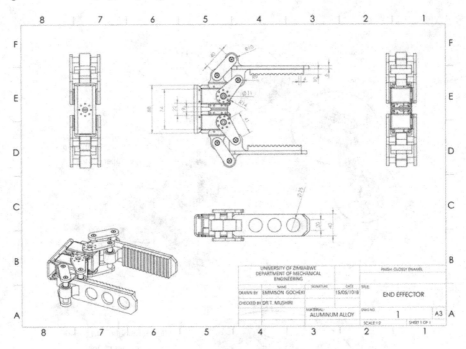

Figure 58. Manipulator details

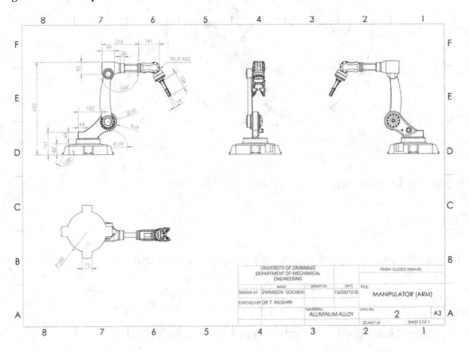

Bill of Materials

End Effector

The end effector of bill materials can be seen by Figure 59.

Manipulator

The manipulator of bill materials can be seen by Figure 60.

COSTING

The total manufacturing cost of the Garbage Collection Robot (GCR) was calculated. Only two classes of costs were considered, that is costs of components (ready-made) and processing costs. The processing costs constitute of welding, drilling, milling, and finishing costs. The grant cost was evaluated in Table 11.

Figure 59. Bill of materials for end effector

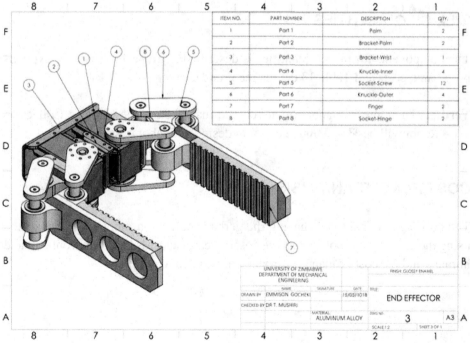

ITEM NO.	PART NUMBER	DESCRIPTION	QTY.
1	Part 1	Palm	2
2	Part 2	Bracket-Palm	2
3	Part 3	Bracket-Wrist	1
4	Part 4	Knuckle-Inner	4
5	Part 5	Socket-Screw	12
6	Part 6	Knuckle-Outer	4
7	Part 7	Finger	2
8	Part 8	Socket-Hinge	2

UNIVERSITY OF ZIMBABWE
DEPARTMENT OF MECHANICAL ENGINEERING

FINISH: GLOSSY ENAMEL

DRAWN BY: EMMISON GOCHEKI
CHECKED BY DR T. MUSHIRI
DATE 15/05/1018
TITLE: END EFFECTOR

MATERIAL: ALUMINUM ALLOY
DWG NO. 3
A3
SCALE 1:2
SHEET 3 OF 1

Figure 60. Bill of materials for manipulator

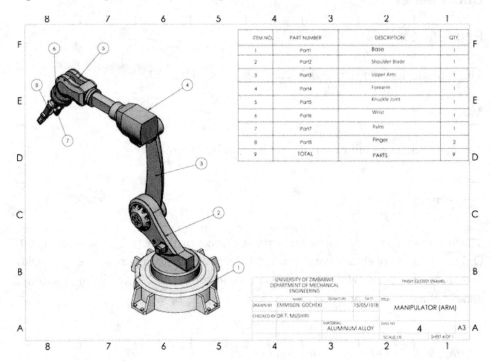

ITEM NO.	PART NUMBER	DESCRIPTION	QTY.
1	Part1	Base	1
2	Part2	Shoulder Blade	1
3	Part3	Upper Arm	1
4	Part4	Forearm	1
5	Part5	Knuckle Joint	1
6	Part6	Wrist	1
7	Part7	Palm	1
8	Part8	Finger	2
9	TOTAL	PARTS	9

UNIVERSITY OF ZIMBABWE
DEPARTMENT OF MECHANICAL
ENGINEERING

DRAWN BY EMMISON GOCHEKI   15/05/1018   MANIPULATOR (ARM)
CHECKED BY DR T. MUSHIRI

MATERIAL ALUMINUM ALLOY

BILL OF QUANTITIES

The components for bill of quantities is described in Table 11. Table 12 explains processing costs and Table 13 details the grand cost.

Verdict: Therefore, the approximate cost to manufacture the Garbage Collection Robot will be **$US 3 000**, excluding design and labor costs.

COST BENEFIT ANALYSIS

The cost benefit analysis of having the garbage collection robot will be analyzed using the cost of the robot against the cost of health hazards to the environment and humans and the cost of manual garbage collection and sorting.

Table 11. Bill of quantities for components

COMPONENT	UNIT PRICE ($US)	QUANTITY	TOTAL COST ($US)
Chassis	70	1	70
Gripper	20	1	20
Forearm	50	1	50
Upper Arm	50	1	50
Shoulder Motor	100	1	100
Elbow Motor	100	1	100
Wrist Motor	100	1	100
Arduino	100	1	100
Battery	100	2	200
Keys	5	4	20
Shafts	10	4	40
Encoder	100	2	200
Proximity sensor	20	5	100
Camera	50	4	200
Tracking System	-	-	800
Wiring	-	-	50
Wheels	50	4	200
Control systems	-	-	80
TOTAL			**2 580**

Table 12. Processing Costs

Operation	Cost
Drilling	75
Milling	90
Welding	80
Finishing	85
Sundry expenses	90
Total	**375**

Table 13. Grand cost

Description	Cost
Components	2580
Process	420
Grand total	**3 000**

Disease Outbreak and Fatality

According to the World Health Organization (WHO, 2016), there is a probability of 37% that there could be an outbreak of typhoid and/ or cholera in a medium density residential area where garbage is manually collected and spend at least 3 days lying idle at dumping site (Zimbabwe National Statistics Agency, 2016). This was attributed to approximately 25% new cases of infection per day where other factors like poor hygiene take a considerably higher contribution to the spread of the bacteria. The healthcare approximately spends $50000 per week in carrying out disease prevention campaigns and in treating the infected fraction of society. There was found to be approximately 20% chance of death per infected individual within a window of 2 weeks, and the chance could go to as high as 67% if the patient stays medically unattended within the window period.

By making an assumption that the municipal authority deploys 2 such garbage collection robot in the aforementioned society, there could be only about 15-minute delay whilst garbage lies idle, and this could account of negligible disease outbreak according the statistics at hand. To do such an efficient job, the operational costs for each robot estimates to not more than $500 per week. Adding the cost of each robot to the operational costs gives $7 000. It follows that the municipal authority would opt to risk $7 000 as compared to $50 000 that they could have gotten lost. Such a garbage collection robot would also save a lot in cutting down the probability of potential death count due to health hazards imposed by mismanagement of garbage.

Depreciation of the Robot

The depreciation rate of the robot over 4 years of its lifespan was evaluated by the Reducing Balance Method with the formula:

$$R = 1 - \frac{S}{C}^N,$$

where: R=rate of depreciation, S=net residual value=$2000, C=cost of robot=$3000, N=number of years=4,

$$R = 1 - \left(\frac{2000}{3000}\right)^{\frac{1}{4}} = 0.0964,$$

Therefore the Depreciation rate **R**=10%

The value of r is seemingly small due to the machine not being in use all the time. This means the robot depreciates more due to its surroundings like temperature, moisture (causing rust) and general aging as compared to wear and tear due to operation.

This type of approach models the failure rate depicted by the bath tub where the higher failure rates take place during the infantry stage of use but then get constant with time. The Reducing Balance Method was opted and deemed to best fit the calculation with the $2000 taken as the net residual value. This method allows for the assignment of cost for each year with regard to the benefits brought by the machine during the very same year. It could be evidenced, through this method, that the depreciation is higher in the few infantry years of use but then noticeably drops down with time.

The value of R is pretty large since the robot is working almost all of the time. Besides other possible minor causes of depreciation such as moisture, temperature and aging, the major factor that depreciates the Garbage Collection Robot (GCR) is wear and tear owing to operation.

The obtained results have shown that the GCR will still remain with a value amounting to $1968.30 (after a 4-year period) which is slightly lower than the net residual value of $2000. It follows that the GCR will still have a considerable value by the time of discard. Due to the increasing adoption of Kaizen Philosophy in the industry, it is advisable to replace the robot after 3 years of continuous operation in order to engulf new technologies into the waste management sector.

Table 14. Depreciation of GCR over 4 years

YEAR	DESCRIPTION	USD $
-	Cost of GCR	3 000.00
1	Depreciation at the rate of 10% of $3 000.00	300.00
	Cost not yet allocated at the end of 1st year	2 700.00
2	Depreciation at 10% of $2700.00	270.00
	Cost not yet allocated at the end of 2nd year	2 430.00
3	Depreciation at 10% of $2 430.00	243.00
	Cost not yet allocated at the end of 3rd year	2 187.00
4	Depreciation at 10% of $2 187.00	218.7
	Cost not yet allocated at the end of 4th year	1968.30

RECOMMENDATIONS AND CONCLUSION

This chapter gives the recommendations pertaining the Garbage Collection Robot (GCR) and the conclusions for the project.

RECOMMENDATIONS

This section will elaborate recommendations pertaining the use and maintenance of the Garbage Collection Robot. Whilst on its mission, the GCR takes as a first priority, its programmed rules on safety precautions, so as not to injure humanity by any means.

Safety Precautions

Relating to Humanity, Environment, and the GCR

- On combating garbage, the GCR detects the four color codes of different garbage bins so in order to prevent confusion, humanity should not spoil the colors on the garbage containers as such an action would misguide the robot.
- The reflectors and indicators around the robot body should be kept intact and full functioning as these help the other traffic to notice the GCR and hence avoiding accidents.
- Humanity should exercise good habits of placing the litter in their respective containers to avoid mixing up of the garbage.

Relating to Operators, Maintenance Team and the GCR

- Batteries should be properly recharged, and also checked often for damage so that the robot would not fail when on mission.
- The GCR should undergo a scheduled plan on preventive maintenance.
- The machine should be kept in a moisture-free environment to avoid corrosion and deterioration of the electronics.
- To maintain the correct precision and machine accuracy, the sensors should be calibrated at a 3-months interval.
- A full device service should be performed at every 3-month mark to ensure a health navigation and stability of the device.
- All moving components should be checked for wear and tear after by means of a condition monitoring technology.

- The fail safe mechanism should be routinely checked for faults. Such mechanism should ensure that the garbage container stays gripped in case the robot fails during the time when the container is lifted.

CONCLUSION

The work of garbage pickup is a physically demanding task and worse over exposes the workers to occupational hazards. This project was designed to bring a fulfilment to the task of garbage collection from designated pickup points, sorting according to garbage type, and then its disposal at a single temporary dump site from which the waste will then be collected for permanent disposal or for a recycling process. The project aims to build an autonomous garbage collection robot namely The Scavenger, using the Arduino microcontroller. The idea of The Scavenger was fulfilled by the construction of a well-sized prototype. This significantly reduces the requirement of strenuous manual collection of solid waste.

REFERENCES

Akkar & Najim (2016). Kinematics Analysis and Modeling of 6 Degree of Freedom Robotic Arm from DFROBOT on Labview. *Research Journal of Applied Sciences, Engineering and Technology*, *13*(7), 569–575. doi:10.19026/rjaset.13.3016

American, S., America, N., & American, S. (2017). *Robot Systems*. Nature America, Inc. Retrieved from http://www.jstor.org/stable/24950284

Apoorva, S., Prabhu, R. S., Shetty, S. B., & Souza, D. D. (2017). *Autonomous Garbage Collector Robot*. doi:10.5923/j.ijit.20170602.06

Chandrasekara, C., Rathnapriya, U., Handwriting, S., & Recognition, C. (2012). *Automatic Garbage Disposal System*. doi:10.13140/RG.2.2.12124.18566/1

Dautenhahn, K. (2017). *Socially intelligent robots: dimensions of human-robot interaction*. doi:10.1098/rstb.2006.2004

Design, E. (2015). *Group 7's BOE-BOT Stair Climbing Robot*. Academic Press.

Engineering, F. O. F. (2013). *Challenges and opportunities in solid waste management in Zimbabwe's urban councils*. Academic Press.

Hanshar, F. (2013). *Pick and place robot. Dynamic vehicle routing using Genetic Algorithms* (2nd ed.). McGraw Hill.

Hayawi, M. J. (2000). *The Closed Form Solution of the Inverse Kinematics of a 6-DOF Robot*. Academic Press.

I, A. P., Much, R., Faithful, M., You, T., & Palumbo, D. (2017). *International Association for the Fantastic in the Arts Alex Proyas's I, Robot : Much More Faithful to Aslmov Than You Think*. Academic Press.

Jamelske, E. M. (2005). Assessing the support for the switch to automated collection of solid waste with single stream recycling in Madison. doi:10.1177/1087724X05283676

Liqing, N., & Qingjiu, H. (2012). Inverse Kinematics for 6-DOF Manipulator by the Method of Sequential Retrieval. *Proceedings of the 1st International Conference on Mechanical Engineering and Material Science, 1*, 255–258. 10.2991/mems.2012.157

Loughery, J. (2018). The Hudson Review. *Inc, 48*(2), 301–307.

Machale, S. (2015). Smart garbage collection system in residential area. *Solid Waste Management and Monitoring, 5*(11), 13–17.

Nurlansa, O., Istiqomah, D. A., Astu, M., & Pawitra, S. (2014). *AGATOR (Automatic Garbage Collector) as Automatic Garbage Collector Robot Model*. doi:10.7763/IJFCC.2014.V3.329

Özturan, P. M., Bozanta, A., Basarir-Ozel, B., Akar, E., & Coşkun, M. (2015). A roadmap for an integrated university information system based on connectivity issues: Case of Turkey. *The International Journal of Management Science and Information Technology, 17*(17), 1–23. doi:10.14313/JAMRIS

Polytechnic, B. S. P. (n.d.). *Pick and place robot*. Academic Press.

Rakesh, N., A, P. K., & Ajay, S. (2013). *Design And Manufacturing Of Low Cost Pneumatic Pick And Place Robot*. Academic Press.

Rakesh, N., A, P. K., & Ajay, S. (2015). Design And Manufacturing Of Low Cost Pneumatic Pick And Place Robot. *Evolutionary Optimization in Uncertain Environments, 2*(8), 131–133.

Sajjad, M., Talpur, H., & Shaikh, M. H. (2012). *Automation of Mobile Pick and Place Robotic System for Small*. Academic Press.

Salmador, A., Cid, J. P., & Novelle, I. R. (1989). *Intelligent Garbage Classifier*. Academic Press.

Saravana, G., Sasi, S., Ragavan, R., & Balakrishnan, M. (2016). *Automatic Garbage Separation Robot Using Image Processing Technique*. Academic Press.

Schultz, J., & Miner, J. (n.d.). *Wall-E Robot Final Report*. Academic Press.

Shepherd, R. F., Ilievski, F., Choi, W., Morin, S. A., Adam, A., Mazzeo, A. D., … Whitesidesb, G. M. (2017). *Multigait soft robot Linked references are available on JSTOR for this article : Multigait soft robot.* doi:10.1073/pnas

Torres-garcía, A., Rodea-aragón, O., Longoria-gandara, O., Sánchez-garcía, F., & González-jiménez, L. E. (2015). *Intelligent Waste Separator.* doi:10.13053/CyS-19-3-2254

UNEP. (2016). Waste Management Criteria. *Cleaner Production, 4*(5), 12–23.

Watanasophon, S., & Ouitrakul, S. (2014). *Garbage Collection Robot on the Beach using Wireless Communications.* . doi:10.7763/IPCBEE

Weiland, M., Boekhoff, A., & Staloch-schultz, T. (2017). n *Strategies.* Academic Press.

Zimbabwe National Statistics Agency. (2016). Zimbabwe Demographic and Health Survey. *Population of Zimbabwe Urban Areas, 3*(17), 13–24.

APPENDIX

Figure 61. Servo Motors for SIMOVERT MASTERDRIVES: Asynchronous Servo Motors

1PH7 Motors

Selection and Ordering Data with SIMOVERT MASTERDRIVES Vector Control

Rated Rotational Speed n_{rated} rpm	Shaft Height SH	Rated Output P_{rated} kW (HP)	Rated Torque M_{rated} Nm (lb$_f$-ft)	Rated Current I_{rated} A	Rated Voltage V_{rated} V	Speed during Field Weakening[1] n_1 rpm	Max. Permissible Continuous Speed[2] n_{S1} rpm	Max. Speed[3] n_{max} rpm	1PH7 Asynchronous Motors Order No.
Supply voltage 3-ph. AC 400 V for SIMOVERT MASTERDRIVES Vector Control									
400	180	16.3 (21.85)	390 (287.7)	51	271	2000	2000	2000	1PH7 184 – ▓ ▓ B ▓ ▓ – ▓ …
		21.2 (28.42)	505 (372.5)	67	268	2000	2000	2000	1PH7 186 – ▓ ▓ B ▓ ▓ – ▓ …
	225	30.4 (40.75)	725 (534.8)	88	268	2000	2000	2000	1PH7 224 – ▓ ▓ B ▓ ▓ – ▓ …
		39.2 (52.55)	935 (689.7)	114	264	2000	2000	2000	1PH7 226 – ▓ ▓ B ▓ ▓ – ▓ …
		48 (64.34)	1145 (844.6)	136	272	2000	2000	2000	1PH7 228 – ▓ ▓ B ▓ ▓ – ▓ …
1150	180	44 (58.98)	366 (270)	89	383	3100	3500[4]	5000	1PH7 184 – ▓ ▓ D ▓ ▓ – ▓ …
		58 (77.75)	482 (355.5)	116	390	3300	3500[4]	5000	1PH7 186 – ▓ ▓ D ▓ ▓ – ▓ …
	225	81 (108.58)	670 (494.2)	160	385	2900	3100[4]	4500	1PH7 224 – ▓ ▓ D ▓ ▓ – ▓ …
		105 (140.75)	870 (641.7)	197	390	2900	3100[4]	4500	1PH7 226 – ▓ ▓ D ▓ ▓ – ▓ …
		129 (172.92)	1070 (789.2)	238	390	2900	3100[4]	4500[4]	1PH7 228 – ▓ ▓ D ▓ ▓ – ▓ …

- **Separate fan:**
  With separate fan — 2
  Without separate fan, for pipe connection — 6
  With separate fan, but with metric cable entries in accordance with EN 50262 — 7
  Without separate fan, for pipe connection,
  but with metric cable entries in accordance with EN 50262 — 8

- **Encoder:**
  Without encoder — A
  Incremental encoder HTL (1,024 pulses/revolution) — H
  Incremental encoder HTL (2,048 pulses/revolution) — J

- **Terminal box arrangement/direction of cable entry (drive-end view):**
  On top/from right — 0
  On top/from drive end — 1
  On top/from non-drive end — 2
  On top/from left — 3

- **Construction type:**
  IM B3 — 0
  IM B3   Hoisting concept for other construction types (IM B6, IM B7, IM B8, IM V5, IM V6) — 1
  IM B35 (only for 1PH7 184 with flange A 400) — 3
  IM B35 (only for 1PH7 184 with flange A 450) — 4
  IM B35 (for 1PH7 186 with flange A 450 and 1PH7 22. with flange A 550) — 3
  IM B35 (only for 1PH7 184 with flange A 400)   Hoisting concept for other construction types (IM V15, IM V36) — 5
  IM B35 (only for 1PH7 184 with flange A 450)   Hoisting concept for other construction types (IM V15, IM V36) — 6
  IM B35 (for 1PH7 186 with flange A 450 and 1PH7 22. with flange A 550)   Hoisting concept for other construction types (IM V15, IM V36) — 5

- **Holding brake with emergency stop function** (suitable for IM B3 coupling drive)[b]:
  Without brake — 0
  With brake (brake includes emergency release screws and microswitch) — 2
  With brake (brake includes manual release and microswitch) — 4

Figure 62. Battery specifications

VOLTAGE AND CAPACITY	LFP-BMS 12,8/60	LFP-BMS 12,8/90	LFP-BMS 12,8/160	LFP-BMS 12,8/200	LFP-BMS 12,8/300
Nominal voltage	12,8V	12,8V	12,8V	12,8V	12,8V
Nominal capacity @ 25°C*	60Ah	90Ah	160Ah	200Ah	300Ah
Nominal capacity @ 0°C*	48Ah	72Ah	130Ah	160Ah	240Ah
Nominal capacity @ -20°C*	30Ah	45Ah	80Ah	100Ah	150Ah
Nominal energy @ 25°C*	768Wh	1152Wh	2048Wh	2560Wh	3840Wh
*Discharge current ≤1C					
CYCLE LIFE (capacity ≥ 80% of nominal)					
80% DoD			2500 cycles		
70% DoD			3000 cycles		
50% DoD			5000 cycles		
DISCHARGE					
Maximum continuous discharge current	180A	270A	400A	500A	750A
Recommended continuous discharge current	≤60A	≤90A	≤160A	≤200A	≤300A
Maximum 10 s pulse current	600A	900A	1200A	1500A	2000A
End of discharge voltage	11V	11V	11V	11V	11V
OPERATING CONDITIONS					
Operating temperature	-20°C to +50°C (maximum charge current when battery temperature < 0°C: 0,05C, i.e. 10A in case of a 200Ah battery)				
Storage temperature	-45°C to +70°C				
Humidity (non-condensing)	Max. 95%				
Protection class	IP 54				
CHARGE					
Charge voltage	Between 14V and 15V (<14,5V recommended)				
Float voltage	13,6V				
Maximum charge current	180A	270A	400A	500A	750A
Recommended charge current	≤30A	≤45A	≤80A	≤100A	≤150A
OTHER					
Max storage time @ 25°C*	1 year				
BMS connection	Male + female cable with M8 circular connector, length 50cm				
Power connection (threaded inserts)	M8	M8	M10	M10	M10
Dimensions (hxwxd) mm	235x293x139	249x293x168	320x338x233	295x425x274	345x425x274
Weight	12kg	16kg	33kg	42kg	51kg
*When fully charged					

Chapter 5
Matrix Models of Cryptographic Transformations of Video Images Transmitted From Aerial–Mobile Robotic Systems

Vladimir G. Krasilenko
Vinnytsia National Technical University, Ukraine

Alexander Lazarev
Vinnytsia National Technical University, Ukraine

Diana Nikitovich
Vinnytsia National Technical University, Ukraine

ABSTRACT

In this chapter, the authors consider the need and relevance of cryptographic transformation of images and video files that are transmitted from unmanned aircraft, airborne robots. The authors propose and consider new multifunctional matrix-algebraic models of cryptographic image transformations, the variety of matrix models, including block parametrical and matrix affine permutation ciphers. The authors show the advantages of the cryptographic models, such as adaptability to various formats, multi-functionality, ease of implementation on matrix parallel structures, interchangeability of iterative procedures and matrix exponentiation modulo, ease of selection, and control of cryptographic transformation parameters. The simulation results of the proposed algorithms and procedures for the direct and inverse transformation of images with the aim of masking them during transmission are demonstrated and discussed in this chapter. The authors evaluate the effectiveness and implementation reliability of matrix-algebraic models of cryptographic image transformations.

DOI: 10.4018/978-1-5225-9924-1.ch005

INTRODUCTION

In various types of industrial activity of a person, as well as in his daily life, photo, static and dynamic images of various formats, video information about various surrounding objects are widely used today. A characteristic feature of modern digital video surveillance systems that are used in unmanned aircraft, when analyzing the traffic situation, remote monitoring and demonstration of emergency or other situations, in security activities, recording events in places of public events, is their distribution. Video information transmitted in such systems, although it is not always secret and has a small period of actuality, is often undesirable for mass distribution and use. In the above-mentioned video systems, especially, such as security, telemedicine and special-purpose systems, in intelligent robotic complexes, not only the tasks of perception, accumulation and transmission of digital video images, but also their protection from unauthorized access, problems of distortion, substitution of information and verification of the integrity of video files are actual tasks. Transmission of video information over open communication channels, IP-networks, and widespread use of wireless technologies for these video systems makes it possible to access information to unauthorized users. The above-mentioned tasks are of particular relevance for mobile robotic and distributed systems implemented on the basis of embedded-class IP modules, for which there are limitations on the computation speed and free computational resource. The specificity of the above systems is that in most cases the transmitted video information is relevant for a short period of time and the use of complex well-studied and widely used cryptographic methods of protection, and especially those requiring significant computational resources, is not required. The analysis showed that in embedded class systems, which include IP-modules of distributed or airmobile video systems, standard cryptographic algorithms are limited, and more often, simpler cryptographic primitives and masking methods are used. The masking information is meant the process of converting digital visual information to a noise-like view in order to protect against unauthorized access, and unmasking is the process of reversely converting masked visual information into restored (outgoing) by applying operations that are inverse to the direct masking procedures. Masking transformations are one of the alternatives to cryptographic methods of photo and video information protection. In the authors' opinion, masking is a special case of some transformations, which are not always cryptographic standard ones. Besides, it is necessary to distinguish matrix masking, as transformation processes using matrixes and matrix procedures follow only when matrices and operations on them appear in the corresponding models. In some cases, by cryptographic masking, authors imply direct and inverse image transformations in which elements of cryptographic methods are used, and the result of masking is the destruction of images to a form that is visually perceived as noise. We stand on the

171

view that reliable and cryptographic protection requires cryptographic procedures and matrix models that transform the original video image into not only noise, but also to provide some important entropy characteristics. A feature of entropy, as a generalized concept of measurement of the uncertainty of processes, is the fact that it reduces to some numerical values, which can be operated as a relative value, and to characterize with it the quality of cryptographic transformations or masking. The entropy of the uniform distribution law (white noise) is an idealization that is maximized and has the greatest disinformation action. The use of masking as a method of protecting video information with a short time of relevance is associated with solving the problems of generating masked data structures, their presentation, exchange between the receiver and the transmitter, and storing and unmasking information. Often, masking takes into account the specific structure of video frames (photos), algorithms for their compression and transmission protocols. It is known that frames are represented as matrix arrays of pixels, the values (intensities) of elements of which are displayed by digital codes, and therefore the matrix apparatus and the operations for converting them are natural. The main types of images that need to be perceived, cryptographically transformed and transmitted in intellectual robotic video surveillance subsystems are half-tone, binary and full-color images, although many more and more hyper-spectral images are being used. The main format of digital images, which directly stores the values of pixel intensities of images obtained from the video-matrix is a Bitmap Picture (BMP), providing storage of images of various sizes and depths. Matrix operations and the matrices themselves are widely used for mathematical modeling of various processes and systems. Matrices are the basic apparatus for most engineering and scientific calculations. Computations over matrices, although laborious, are focused on parallel computing, on significant increases in computational performance, and are a classic example and direction for the further development of more intelligent computer architectures of parallel action. The emerged multi-core processors, graphics accelerators, digital signal processing (DSP) processors, structures on the FPGA, essentially support vector and matrix calculations and increase the speed and performance of the latters. Therefore, it is precisely for more modern hardware implementations that matrix models are ideally suited for implementations of cryptographic or similar methods for transforming and masking information objects in order to protect them during transmission. Matrix algebra and its operations are well studied, they are structured, regular, and easily mapped to hardware matrix structures that provide parallelization of computations, increasing computational performance and efficiency. In addition, such structures are more efficiently implemented using DSP or FPGA, which is important for systems of embedded classes, especially small and mobile ones. The actual task is to create such cryptographic procedures for direct and inverse transformation of video information, which use fully matrix models and procedures that are easily mapped to the corresponding matrix equipment. At the same time ensuring their simplicity, meeting the requirements of speed and computation performance while ensuring the best entropy characteristics .

Review and Analysis of Publications and Formulation of Problems and Challenges

The necessity of solving theoretical and practical tasks of information security and achieving the necessary level of information protection for state, military, commercial and private content caused the corresponding accelerated development of cryptography and related new scientific disciplines. In the era of electronic communications, the need to process and transmit specific text and graphic documents (TGDs) in the form of digital, table data, drawings, charts, diagrams, signatures, visas, resolutions, etc., has essentially increased, and the data are essentially 2D arrays (images) of significant dimension. In addition, the sharing of new tasks in which cryptographic transformations over multidimensional signals is required, among which a variety of semi-tones, color multispectral images, 2-D, 3-D, and even 4-D arrays (Yemets, 2003; Khoroshko, 2003; Korkishko, 2003; Kovalchuk, 2009; Rashkevich, 2009; Deergha, 2011; Han Shuihua, 2005; Chin-Chen, 2001) occupy an important place. In recognition, identification, biometric, navigation monitoring systems, robotics, intelligent management, when deciding, it is necessary to process and transmit a large number of various images in encrypted form, for example, fingerprints, photographs of persons, images of moving objects, iris eye retina, etc. Expansion of the spectral range that is perceived by modern multisensory remote sensing and monitoring systems has necessitated the processing of large arrays of large-scale multi-spectral images. Since this information is often confidential, there is an urgent need for cryptographic transformations to protect against unauthorized access. Many TGDs contain restricted access information that should be reported to tax and other government agencies, in a timely manner and in encrypted form, transmitting over communication channels and providing only authorized access, to certify their digital signatures. Authorized access many information resources such as library, archival and book funds, scientific publications, patent documents, which are formed in the process of activities of information actors, can be provided with appropriate technologies of cryptography and measures with the issuance of permits, certificates and access keys.

For such information security purposes, methods and tools for cryptographic transformations (CTs) of information arrays or images (Yemets, 2003; Khoroshko, 2003; Korkishko, 2003; Kovalchuk, 2009; Rashkevich, 2009; Deergha, 2011; Han Shuihua, 2005; Chin-Chen, 2001; Krasilenko, 2004; Krasilenko, 2006), procedures and protocols for the formation of keys and their exchange (Yemets, 2003; Krasilenko,

2012; Krasilenko, 2008) are used. Among their great variety (Yemets, 2003; Khoroshko, 2003; Korkishko, 2003; Kovalchuk, 2009; Rashkevich, 2009; Deergha, 2011; Han Shuihua, 2005; Chin-Chen, 2001; Krasilenko, 2004; Krasilenko, 2006) most of them focused on sequential scalar processing of TGD blocks transformed into digital formats, and only a small part is devoted to methods and algorithms oriented on matrix models (Krasilenko, 2012; Krasilenko, 2012; Krasilenko, 2011; Krasilenko, 2009; Krasilenko, 2012; Krasilenko, 2013; Krasilenko, 2013; Krasilenko, 2014; Krasilenko, 2010) and matrix specialized algorithms and tools. At the same time, the emergence of parallel algorithms, and especially matrix multiprocessor, matrix linear-algebraic, specialized multi-core, parallel and matrix (image-type) processors (Korkishko, 2003; Krasilenko, 2004) contributed to the reorientation in the study of image CTs on these new tools and the creation and corresponding models of matrix type (MT) (Krasilenko, 2012; Krasilenko, 2012; Krasilenko, 2011; Krasilenko, 2009). In addition, the urgency of the problem of creating new high-performance models, algorithms, protocols for processing and cryptographic transformations of images is confirmed by the significant increase in the number of works devoted to encryption and decoding of images in recent years (Kovalchuk, 2009; Rashkevich, 2009; Deergha, 2011; Han Shuihua, 2005; Chin-Chen Chang, 2001; Krasilenko, 2012; Krasilenko, 2009; Krasilenko, 2012; Krasilenko, 2013; Krasilenko, 2013; Krasilenko, 2014; Krasilenko, 2010; Krasilenko, 2016; Krasilenko, 2016; Krasilenko, 2016). That is why the search and research of new matrix models (MM) of CT, improvement of existing matrix ciphers and means for their realization are an actual strategic task.

Analysis of recent research and publications. The results of modeling the processes of cryptographic transformations of images on the basis of the proposed work by V.G. Krasilenko and the investigated matrix algorithms and models of cryptographic protection show their advantages. For example in (Krasilenko, 2006; Krasilenko, 2006) matrix algorithms and the implementation on the Delphi language in CryptoFax program were considered. It has been shown that the developed methods of permutations are resistant to the effects of disturbances and various distortions. The disadvantage of the CryptoFax program was that the transformations did not change the histogram of converted ciphered images. Therefore, in order to eliminate this disadvantage and improve the stability of the algorithms of cryptographic transformations of images, generalization of affine ciphers and their expansion into matrix cases (Krasilenko, 2009) were proposed. Experiments in the MathCad environment partially demonstrated the possibilities and advantages for practical applications of matrix algorithms for cryptographic protection on the basis of more generalized matrix affinity ciphers (MACs). In Krasilenko (2012) and Krasilenko (2011) more generalized matrix algorithms for cryptographic transformations of images and so-called matrix affine-permutation algorithms (MAPA) (Krasilenko,

2012) based on modifications of known affine ciphers were proposed and modified. The results of simulation (Krasilenko, 2012; Krasilenko, 2012; Krasilenko, 2011; Krasilenko, 2009) of processes of cryptographic transformations of multi-gradation and color images (Krasilenko, 2010) on the basis of such models and algorithms have shown their significant advantages over traditional scalar affine asymmetric ciphers such as: greater stability, increase in speed, the possibility of parallel computing procedures and processes and implement them using parallel problem-specific tools, matrix processors. In work (Krasilenko, 2011) on the basis of MACs the algorithm and the procedure for creating a digital blind signature (DBS) is proposed on the TGD, and the results of simulation of a developed and practically verified program for the formation and verification of such DBS are presented. Such matrix cryptographic models, algorithms and cryptographic systems based on them are better and more effectively based on completely parallel matrix computing devices, since they are described purely by mathematical matrix models, which significantly increase the processing efficiency during transformations and reduces the time for their execution.

The results of modeling algorithms for creating a 2D key are also known (Krasilenko, 2012; Krasilenko, 2008), the essence of which is the synthesis of known protocols for creating and generating keys on the matrix case, and the formation and description of these protocols using matrix models. Paper (Krasilenko, 2012) is devoted to creation of DBS on TGD, but on the basis of other models of matrix type. One of the main components of the most generalized matrix affine-permutation ciphers or MAPA, proposed and investigated in paper (Krasilenko, 2012), is matrix permutation model (MM_P), which has obvious simplicity. Further application and improvement of matrix-type ciphers based on such MM_P is highlighted in papers (Krasilenko, 2013; Krasilenko, 2013; Krasilenko, 2014; Krasilenko, 2016; Krasilenko, 2016). However, as shown in papers (Krasilenko, 2013; Krasilenko, 2014), the CPs on their basis, without additional operations, do not modify histograms of images or TGDs, and the proposed modified MM_Ps with decomposition of bit sections eliminate this defect, although in some cases they require two vector keys (VK) in addition to two matrix keys (MK). At the same time, for most of the above-mentioned works, there is a common significant disadvantage, especially for work related to MAC (Krasilenko, 2009; Krasilenko, 2010), MAPA (Krasilenko, 2009) and the like (Krasilenko, 2012; Krasilenko, 2012; Krasilenko, 2011; Krasilenko, 2014; Krasilenko, 2016; Krasilenko, 2016; Krasilenko, 2016), which requires the use of at least two MK, if implemented in models MAC, MAPA, MT and multiplicative and additive matrix components. But the kind of MK that is used is of two types: in the form of random images (basically black and white 8-bit) for MAC and square matrix of permutations for the implementation of MM_P and algorithms on them (Krasilenko, 2013; Krasilenko, 2013; Krasilenko, 2014). The first kind is less investigated. Therefore, the search for ways to improve the MAC and especially the multi-step MAC, MAPA (Krasilenko, 2012) in order to reduce the number of MKs to one, while maintaining stability and other characteristics of the matrix models (MM), their experimental verification on various images is a necessary task and which is justified by the above survey of publications.

Formulation of the problem. It is necessary to further modify and improve the well-known MACs with spectral decomposition for the CT over color images in order to simplify, improve and to study the models that implement MAC in different environments, to identify their specific features of specific applications and expansion their functional capabilities. Testing the created models, carrying out experiments with real images of different formats and dimensionalities allow assessing their adequacy, characteristics, indicators and features.

Therefore, the purpose of this work is to study and modify such modifications and enhancements of the MACs in the Mathcad software environment for the purpose of their use in the CT over black-and-white and color images, including large-scale and multi-spectral, in which the number of necessary matrix keys for these transformations would be reduced to one, so-called main or basic, while retaining the same functionality. One of the sub-tasks is an experimental verification of the correctness and quality of the work of such MAC in their work with different types and formats, image sizes to study their impact on indicators, characteristics of ciphers, models and algorithms for their implementation.

PRESENTATION OF THE MAIN MATERIAL AND RESEARCH RESULTS

Theoretical Foundations of Matrix Affine Ciphers

Let's recall some of the simplest theoretical foundations of the matrix affine ciphers (MAC). The encryption and decryption processes on the basis of the MAC for the message of an arbitrary form and size of the matrix **M** and for the created corresponding cryptogram **C** using cryptographic transformations (CT) described by the matrix model (MM), which are expressed by the following matrix formulas (Krasilenko, 2009):

$$\mathbf{C} = \left(\mathbf{M} \underset{N}{\otimes} \mathbf{A} + \mathbf{S} \right); \quad \mathbf{M} = (\mathbf{C} \underset{N}{\otimes} \mathbf{AD} + \mathbf{SD});$$

where **A** and **S** – two keys (multiplicative and additive components) for encryption in the form of matrices, **AD** and **SD** – decryption keys, moreover, **AD** – respectively, the multiplicative component of the matrix affine cipher, and **SD** – additive component of the matrix affine cipher, **N** – matrix, all elements of which are equal to n (simple large number), and components of all matrices are selected from the range 1÷(n-1), in addition, symbols $\underset{N}{\otimes}$ and $\underset{N}{+}$ denote element-wise matrix multiplication and matrix addition by modulo N.

To reduce the number of matrix keys, you can use the following formulas for one-key MAC:

$$C = \left(M \underset{N}{\otimes} A \underset{N}{+} AD \right); \quad M = (C \underset{N}{\otimes} AD \underset{N}{+} (-AD \underset{N}{\otimes} AD)) ;$$

in this case **S= AD, SD=** $(-AD \underset{N}{\otimes} AD)$.

Simulation Results of Matrix Affine Ciphers

Let's consider the essence of MAC algorithms on the MM with spectral decomposition basis. The offered modification of the MM MAC is essentially one matrix key (MK) for the corresponding both multiplicative and additive direct and inverse transformations that are components of one or multi-step MAC and implement cryptographic procedures for black and white or for all spectral components of color images. The idea is that the secret MK, which is selected or generated by known methods in the form of pseudo-random black and white or multi-level image with dimensions equal to the size of the input image. There is always, under the fulfillment of some simple additional conditions, the inverse matrix key that we denote as MK, and its elements are reversed by the corresponding modulo to the MK elements. This idea and its explanation were proposed by Krasilenko V.G. and are covered in our previous works (Krasilenko, 2009; Krasilenko, 2012), so here we note only the fact that when using a simple number 257 as the modulo, the entire range of 0-255 graduations of the 8-bit image which are displaced in the range 1-256, will have unambiguous inverse values in the same range of 1-256, and hence with their inverse shift and in the range of 0-255, that is, have a similar 8-bit representation. Before moving to some the new suggestions and improvements, let's consider the simulation results of the simplest MAC with only one multiplicative component, which is a generalization of the scalar linear cipher to the matrix case. In the first of a series MAC simulation experiments conducted with the Mathcad software, we recreated the direct and inverse cryptographic processes over two different images (C) (256x256) using the MK Key_GC and its associated inverse MKi Key_GD. The Experimental results are shown in Fig. 1 and testify the correct and adequate work of the models. The formulas used for transformations are shown in Fig. 1, especially in the scalar form, and those used for verification in the matrix form (left in the 1st and 2nd rows - cryptograms, in the center are decoded images, they are initial, in the right there are differences (zero) matrix). The inscriptions in the Figure 1 are fuzzy and blurry.

Our second idea is to use the inverse MKi for the second step, namely the additive component, for the direct transformation of the MAC, since the use of the direct MK is primitive. Since the matrix **SD** is a matrix, all elements of which are equal to (-1). And since, in essence, the MK and MKi are secret and interconnected, this leads to the need for the two parties to coordinate or formally create only one MK in the process of creating and transmitting encrypted data. At the same time, it is not desirable to apply the same MK when applying MAC for cryptographic transformations of color images. Let's move on to the application of the second idea and its verification by encrypting and deciphering a color image, using each of its components R, G, B of its MK, that is three random R, G, B components, equivalent

Figure 1. The simulation results of the processes of direct and inverse cryptographic transformations over two images by the matrix affinity cipher: the formulas used for the multiplicative component of the MAC, encrypted, decrypted and difference matrixes

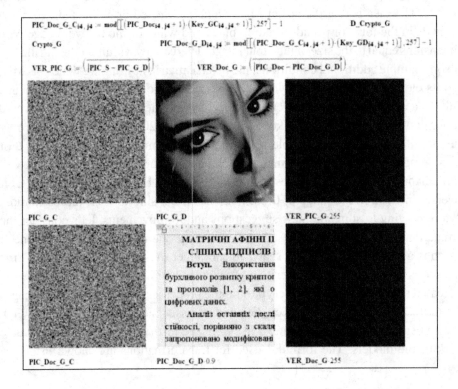

to one MK in color format. Figure 2 shows one of the windows with the formulas that were used to generate keys, direct and inverse to them in modulo 257, encrypt and decrypt each R, G, B_pic component of C (600 x 549), three MK Key_C_ (R, G, B) and Key_D_ (R, G, B) respectively.

Fig. 3 shows the results of a MAC-based CT with only one MK for each component: a color output image, MK (1 row, right), a cryptogram (2 rows, left) and a decoded image. They testify to the correct operation of models for such modification of MAC. We note here that the components of the CT are executed in elementary matrix procedures of multiplication and addition, respectively, by modulo 257 and 256, using practically one corresponding MK, since the inverse key MKi is essentially an additive component of the MAC.

Figure 2. The Mathcad window with formulas for the CT over color image MAC when using spectral components, but only one MK, which performs a multiplicative direct transformation, and the inverse MKi - multiplicative, direct and inverse additive transformations.

$$Key_C_R_{i7,j7} := \mathrm{round}(\mathrm{rnd}(255),0)$$
$$Key_C_G_{i7,j7} := \mathrm{round}(\mathrm{rnd}(255),0)$$
$$Key_C_B_{i7,j7} := \mathrm{round}(\mathrm{rnd}(255),0)$$

$$Key_D_Rv_{i7,j7} := \left| \begin{array}{l} s \leftarrow 0 \\ \mathbf{while}\ \mathrm{mod}\left[\left[(Key_C_R_{i7,j7}+1)\cdot s\right],257\right] \neq 1 \\ \quad s \leftarrow s+1 \end{array} \right.$$

$$Key_D_Gv_{i7,j7} := \left| \begin{array}{l} s \leftarrow 0 \\ \mathbf{while}\ \mathrm{mod}\left[\left[(Key_C_G_{i7,j7}+1)\cdot s\right],257\right] \neq 1 \\ \quad s \leftarrow s+1 \end{array} \right.$$

$$Key_D_R_{i7,j7} := Key_D_Rv_{i7,j7} - 1$$

$$Key_D_Bv_{i7,j7} := \left| \begin{array}{l} s \leftarrow 0 \\ \mathbf{while}\ \mathrm{mod}\left[\left[(Key_C_B_{i7,j7}+1)\cdot s\right],257\right] \neq 1 \\ \quad s \leftarrow s+1 \end{array} \right.$$

$$Key_D_G_{i7,j7} := Key_D_Gv_{i7,j7} - 1$$

$$Key_D_B_{i7,j7} := Key_D_Bv_{i7,j7} - 1$$

$$R_pic_C_{i7,j7} := \mathrm{mod}\left[\left[\left[\mathrm{mod}\left[\left[(R_pic_{i7,j7}+1)\cdot(Key_C_R_{i7,j7}+1)\right],257\right]-1\right]+Key_D_R_{i7,j7}\right]\right],256\right]$$
$$G_pic_C_{i7,j7} := \mathrm{mod}\left[\left[\left[\mathrm{mod}\left[\left[(G_pic_{i7,j7}+1)\cdot(Key_C_G_{i7,j7}+1)\right],257\right]-1\right]+Key_D_G_{i7,j7}\right]\right],256\right]$$
$$B_pic_C_{i7,j7} := \mathrm{mod}\left[\left[\left[\mathrm{mod}\left[\left[(B_pic_{i7,j7}+1)\cdot(Key_C_B_{i7,j7}+1)\right],257\right]-1\right]+Key_D_B_{i7,j7}\right]\right],256\right]$$

$$R_pic_D_{i7,j7} := \mathrm{mod}\left[\left[\left[\left[\mathrm{mod}\left[\left((R_pic_C_{i7,j7}+256-Key_D_R_{i7,j7})\right),256\right]\right]+1\right]\cdot(Key_D_R_{i7,j7}+1)\right],257\right]-1$$
$$G_pic_D_{i7,j7} := \mathrm{mod}\left[\left[\left[\left[\mathrm{mod}\left[\left((G_pic_C_{i7,j7}+256-Key_D_G_{i7,j7})\right),256\right]\right]+1\right]\cdot(Key_D_G_{i7,j7}+1)\right],257\right]-1$$
$$B_pic_D_{i7,j7} := \mathrm{mod}\left[\left[\left[\left[\mathrm{mod}\left[\left((B_pic_C_{i7,j7}+256-Key_D_B_{i7,j7})\right),256\right]\right]+1\right]\cdot(Key_D_B_{i7,j7}+1)\right],257\right]-1$$

$$Ver_R_pic := \left(\overrightarrow{|R_pic - R_pic_D|}\right) \quad Ver_G_pic := \left(\overrightarrow{|G_pic - G_pic_D|}\right) \quad Ver_B_pic := \left(\overrightarrow{|B_pic - B_pic_D|}\right)$$

$$\max(Ver_R_pic) = 0 \quad \min(Ver_R_pic) = 0 \quad \max(Ver_G_pic) = 0 \quad \min(Ver_G_pic) = 0$$
$$\max(Ver_B_pic) = 0 \quad \min(Ver_B_pic) = 0$$

Figure 3. Simulation results (fragment of interface window from Mathcad) of processes of direct and inverse cryptographic transformations by matrix affinity cipher: encryption image, matrix key (three-key set), crypto graph and decoded image in color formats.

R_pic , G_pic , B_pic   Key_C_R , Key_C_G , Key_C_B

R_pic_C , G_pic_C , B_pic_C   R_pic_D , G_pic_D , B_pic_D

Our third suggestion is that it is possible to create matrix keys from one main or basic key for other or all spectral components, not even for color, but for multispectral images or 3-D arrays. The use of scalar keys and procedures of elemental powering by modulo each MK (even one agreed key!) gives the realization of one and multi-step MAC (Krasilenko, 2011; Krasilenko, 2009; Krasilenko, 2012) with only one secret MK, from which other MK are formed. Thus, our third experiment was to develop models and create a procedure for generating a series of MKs, as derivatives from one base in accordance with an agreed sequence of the numerical values that will be taken as degrees in elemental powering by modulo and in an attempt to implement them on the basis of MAC images of different formats. The vector of scalar keys with the dimension equal to the required number of matrix keys taking into account the number of spectral components. This model experiment was performed on the basis of formulas in a matrix form, some of which for sufficient understanding and with allowance for restrictions are shown in Fig. 4. It shows a copy of the fragment of the Mathcad window with formulas, procedures for forming a number of auxiliary lines and matrix keys turned to them, and formulas for the multiplicative and additive components of direct and inverse cryptographic transformations. As can be seen from Fig. 4 a), the keys Key_Cw_Rz (w) are created by a recursive procedure of elemental powering according to the modulus of the previous MK, starting from the base, and the degree of these MK and their corresponding matrix of values depend on the parameter w. If w = 0, then the matrix Key_Cw_Rz (0) is formed that equal to matrix R_C2, all elements of which is "1".

Some copies of the Mathcad windows of these MKs with dimensions of 600x549 corresponding to the sizes of one of a series of images for the CT are shown in Fig. 5 in digital format and correspond exactly to those MKs having the value w such as 1, 2, 7. The agreed secret key is marked as Key_C_R. To display the MK in the format of 8-bit images, the displacement of the values of matrices MK by subtracting from them the matrix R_C2. Note that the check shows the correctness of getting all values of the elements of all matrix keys to the required range. The results of this experiment using the prevailing and shown in Fig. 5 keys at the CT of the color image and its spectral components with such an improved MAC are shown in Fig. 6, 7, 8, 9. They testify to the qualitative correct operation of MAC models when using the correct keys and the impossibility of deciphering without the knowledge of keys and the base (not shown for the wrong keys!).

We have also created and experimentally tested the subroutine, which allows in accordance with the automatically determined sizes of input arrays or images, to form the basis of the Diffie-Helman protocol, generalized on the matrix case, as agreed upon by the parties of the secured data transmission MK (MKi), made on the basis of the results considered in papers (Krasilenko, 2012; Krasilenko, 2008), to verify them and generate keys derived from it. The Created keys are shown in Fig. 10, and the results of the direct and inverse CT of these MKs of a specific color image of the natural scene with fragments of the same intensity values using the improved MAC are shown in Fig. 11, 12. Similar studies performed in (Krasilenko, 2012; Krasilenko, 2009; Krasilenko, 2012; Krasilenko, 2014) histogram and entropy analyzes also showed good indexes of formed cryptograms and increase their entropy to almost 90-95% of the maximum possible. In more detail, we discuss these issues below and show some histograms.

Figure 4. A fragment of Mathcad window with formulas, procedures of forming a number of auxiliary direct and inverse matrix keys and formulas for the multiplicative and additive components of direct and inverse cryptographic transformations: a) the R-spectral component and b) the B-spectral component of the color image

$$Key_C\omega_R(\omega) := Key_C\omega_Rz(\omega) - R_C2$$

$$Key_D\omega_R(\omega) := Key_D\omega_Rz(\omega) - R_C2$$

Crypto_ω

$$R_pic_C\omega M := \left(\overrightarrow{mod\left(\overrightarrow{\left[(B_pic + R_C2)\cdot(Key_C\omega_R(5) + R_C2)\right]}, m1\right)}\right) - R_C2$$

$$R_pic_C\omega MA := \left(\overrightarrow{mod(R_pic_C\omega M + Key_D\omega_R(4), m2)}\right)$$

D_Crypto_ω

$$R_pic_D\omega A := \left(\overrightarrow{mod(R_pic_C\omega MA + R_C2\cdot 256 - Key_D\omega_R(4), m2)}\right)$$

$$R_pic_D\omega AM := \left(\overrightarrow{mod\left(\overrightarrow{\left[(R_pic_D\omega A + R_C2)\cdot(Key_D\omega_R(5) + R_C2)\right]}, m1\right)}\right) - R\ C2$$

$$Ver_\omega_pic := \left(\overrightarrow{|B_pic - R_pic_D\omega AM|}\right)$$

$$min(Ver_\omega_pic) = 0 \qquad max(Ver_\omega_pic) = 0$$

a)

$$Key_C\omega_R(\omega) := Key_C\omega_Rz(\omega) - R_C2$$

$$Key_D\omega_R(\omega) := Key_D\omega_Rz(\omega) - R_C2$$

Crypto_ω

$$R_pic_C\omega M := \left(\overrightarrow{mod\left(\overrightarrow{\left[(B_pic + R_C2)\cdot(Key_C\omega_R(5) + R_C2)\right]}, m1\right)}\right) - R_C2$$

$$R_pic_C\omega MA := \left(\overrightarrow{mod(R_pic_C\omega M + Key_D\omega_R(4), m2)}\right)$$

D_Crypto_ω

$$R_pic_D\omega A := \left(\overrightarrow{mod(R_pic_C\omega MA + R_C2\cdot 256 - Key_D\omega_R(4), m2)}\right)$$

$$R_pic_D\omega AM := \left(\overrightarrow{mod\left(\overrightarrow{\left[(R_pic_D\omega A + R_C2)\cdot(Key_D\omega_R(5) + R_C2)\right]}, m1\right)}\right) - R\ C2$$

$$Ver_\omega_pic := \left(\overrightarrow{|B_pic - R_pic_D\omega AM|}\right)$$

$$min(Ver_\omega_pic) = 0 \qquad max(Ver_\omega_pic) = 0$$

b)

To test the influence of sizes, number of spectral components, statistical characteristics, structural and texture peculiarities of images subject to cryptographic transformations on some of the performance indicators of the proposed improved MAC, and especially on the histogram-entropy and visual characteristics of the obtained cryptograms, we have performed a group of other experiments. The results of these model experiments with other images, including video streams, large-scale (640x1024) multispectral (100 spectral channels) images and their constituents, text documents in color format, etc., are shown in Fig. 13-17 and also confirm the correct functioning of the MAC with a reduced number of keys. They showed

Figure 5. Results (Mathcad window interface) of the the base MK formation and its elemental powers modulus as auxiliary keys with scalar keys

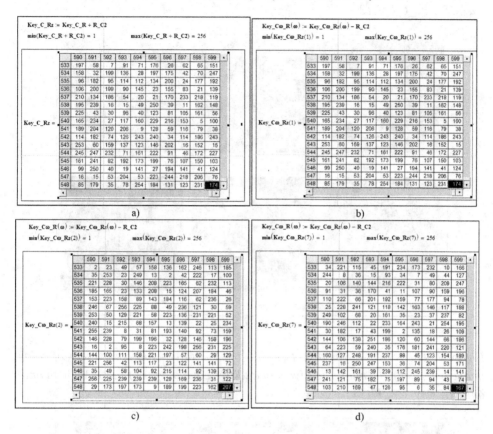

a)   b)

c)   d)

that the duration of CT procedures does not exceed a few seconds, even for large-scale (640x1024) color images when they are modeled in the Mathcad and the use of medium-class PCs. Of course, the total duration of the required procedures is influenced by both the number of spectral channels, the way of writing software modules when emulating models, and the use of vector parallel computing, and therefore making more detailed estimates inappropriate, and here we note only the fact that our matrix models MAC with decomposition, have internal parallelism, are more easily structurally reflected on hardware matrix means.

The visual analysis of the obtained cryptograms in the simulation and shown in Fig. 13 - 17 shows the qualitative encryption and the correct operation of the MM of direct and inverse cryptographic transformations on the basis of MAC in all cases. For a more accurate analysis, we determined the entropy of images, keys and cryptogram and their histograms were constructed using the Mathcad tools.

Figure 6. The results of the direct and inverse CT MAC of R component of color image with using the multiplicative and additive components of the MAC with only one MK Key_Cw_R (2).

Figure 7. The results of the direct and inverse CT MAC of G component of color image with using the multiplicative and additive components of the MAC with only one MK Key_Cw_R (7).

Figure 8. The results of the direct and inverse CT MAC of B component of color image with using the multiplicative and additive components of the MAC with only one MK Key_Cw_R (5).

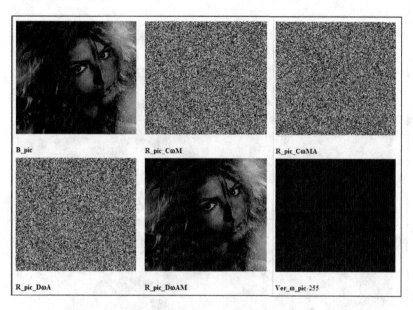

They are shown in Fig. 18-19. From the drawings it can be seen that despite the very specific histograms of the composite images, the histograms of the components of the cryptogram have changed significantly, and do not allow them to guess, as they are, to recognize the possible appearance of the image or read the message or understand the TGD.

To evaluate the quality of encrypted images or documents, we also used, developed and covered in paper (Krasilenko, 2011) a subroutine in MathCad, which allows to calculate the average entropy by 1 pixel of specific images. As shown in Fig. 18-19 and with histogram distributions R, G, B components of explicit color image and confirmed by the definition of entropy, the entropy of the initial explicit image, namely its 8-bit components, is within 3-4 bits per pixel, and the entropy of the cryptogram (its components) for various experiments fluctuated within 7.5-7.8 bits per pixel, which is very close to the maximum possible value of 8. Similar values of entropy also have 8-bit components of MK, see Fig. 18. Note, that the comparison (from Figure 19) of histograms and entropy of the components of cryptograms obtained after **multiplicative and multiplicative** with subsequent additive transformations, allows us to conclude that they are insignificant. We also found that the multichannel MAC-based CT practically improves the histogram-entropy characteristics if the base agreed key is correctly selected or generated, meets the necessary requirements, and is

Figure 9. Results of the direct and inverse CT MAC of color image with using the multiplicative and additive components of the cipher for spectral components transformation with only derivatives from the base MK and parametric scalar keys

also close to the maximum possible entropy. The greater the entropy of cryptograms, the greater the degree of uncertainty of the corresponding image and the more difficult it is to attack this algorithm. The modeling established by the fact that all derivative keys have the required histogram distribution, which is practically close to the uniform distribution, and therefore they are close to the maximum entropy. This allows even by the visual appearance of the histogram to evaluate the quality of both keys and cryptogram.

We note that the use of spectral decomposition and recursive procedures for forming a set of MK, the consistency between the sizes of matrices, displacements of the ranges of values of matrix elements make possible to use for the CT on the basis of the improved MAC single secret matrix key, which can be easily represented as image. And having only one such MK it is possible to implement all the necessary procedures for specific applications and for different types of data, reliable CT

Figure 10. View of one of the basic matrix keys (Key_OC), the inverse MKi (Key_OD), auxiliary (Key_Cw_R (5)) with parameter 5 and the verifying matrix (Ver_O_CD) formed in accordance with the selected parameters and the dimension of the encrypted images.

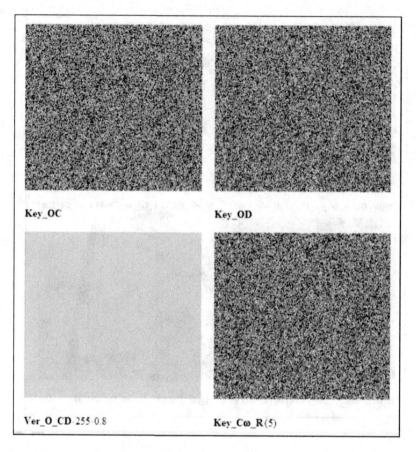

procedures of MAC. All the experiments performed and the results presented here have confirmed the correct work of the proposed models and their modifications, the convenience of their adaptation to the size of the images or their fragments-blocks for cryptographic transformations, the convenience and ease of choosing the necessary keys. Some effective procedures, secret key negotiation protocols and their exchange, updates, were partly considered in papers (Krasilenko, 2012; Krasilenko, 2008) for some more general types.

The section conclusions: Based on the review and analysis of publications, the prospects and necessity of further research and improvement of matrix affine ciphers and their derivatives have been substantiated. The ways of perfection are proposed and the results of the simulation of the matrix affine advanced ciphers for

cryptographic transformations of black-and-white and color images with a reduced number of matrix keys are given. Modified models and algorithmic procedures of keys formation, direct and inverse cryptographic transformations, reduced to matrix-matrix elemental operations by modulo, are developed. It was shown that the use of decomposition of color and multispectral images on their black and white components allowed unifying the transformation procedures, using only one agreed matrix key and expanding the types, data formats and the range of the cipher applications. It is suggested and confirmed experimentally that as an auxiliary derivative key you can use the power of the master key by modulo. Based on a series of experiments in Mathcad with different multi-gradation and color images to encrypt and decrypt them with proposed models, it is shown that the proposed improvements to such ciphers are correct, adequate, easy to use, have advantages and even allow to increase their

Figure 12. A fragment of the interface window that shows the process of direct and inverse cryptographic transformations of one, namely G, of the spectral component of the color image

functionality. The histogram-entropy characteristics of the cryptograms obtained with MAC are determined and evaluated, which also testify to their cryptographic properties and stability.

Multi-Functional Parametric Matrix-Algebraic Models (MAM) of Cryptographic Transformations (CT) with Operations by Modulo and Their Modeling

Modifications of the above mentioned models allow the CT to check the integrity of the cryptogram and their distortions, as shown in paper (Krasilenko, 2016; Krasilenko, 2016), for both black and white and color images. However, as experiments have shown, some specific TGDs, for example scanned documents, have a sizeable area of almost the same intensity of pixels, a small number of graduations and very characteristic histograms, which requires their CT to increase cryptostability by seeking improvements to MAM, including and by expanding their functionality while maintaining unified matrix operations and procedures ((Krasilenko, 2016). Thus, the purpose of this section is development and further modification, universalization

Figure 13. A fragment of a window with cryptograms and decoded images demonstrating the process of direct and inverse cryptographic transformations of components of a large-scale multispectral image obtained from a remote monitoring aircraft and used for model experiments

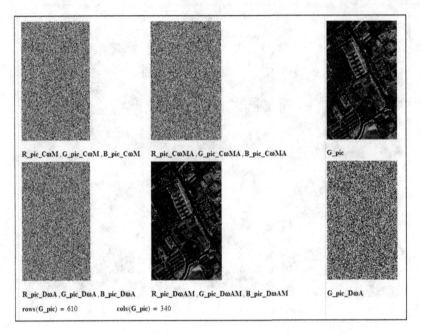

and generalization of MAM for the CT in order to improve their characteristics, sustainability, simulation **and** testing the created models on real information objects (IO) that will allow evaluating their parameters, possibilities and application features.

The essence of the proposed MAM for CT is to apply matrix multiplication procedures to the corresponding 8-bit MK of the same dimension (KLC256, KLD256) for matrix size NxN, as sets of bytes or 8-bit images (PIC_S, PIC_Doc, see Fig. 20) using multiplication and add operations by modulo. As can be seen from Fig. 20 - 24, the simulation results of the processes of direct and inverse CT TGD with a dimension of 256x256 confirmed the correct operation of models when applying the correct (Fig. 23) and wrong (Fig. 24) keys. MK had a hierarchical structure, the dimension of 256x256 consisted of a block matrix of 16x16 units with each unit size of 16x16, and each of the blocks (KLC16, KLD16) had 4 sub-blocks of 4x4 elements. Using matrices of permutations **P** of types K, KP16V1, KP16V2, allow arbitrary permutations of blocks and sub-blocks, as shown in Fig. 20. Blocks KLC, KLD and full keys are mutually inverse matrices when multiplying them by the corresponding modulo.

190

Figure 14. A fragment of the window with cryptograms and decoded images (all color!) that demonstrates the process of direct and inverse cryptographic transformations based on the improved MAC of a large-scale image (640x1024 elements) that was used for experiments

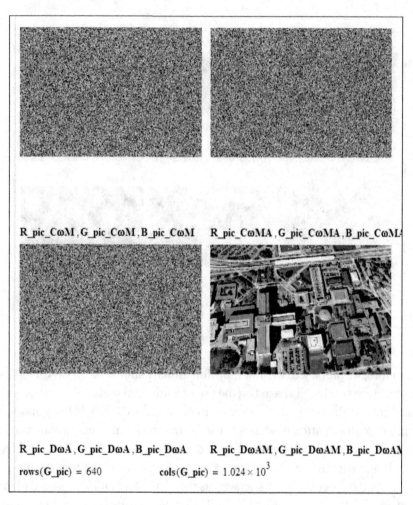

The essential difference between the proposed MKs is that both the blocks themselves in the entire matrix and sub-blocks, and elements in them can be mixed, and their structures are similar to the permutations matrix. Thus, the cryptographic block processing is accompanied by simultaneous mixing blocks and sub-blocks, as well as their elements (Fig. 21 - 23). But the analysis of entropy, histograms of TGD and their cryptogram (Fig. 20) **shows** that for TGD, in contrast to an image of a person, even several iterative multiplications of the data matrix (DM) by the MK may not be sufficient, more when applying the same MK.

Figure 15.A Fragment of the interface window, which shows the details of the processes of direct and inverse cryptographic transformations and their verification of one, namely G, of the spectral component of the color image shown in Fig. 14 and was used to simulate the improved MAC with a reduced number (one base!) of MK

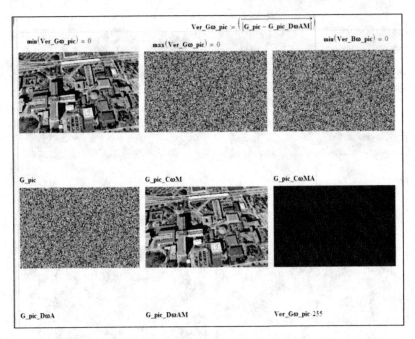

Therefore, we proposed two new multifunctional parametric MAM CTs, the main conceptual idea of which is based on the use of additional scalar or vector keys (VK) as parameters influencing the power of matrices of MD and MK by modulus in their matrix multiplication models and the degree and form of permutation matrices blocks or elements. At each iterative step, depending on the VK, different MKs are formed. Fragments of the simulation of the processes of formation of matrices P, cyclic MK and their components, as well as the MAM formula for direct and inverse CT and verification using parametric MK are **shown** in Fig. 21. Fig. 22 **shows** the appearance of some parametric MK, and Fig. 23 and 24 **show** the results of modeling the CT TGD on the basis of parametric MAM and MK for cases of correct and accordingly, incorrect MK. The appearance of the initial histograms and after the CT confirms that even for the selected TGD-specific histogram, the proposed models give better results. The power of the set of possible keys has increased by an order of magnitude (more than 10^{300}), and as the estimates show, only the power of a plurality of mini-blocks (8x8 8-bit) is of the order of more than 10^{150}. Thus, the stability of the models has increased significantly.

Figure 16. Mathcad interface window (fully) with tools and a text-graphical document (TGD) displayed in its window that was used to simulate advanced MACs

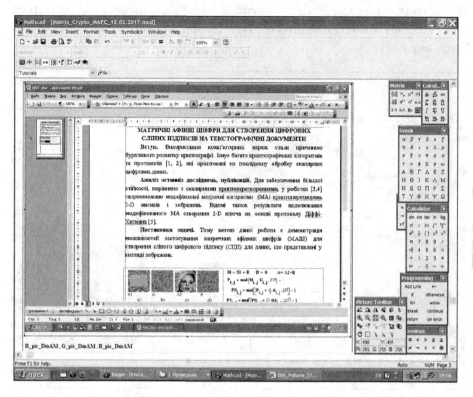

Without the knowledge of keys, it is impossible to restore MD and, as shown in papers (Krasilenko, 2011; Krasilenko, 2012), even with the dimension of MK, equal to 32x32, the stability of models is ensured and we have the keys of 256x256 8-bit elements, which gives a substantial strength!

The section conclusions: New models with modular operations for MD, including images, are proposed and considered. The results of their simulation are presented on the example of direct and inverse CT over images, which testify to their correct work, convenience (only 1 matrix procedure and one in essence MK!), adaptability to formats, multi-functionality (combination of operations of matrix block replacements with permutations, interchangeability of cyclic iterations procedures and matrix substitutions in modulus with convenient choice of parameters and management of transformations and key shapes) and efficiency (orientation to matrix processors). The aspects of matrix algebraic procedures and operations by modulo and creation of MK are considered. The results of simulation of direct and inverse CT, their verification confirmed the adequacy of parametric generalized MAM, their convenience, multi-functionality, efficiency for use. They both are

Figure 17. A Fragment of the window with fragments of cryptogram and decoded images of TGD from Fig. 16 (both colored in the center and left, and black and white corresponding spectral components!), which demonstrates the correctness of the processes of direct and inverse cryptographic transformations of TGD with the improved MAC

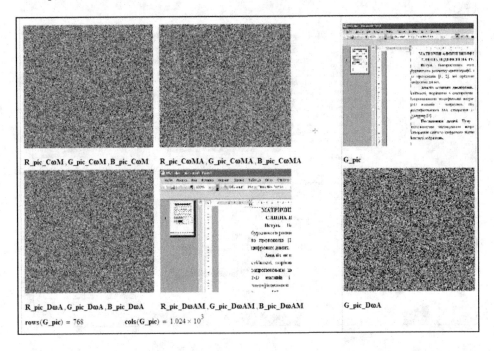

implemented programmatically and with matrix processors, have high speed and stability of transformations and adapt to the CT over image of different formats.

Models of Block Matrix Affine-Permutation Ciphers (MAPCs) for Cryptographic Transformations and Their Research

The emergence of parallel algorithms and especially the matrices of multiprocessor means, requires the creation of appropriate matrix-algebraic models (MAM), matrix-type systems (MT) for CT. Advantages of the TGD, black and white, color images by generalized matrix affine and affine-permutation ciphers (MAPCs), including the creation of blind digital signatures, were demonstrated in works (Krasilenko, 2011; Krasilenko, 2009; Krasilenko, 2012) . Their basic operations are elemental multiplication, matrix addition and matrix permutation models (MM_P) with multiplication matrices. But the disadvantage of these works is the large size of the matrix keys (MK) and the lack of demonstration of their effective work with

Figure 18. Histograms of three formed from the main matrix keys (left) and corresponding R, G, B spectral components (right) of the color image over **which** *the CT based on the MAC* **which** *is shown in Fig. 9 on the right in the bottom row*

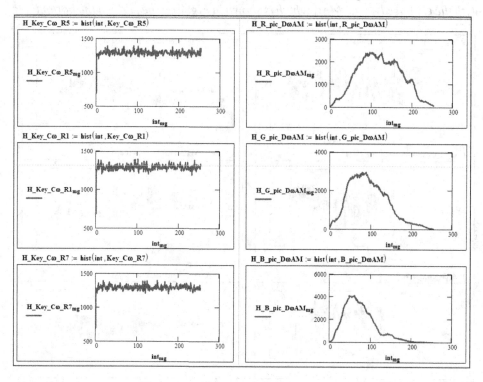

blocks in the form of matrices, which split multi-page data. Some MAMs based on MM_P require decomposition of bit-sections and in addition to the 2 MKs, there are two vector keys (VK) for increasing entropy and the change of histograms with the CT (Krasilenko, 2014; Krasilenko, 2010; Krasilenko, 2016). The promise of the MAM and its modifications for the CT is evidenced by the ability to check the integrity of the cryptogram of the images and the presence of distortions in them, see papers (Krasilenko, 2016; Krasilenko, 2016), increasing the crypto-stability and expanding their functionality while maintaining unified matrix operations, procedures, even for very specific characteristic histograms) of scanned TGDs, as experimentally shown in paper (Krasilenko, 2016). The generalization of the AM to a matrix-block view is necessary in terms of the versatility of block algorithms and independence on data volumes. Therefore, the improvement of the MAPCs, aimed at reducing the number of MK while maintaining the stability and other characteristics of the matrix models (MMs), their experimental testing on various images is also an urgent task. Thus, the actual purpose of this section is the development of block

Figure 19. Histograms of three R, G, B spectral components (left) of the received cryptogram after the multiplicative CT and the corresponding spectral components R, G, B of the obtained cryptogram after the second additive CT over the color image with the CT based on the MAC which is shown in Fig. 9 on the right in the bottom row

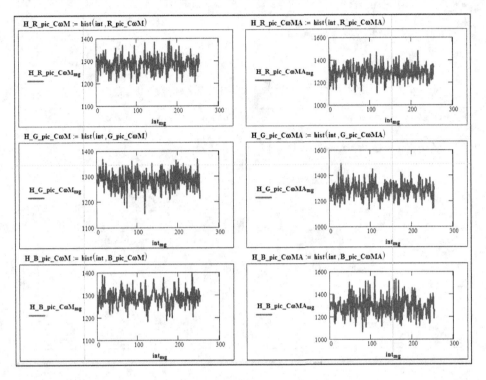

modifications of the MAPCs with a minimum length of 2048 bits, with the possibility of choosing its parameters and cyclic or block keys of similar length, their simulation on real information objects (IO) and demonstration, evaluation of their advantages, characteristics and durability, application possibilities.

Presentation of section material and research results. The proposed CT algorithm for encryption consists of the following steps: 1) the partition of IO into blocks in the form of matrices with a dimension $2^m \times 2^m$, where $m = 4, 5, 6, ...$ and with element-bytes in a digital format that $m = 4$ is equivalent to the length of the block $256 \times 8 = 2024$ bits; 2) the permutation of the bytes of each current block using the current key, which is formed synchronously as the power of the main according to the parametric model, the argument of which is index block, 3) matrix affine or affine-permutation transformations (MAPTs) of matrix of bytes of current keys, the same as on stage 2 or similar, but according to another parametric model, 4) concatenation of the received blocks for the formation of cryptogram of IO. The decryption process

Figure 20. Fragments of Mathcad windows with the results of MK formation and simulation of MAM CT

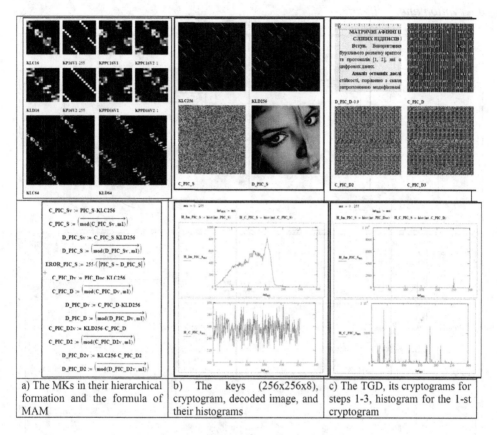

a) The MKs in their hierarchical formation and the formula of MAM	b) The keys (256x256x8), cryptogram, decoded image, and their histograms	c) The TGD, its cryptograms for steps 1-3, histogram for the 1-st cryptogram

has the following steps: 1) decomposing the cryptograms on blocks, 2) reversing the MAPT blocks based on the reversed current keys; 3) reversing bytes of blocks by current keys (vectors); 4) concatenating the transformed blocks into the restored IO. Blocking MAPS modeling was done with Mathcad using black and white and color images of different dimensions for visual demonstration. Mathcad windows with formulas for modeling the CT of the image by the algorithm of block MAPTs for two black and white images (256x256 elements) with the M-key M_V are shown in Fig. 25. Fig. 26 shows the results of the CT and the form of keys, blocks before and after the CT, the difference verification matrix blocks. Fragments of Mathcad windows with modules for the MK formation and the CT formulas are shown in Fig. 27, 28. Results of these CT are shown in Fig. 29.

The random bitmap KPX (256x256x1) of permutations formed in any way is used for permutations of bytes in each kp-th block (256 component vector VID (C_VID) or matrix C_M_V (16x16) with 8-bit numbers). It can be uniquely represented in

*Figure 21. Fragments of modeling of the processes of forming matrices **P**, cyclic parametric **MK**, their constituents, as well as MAM formulas for encryption, decryption and verification*

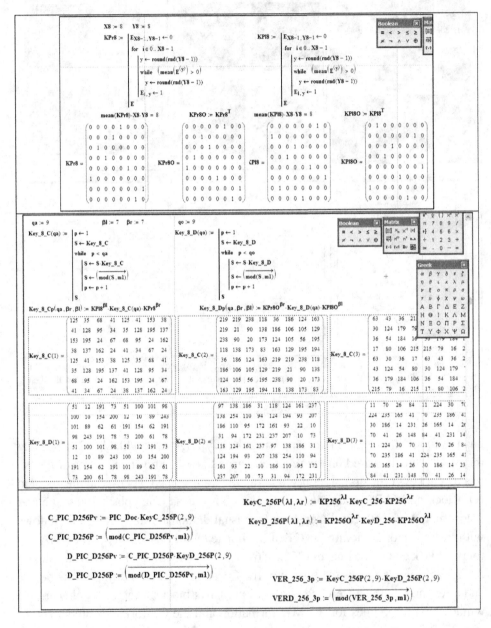

Figure 22. The appearance of some parametric MKs, their component hierarchical blocks, and the unity- matrix (at checking) in different formats (2D, 3D, and digital)

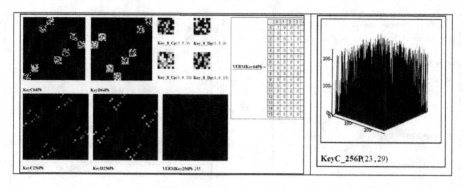

Figure 23. Results of simulation of CT TGD on the basis of parametric MAM and MK with the correct keys (1 experiment) and histogram TGD and cryptograms (right)

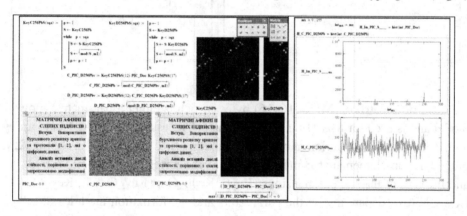

Figure 24. Results of simulation of CT TGD on the basis of parametric MAM and MK with wrong keys (2 experiments) and TGD histograms and cryptograms (right)

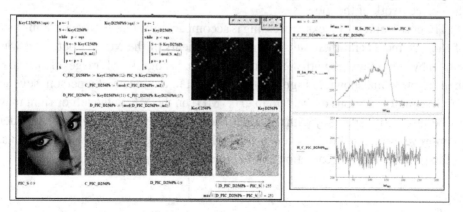

Figure 25. Fragments of Mathcad windows with formulas for forming (concatenate) blocks, encrypting, decoding images with block algorithm MAPTs and verification

PIC_Sv := READBMP("D:\TatoD\tato\2\Ris\g_1.bmp")

PIC_S := submatrix(PIC_Sv, 110, 365, 220, 475)   rows(PIC_Sv) = 549

PIC_Docv := READBMP("D:\tatpic\Doc_1204.bmp")   cols(PIC_Sv) = 600

PIC_Doc := submatrix(PIC_Docv, 110, 365, 300, 555)

rows(PIC_S) = 256   cols(PIC_S) = 256   rows(PIC_Docv) = 768

cols(PIC_Docv) = 1.024 × 10³

PIC_SD := PIC_S

kp := 0..255   ki := 0..15

V1D_{kp} := submatrix(PIC_SD, kp, kp, 0, 255)

Rkl := submatrix(R1, 0, 15, 0, 15)   C_V1D_{kp} := V1D_{kp}·KPX

CP_V1D_{kp,ki} := submatrix(C_V1D_{kp}, 0, 0, ki·16, ki·16 + 15)

$$C\_M\_V_{kp} := \begin{vmatrix} VC0 \leftarrow CP\_V1D_{kp,0} \\ \text{for } ki \in 1..15 \\ \quad VC0 \leftarrow stack(VC0, CP\_V1D_{kp,ki}) \\ VC0 \end{vmatrix}$$

Key0 := M_V + Rkl

min(Key0) = 1

$$CC\_M\_VM_{kp} := \overrightarrow{\left[(C\_M\_V_{kp} + Rkl)·(Key0)\right]}$$

$$CC\_M\_V_{kp} := \overrightarrow{\left(mod(CC\_M\_VM_{kp}, 257)\right)} - Rkl$$

min(CC_M_V25) = 0   max(CC_M_V25) = 255

min(C_M_V25) = 41   max(C_M_V25) = 179

RM_V := 16

$$CV\_V1D_{kp} := \begin{vmatrix} VR1 \leftarrow \left(C\_M\_V_{kp}^T\right)^{(0)^T} \\ \text{for } i \in 1..RM\_V - 1 \\ \quad VR1 \leftarrow augment\left(VR1, \left(C\_M\_V_{kp}^T\right)^{(i)^T}\right) \\ VR1 \end{vmatrix}$$

CVd_V1D_{kp} := CV_V1D_{kp}·KPXO

$$PIC\_SDV := \begin{vmatrix} VC0 \leftarrow CVd\_V1D_0 \\ \text{for } kp \in 1..255 \\ \quad VC0 \leftarrow stack(VC0, CVd\_V1D_{kp}) \\ VC0 \end{vmatrix}$$

$$C\_V1D\_M := \begin{vmatrix} VC0 \leftarrow CV\_V1D_0 \\ \text{for } kp \in 1..255 \\ \quad VC0 \leftarrow stack(VC0, CV\_V1D_{kp}) \\ VC0 \end{vmatrix}$$

$$CCV\_V1D_{kp} := \begin{vmatrix} VR1 \leftarrow \left(CC\_M\_V_{kp}^T\right)^{(0)^T} \\ \text{for } i \in 1..RM\_V - 1 \\ \quad VR1 \leftarrow augment\left(VR1, \left(CC\_M\_V_{kp}^T\right)^{(i)^T}\right) \\ VR1 \end{vmatrix}$$

$$C2\_PIC\_SD := \begin{vmatrix} VC0 \leftarrow CCV\_V1D_0 \\ \text{for } kp \in 1..255 \\ \quad VC0 \leftarrow stack(VC0, CCV\_V1D_{kp}) \\ VC0 \end{vmatrix}$$

the form of a matrix of M_V (16x16) bytes, which is either a parametric (power) model and is used for MAPTs in the next stage. The essence of MAPT is to apply to matrices-B, as a collection of bytes (8-bit images (PIC_S, PIC_Doc, see Fig. 25), procedures on-element matrix multiplication by the corresponding 8-bit MKs (direct and inverse) of the same dimensions (Key0, Key0_O or Key_C (qa), Key_C (qo), Key_CN (qs), depending on the parameters and the formation modules of which are shown in Fig. 27, 28) using the multiplication and modulo operations. As can be seen from Fig. 29 and 30, the simulation results of the processes of direct and reverse CT TGDs and images with the dimension of 256x256 elements are confirmed the correct work of the models.

The cryptographic blocks processing is accompanied by the simultaneous mixing of their elements and their subsequent replacements of the MAPT, but, as it was shown by our researches with the entropy analysis, histograms of images, TGDs and their cryptogram, shown in Figure 29, in contrast to the image of a person, several iterative multiplications of the data matrix (MD) on the MK to the left or right may not be sufficient, especially with the application of the same MK. Therefore, in order to improve the algorithm, we propose to apply different current MKs to the blocks, as the process of their generation can be reduced to simple parametric models.

Figure 26. The results of the CT and the form of the current keys and blocks before and after the CT, the difference verification matrix-blocks: left - for the 1-st image, right - for the TGD

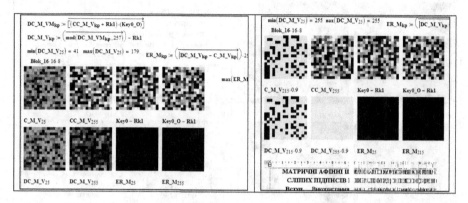

Figure 27. Fragments of Mathcad windows with modules of MK formation

Key_C(qa) :=	$p \leftarrow 1$
	$S \leftarrow Key0$
	while $p < qa$
	   $S \leftarrow \overrightarrow{[(S) \cdot (Key0)]}$
	   $S \leftarrow \overrightarrow{(mod(S, m1))}$
	   $p \leftarrow p + 1$
	S
		Key_C_O(qo) :=	$p \leftarrow 1$
			$S \leftarrow Key0\_O$
			while $p < qo$
			   $S \leftarrow \overrightarrow{[(S) \cdot (Key0\_O)]}$
			   $S \leftarrow \overrightarrow{(mod(S, m1))}$
			   $p \leftarrow p + 1$
			S

Module of current MK generation		Generation of current reversed MK	

Figure 28. Fragments of Mathcad windows with modules for the MK formation and the CT formulas

Key_CN(qs) :=	$p \leftarrow 1$
	$S \leftarrow Key0N$
	while $p < qs$
	   $S \leftarrow \overrightarrow{[(S) \cdot (Key0N)]}$
	   $S \leftarrow \overrightarrow{(mod(S, 257))}$
	   $p \leftarrow p + 1$
	S
		$\mu_{kp} := mod(kp, 5) + 3$

		$CC\_M\_VM_{kp} := \overline{\left[(C\_M\_V_{kp} + Rk1) \cdot \overrightarrow{(Key\_C(\mu_{kp}))} \right]}$ $CC\_M\_V_{kp} := \overrightarrow{\left(mod(CC\_M\_VM_{kp}, 257) \right)} - Rk1$
		Formulas for direct CT with parametric MK
		$DC\_M\_VM_{kp} := \overline{\left[(CC\_M\_V_{kp} + Rk1) \cdot \overrightarrow{(Key\_C\_O(\mu_{kp}))} \right]}$ $DC\_M\_V_{kp} := \overrightarrow{\left(mod(DC\_M\_VM_{kp}, 257) \right)} - Rk1$ $ER\_M_{kp} := \left(\overrightarrow{|DC\_M\_V_{kp} - C\_M\_V_{kp}|} \right) \cdot 255$
Module for generating MK with CTX		For the inverse CT with parametric MK

Figure 29. Fragments of Mathcad windows with the results of CT modeling the block MAPC

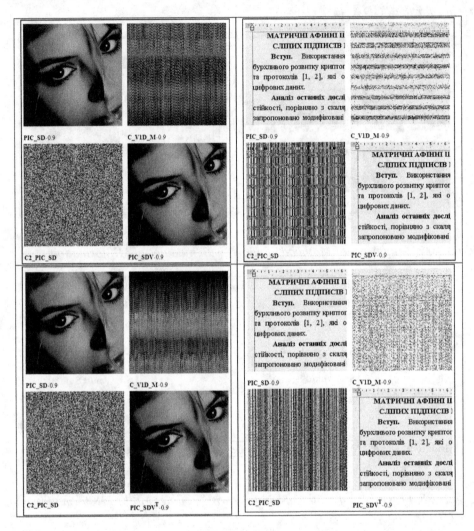

Parametric block MAPC CT, the idea of which is based on the use of dependencies on the indexes of blocks and additional scalar-vector keys (VK) and as parameters influencing the power of matrices MD and MK by modulus in models of their matrix multiplication and the degree and form of permutation matrices. For different blocks and iterative steps different MKs are taken.

The analysis of histograms before and after the KT confirms that the proposed models give better results. The entropy of the TGD was 0.738, and the entropy of the cryptogram of the TGD increased 10.62 times and became equal to 7.837. The

Figure 30. The form of the parametric current MK (right) and the CT of a color image (left)

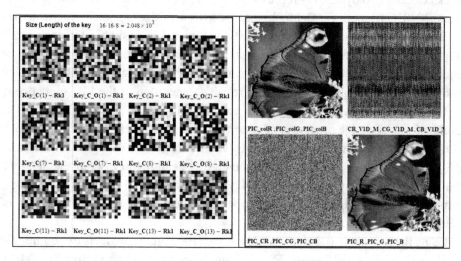

entropy of the image cryptogram has become almost equal to 8 bits per element: 7,997 (-0.04%!). Without knowledge of MK it is impossible to restore MD and, as was shown in (Krasilenko, 2011; Krasilenko, 2012), already with a dimension of 32x32 MK of type **P** The stability of models is ensured and with keys 16x16 8-bit elements, gives a substantial strength. The power of the set of possible keys has increased by an order of magnitude more than 10^{300} . Therefore, the stability of the models has increased significantly.

The section conclusions: New parametric matrix-algebraic models (MAMs) of block MAPC for CT are proposed and modulated. The results of their simulation are presented on the example of direct and inverse CT over images, which testify to their correct performance and efficiency. Considered aspects of creating current MK, models can be implemented with software or hardware matrix processors, and have high speed and stability of transformations.

Modeling and Study of the Generation Method of the Matrix Keys Flow and Their Quality

For the MAM there is an urgent need to form a whole range of permutation matrixes (MPs) from the main MK, which would satisfy a number of requirements. Since in papers (Krasilenko, 2017; Krasilenko, 2017) only the main MK of the general type, but not the series (flow) of the MP was considered, the purpose of this section is to model and study the processes for forming the flow of MP for MAM CT, checking the statistical and correlation properties of a series of generated MP.

Presentation of the main material of the section. Consider the situation for blocks of 256x256 bytes long representing a black-and-white image matrix or 256 bytes (2048 bits) length vector blocks that use MPs with size of 256x256 (Krasilenko, 2012; Krasilenko, 2014; Krasilenko, 2016). Since it is desirable for each block to have a number of MKs generated from the master key, taking into account the requirements for the cryptographic and statistical characteristics of MK, the urgent task of studying the processes of rapid reliable generation of the MK sequence in the form of MP is established, assuming that their number is also equal to 256. The results of modeling the processes of generating a series of MPs for such a situation with the Mathcad formulas and matrices of MP are shown in Fig 31. If the main MK is a random MP of KPX (Fig. 31), then it is unambiguously represented by a 256-component permutation (vector) of V_KPX and also in the form of an image or matrix of bytes (MB) of 16× 16 size with the peculiarity that all 256 grades of intensity are different.

Using the coordinated scalars xa and xm, as KPX degrees, we form two additional matrices C_MKa, C_MKm, see Fig. 31, and the corresponding vectors V_CMKa, V_CMKm, which together with the vector V_KPX (vector representation KPX) are shown in Fig. 32. The histograms of all these vectors are horizontal lines, see Fig. 33, as well as all vector representations of generated permutations that are formed from V_KPX, as its i-th cryptograms, using the affine cipher and the pair of their components of vectors V_CMKa, V_CMKm (additive and multiplicative components). These cryptograms are i-th current permutations (vectors) of KeyCma, which can be uniquely represented in the form of bit matrices KeyCmaR dimension (256x256), for example, KeyCmaP1-254, Fig. 31. Fragments of Mathcad windows are shown in Fig. 34.

Since the histograms of all MPs (their vectors) are horizontal lines, and their entropy is 8 bits, then cryptanalysis on their basis is impossible. In addition, the main and two subsidiary MKs are secret, which allows only the CT parties to create or have this series of MK (MP). In principle, only the main and aforementioned xa and xm scalar keys can be secret or negotiated.

To study the quality of MK (MP), their properties **were investigated**, we calculated all of their possible correlation and equivalent normalized functions, which are represented as fragments of Mathcad windows (Fig. 35-37) and confirm the achievements of surprisingly beautiful properties. We note that the obtained results and their comparison also show that the mutually-equivalently normalized functions are better than mutually correlated ones.

For better perception and more efficient transmission of basic MK (MP) and sequence of created MP, the latter using software modules are converted to colored or black and white images, shown in Fig. 38 and can go as video stream frames (colored images corresponds to three basic MK).

Figure 31. Results of modeling processes for generating an array of MK (MP)

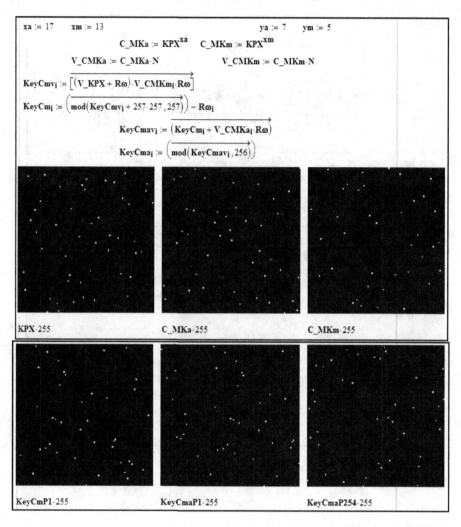

As can be seen from Fig. 36-37, for one MP (in the 200-th experiment) there is a similarity to another key, but this is due to the fact that xm is equal to "1". This can easily be eliminated if the number of MPs in a sequence decreases from 256 to 255 for the one selected in the simulation and the situation described here.

The section conclusions. A method for generating a series of MK (MP) for multipage, block, matrix affine-permutation algorithms and MAM CT is proposed and modulated with Mathcad. The properties of a series of MK (MP) with the mutually equivalent normed functions that are more effective than correlations are investigated, and the adequacy and stability of the method is confirmed.

Figure 32. Vector representations of base MK for generating an array of MK (MP)

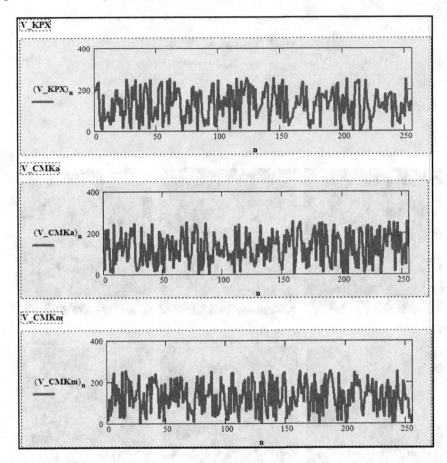

Figure 33. Histograms of vector representations of the base (left) and some (first, second) generated (right) MK (MP)

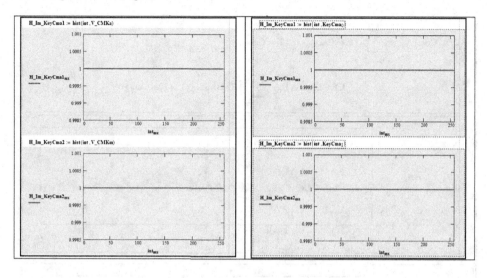

Figure 34. Fragments of Mathcad windows: one of the keys forming procedures (left) and vector representations of some (zero, first, 255th) generated (right) MK (MP)

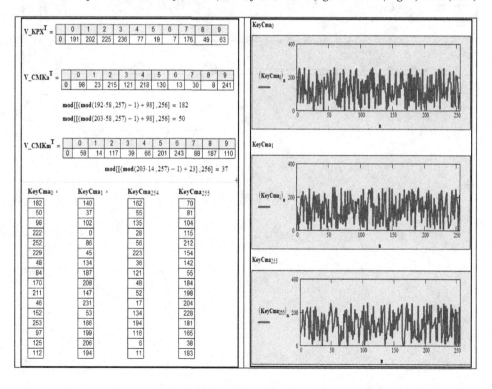

Figure 35. Formulas and the auto-correlation CFa_Cma and mutual-correlation CFv_Cma functions, depending on cyclic shift, displacement of elements of vectors MP

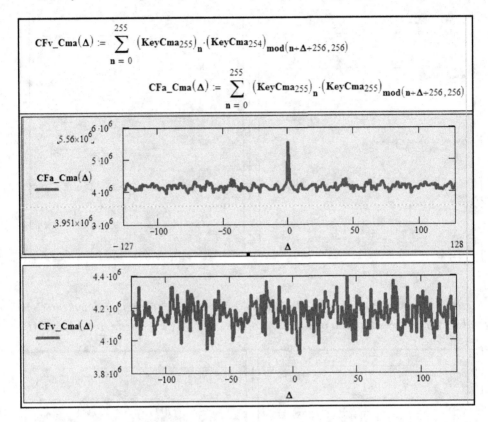

Figure 36. Formulas and the mutually-equivalent CFv_CmaG functions depending on the number of MP (i) and cyclic shift, bias of the elements of t vectors MP

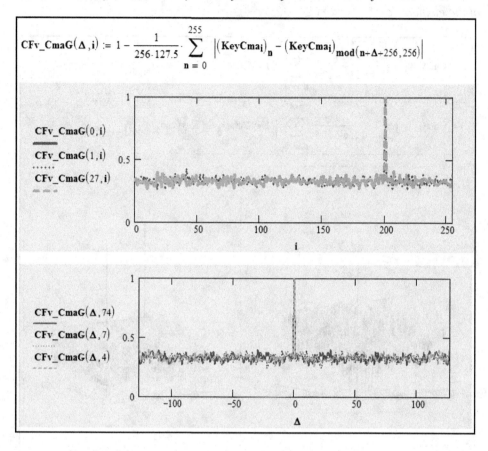

Figure 37. Formulas and appearance (3D) of the mutually-equivalent CFv_CmaG functions depending on the numbers of the MP (i, j) for the "0" and "1st" displacements of the elements of the vectors

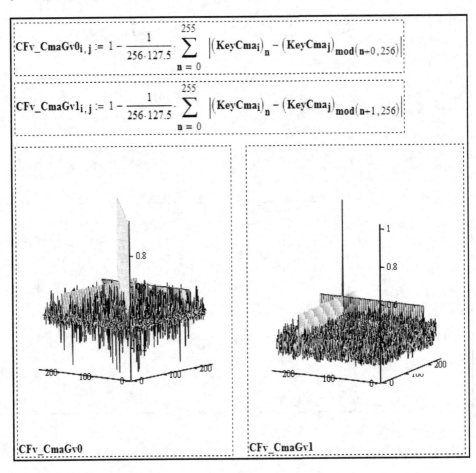

$$CFv\_CmaGv0_{i,j} := 1 - \frac{1}{256 \cdot 127.5} \cdot \sum_{n=0}^{255} \left| \left(KeyCma_i\right)_n - \left(KeyCma_j\right)_{mod(n+0,256)} \right|$$

$$CFv\_CmaGv1_{i,j} := 1 - \frac{1}{256 \cdot 127.5} \cdot \sum_{n=0}^{255} \left| \left(KeyCma_i\right)_n - \left(KeyCma_j\right)_{mod(n+1,256)} \right|$$

CFv_CmaGv0

CFv_CmaGv1

Figure 38. Matrix representation of the basic MK and a number of MPs.

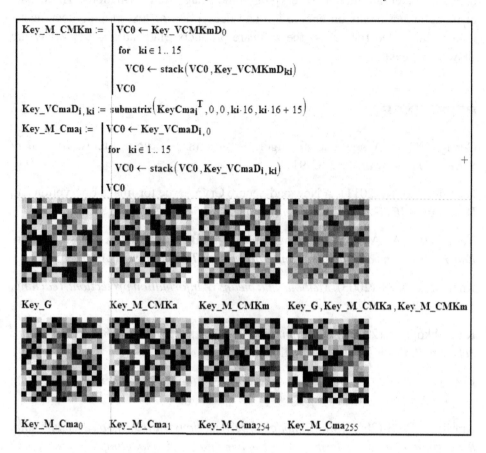

Key_M_CMKm := VC0 ← Key_VCMKmD$_0$

for ki ∈ 1 .. 15

VC0 ← stack(VC0, Key_VCMKmD$_{ki}$)

VC0

Key_VCmaD$_{i, ki}$:= submatrix(KeyCma$_i^T$, 0, 0, ki·16, ki·16 + 15)

Key_M_Cma$_i$:= VC0 ← Key_VCmaD$_{i, 0}$

for ki ∈ 1 .. 15

VC0 ← stack(VC0, Key_VCmaD$_{i, ki}$)

VC0

Key_G   Key_M_CMKa   Key_M_CMKm   Key_G, Key_M_CMKa, Key_M_CMKm

Key_M_Cma$_0$ Key_M_Cma$_1$ Key_M_Cma$_{254}$ Key_M_Cma$_{255}$

CONCLUSION

In the four sections of the chapter, the authors propose and consider new multifunctional matrix-algebraic models of cryptographic image transformations, the variety of matrix models, including matrix affine ciphers, block parametrical and matrix affine permutation ciphers. The algorithms and protocols for generating the necessary matrix keys are discussed in the chapter. The authors show the advantages of the cryptographic models, such as: adaptability to various formats, multi-functionality, ease of implementation on matrix parallel structures, interchangeability of iterative procedures and matrix exponentiation modulo, ease of selection and control of cryptographic transformation parameters. The simulation results of the proposed algorithms and procedures for the direct and inverse transformation of images, with the aim of masking them during transmission, are demonstrated and discussed in this chapter. The authors evaluate the effectiveness and implementation reliability of matrix-algebraic models of cryptographic image transformations. The results of model experiments on encryption and decryption of text-graphic documents, images and video files using the software products Mathcad and LabVIEW are shown by the authors.

REFERENCES

Chang. (2001). A new encycle algorithm for image cryptosystems. *Journal of Systems and Software, 58*, 83-91.

Deergha Rao, K. (2011). A New and Secure Cryptosyce for Image Encryption and Decryption. *IETE Journal of research, 57*(2), 165-171.

Han. (2005). An Asymmetric Image Encryption Based on Matrix Transformation. *Ecti Transactions on Computer and Information Technology, 1*(2), 126-133.

Khoroshko, V. O. (2003). *Methods and means of information protection: Teaching.* Academic Press.

Korkishko, T. A. (2003). *Algorithms and Processors of Symmetric Block Encryption: Scientific Edition.* Baku: V.A. Melnik. - Lviv.

Kovalchuk A. (2009). *Increasing the stability of the RSA system when encrypting images.* Academic Press.

Krasilenko, V. G. (2013). *Matrix models of permutations with matrix-bit decomposition for cryptographic transformations of images and their modeling. In Science and educational process: a scientific and methodical collection of materials of the NPC of all the Universities "Ukraine"* (pp. 90–92). Vinnytsya: Vinnytsia Socio-Economic Institute of the University of Ukraine.

Krasilenko, V. G. (2006). A noise-immune crptographis information protection method for facsimile information transmission and the realization algorithms. *Proc. SIEE, 6241*, 316-322.

Krasilenko, V.G. (2004). Algorithms and architecture for high-precision matrix-matrix multipliers based on optical four-digit alternating arithmetic. *Measuring and computing engineering in technological processes, 1*, 13-26.

Krasilenko, V.G. (2012). Algorithms for the formation of two-dimensional keys for matrix algorithms of cryptographic transformations of images and their modeling. *Systems of information processing, 8*, 107-110.

Krasilenko, V. G. (2008). Simulation of the modified algorithm for creating 2-D keys in cryptographic applications. *Scientific-methodical collection of the scientific-practical conference "Science and educational process"*, 107-109.

Krasilenko, V. G. (2006). Development of the method of cryptographic protection of information text-graphic type. Science and educational process: a scientific and methodical collection of scientific and practical conference, 73-74.

Krasilenko, V. G. (2012). Simulation of Blind Electronic Digital Signatures of Matrix Type on Confidential Text-Graphic Documentation. *International Scientific-Methodical Conference*, 103-107.

Krasilenko, V. G. (2012). Modifications of the RSA system for creation of matrix models and algorithms for encryption and decryption of images on its basis. Systems of information processing, 8, 102-106.

Krasilenko, V.G. (2011). Matrix Affine Ciphers for the Creation of Digital Blind Signatures for Text-Graphic Documents. *Systems of information processing, 7*(97), 60 - 63.

Krasilenko, V.G. (2009). Modeling of Matrix Cryptographic Protection Algorithms. *Bulletin of the National University of Lviv Polytechnic "Computer Systems and Networks", 658*, 59-63.

Krasilenko, V.G. (2012). Matrix affine and permutation ciphers for encryption and decryption of images. *Systems of information processing, 3*(101), 53-62.

Krasilenko, V. G. (2013). Matrix models of cryptographic transformations of images with matrix-bit-map decomposition and mixing and their modeling. Materials of 68 NTC "Modern Information Systems and Technologies. Informational security", 139-143.

Krasilenko, V. G. (2014). Cryptographic transformations of images based on matrix models of permutations with matrix-bit-map decomposition and their modeling. *Bulletin of Khmelnitsky National University. Technical sciences, 1*, 74-79.

Krasilenko, V. G. (2010). Modeling of Matrix Affine Algorithms for the Encryption of Color Images. Computer technologies: science and education: abstracts of reports v VseUkr. sci. conf., 120-124.

Krasilenko, V. G. (2016). Modeling and research of cryptographic transformations of images based on their matrix-bit-map decomposition and matrix models of permutations with verification of integrity. In *Electronics and Information Technologies: a collection of scientific works*. Lviv: Lviv Ivan Franko National University. Retrieved from http://elit.lnu.edu.ua/pdf/6_12.pdf

Krasilenko, V. G. (2016). Simulation of cryptographic transformations of color images based on matrix models of permutations with spectral and bit-map decompositions. *Computer-integrated technologies: education, science, 23*, 31-36. Retrieved from http://ki.lutsk-ntu.com.ua/node/132/section/9

Krasilenko, V. G. (2016). Modeling cryptographic transformations of color images with verification of the integrity of cryptograms based on matrix permutation models. *Materials of the scientific and practical Internet conference "Problems of modeling and development of information systems"*, 128-136. Retrieved from http://ddpu.drohobych.net/wp-content/uploads/2016/04/material_konf.pdf37

Krasilenko, V. G. (2016). Cryptographic transformations (CTs) of color images based on matrix models with operations on modules. In *Modern methods, information and software management systems for organizational and technical complexes: a collection of abstracts of reports of the All-Ukrainian scientific and practical Internet conference*. Lutsk: RVB of Lutsk National Technical University.

Krasilenko, V. G. (2017). Modeling Protocols for Matching a Secret Matrix Key for Cryptographic Transformations and Matrix-type Systems. *Systems of information processing, 3*(149), 151-157.

Krasilenko, V. G. (2017). Modeling of multi-stage and multi-protocol protocols for the harmonization of secret matrix keys. *Computer-integrated technologies: education, science, production: scientific journal, 26*, 111-120. Retrieved from http://ki.lutsk-ntu.com.ua/node/134/section/27

Rashkevich, Y.M. (2009). Affine transformations in modifications of the RSA image encryption algorithm. *Automation. Electrotechnical complexes and systems, 2*(24), 59-66.

Yemets, V. (2003). Modern cryptography. Lviv: Baku.

Chapter 6
Digital Control Theory Application and Signal Processing in a Laser Scanning System Applied for Mobile Robotics

Miguel Reyes-Garcia
Universidad Autónoma de Baja California, Mexico

Cesar Sepulveda-Valdez
Universidad Autónoma de Baja California, Mexico

Oleg Sergiyenko
iD https://orcid.org/0000-0003-4270-6872
Universidad Autónoma de Baja California, Mexico

Moisés Rivas-López
Universidad Autónoma de Baja California, Mexico

Julio Rodríguez-Quiñonez
Universidad Autónoma de Baja California, Mexico

Wendy Flores-Fuentes
iD https://orcid.org/0000-0002-1477-7449
Universidad Autónoma de Baja California, Mexico

Daniel Hernandez-Balbuena
Universidad Autónoma de Baja California, Mexico

DOI: 10.4018/978-1-5225-9924-1.ch006

Juan-Ivan Nieto-Hipolito

iD https://orcid.org/0000-0003-0105-6789
Universidad Autónoma de Baja California, Mexico

Fabian N. Murrieta-Rico

iD https://orcid.org/0000-0001-9829-3013
Universidad Autónoma de Baja California, Mexico

Lars Lindner

iD https://orcid.org/0000-0002-0623-6976
Universidad Autónoma de Baja California, Mexico

Mykhailo Ivanov

Universidad Autónoma de Baja California, Mexico

ABSTRACT

Positioning technologies are useful in a great number of applications, which are oriented for pick and place robots, manipulation of machine tools, especially on machines oriented for artificial vision and detection systems, such as vision-guided robotic systems and object existence in a limited environment. Due to the high demand of those applications, this chapter presents digital control theory application using a laser positioner, which obtains 3D coordinates in a defined field of view. Using the LM629N-8 motion controller, representing the main digital controller for the motion task of the laser positioner as an active element, which is analyzed via modeling and simulation using Matlab-Simulink. Additionally, this chapter focuses on some of the principal sources of uncertainties that exist in a laser scanning system and mainly on the receptive part of such system, which is driven by a brushed DC motor. The processed signal will be analyzed in different environmental conditions to analyze how it is affected by the instability characteristics of this main actuator.

INTRODUCTION

In accordance to positioning technologies which nowadays represent a principal part for vision-guided mobile robotics, this chapter explains methodologies and techniques to obtain three-dimensional coordinates using a vision system. The conception of a single-sensor high speed and low cost scanning system called Technical Vision System (TVS) to obtain data of space coordinates of any surface of an object under observation using a Laser Positioner (LP), which defines accuracy measurements when positioning a laser beam in the mobile robots. Especially, this chapter presents detailed information about the Technical Vision System (TVS), which has two principal parts, namely the Laser Positioner (LP) as the active element and the Scanning Aperture (SA) as the passive element. Both systems collectively obtain data about the limited surrounding environment in a field-of-view of TVS using 3D laser mapping. The TVS has different applications like automatic inspection, quality control of industrial processes, health monitoring of crops and structures.

Thus, this chapter focuses on the components LP and SA. The Laser Positioner is constituted for different components particularly, this system works with an implementation in closed-loop (Atkinson, 2012), that is a precision motion control which integrates the embedded digital controller (Wescott, 2006), LM629N-8 made by Texas Instrument. In fact, this controller represents the main device to execute arbitrary angular displacements of the actuator, that is a high quality brushed DC motor of the LP-TVS. Also this controller is commanded by a Graphic User Interface (GUI) developed in LabVIEW using an Arduino Mega as gateway between the LM629N-8 and the GUI. The GUI includes all available commands of the LM629N-8. The Scanning Aperture (SA) principles and elements are then presented as a useful way to receive signals generated by a light emitter sources, such as the LP first presented in this chapter. In addition to the utility of the SA in Laser Scanner Systems, uncertainty factors of the driving conditions in the SA, due to the signal processing characteristics of the system, may affect the resolution of the detected coordinate (angle). Thus, a closed-loop control system is proposed for the SA DC motor. The final part of this chapter shows experimentation with both the LP and the SA. Therefore, this chapter has been written for engineering students, engineers, and new researchers that are interested in machine vision applied for mobile robotics taking into account, the operation of the principal components of an optoelectronic device (Bass, et al., 2009) named Technical Scanner System.

Background

Digital Control Theory Application (DCTA) and Signal Processing (SP) are topics considered for Discrete-Time Systems (DTS), due to those analyze and interpret mathematically physical phenomenon behavior. With DCTA and SP new knowledge about approaches and limitation of DTS is created. Currently, the tools used in DCTA are more capable to perform complex tasks in comparison as they were before. Examples of these tools are some single-board computers such as Raspberry Pi, Jaguar One, Orange Pi, pcDuino4, among others. Furthermore, new Integrated Development Environments (IDE's) maximizes the productivity of the designer during the software development for high-performance microcontrollers such as the Intel® Quark™ microcontroller D2000, Arduino Ethernet, Arduino DUE, Arduino Mega 2560. Due to the powerful features of these processors it is possible to develop complex control algorithms to handle automatic systems. The

SP is necessary for control methodologies, due to signal processing operations main task is to convert physical signals to electric signals which can be fed back to the automatic systems. This can be done by the implementation of different elements such as filters, transducers, analogic-to-digital converters, as well as mathematical, computational algorithms and statistical methods (Madisetti, 2009). The wide variety of tools and methodologies used to SP and DCTA gives the opportunity to develop more specialized, efficient and low-cost systems. Thus, in this chapter the DCTA and SP are applied in the development of the TVS used for Mobile Robotics.

MAIN FOCUS OF THE CHAPTER

This chapter focuses mainly on two optoelectronic devices: Laser Positioner (LP) and Scanning Aperture (SA), both systems are the main elements of a Laser Scanning System (LSS) applied for Mobile Robotic (MR). Furthermore the LSS can be used for other applications (Lindner, et al., Machine vision system errors for unmanned aerial vehicle navigation, 2017), where the LP and SA of LSS are analyzed by means of Digital Control Theory (DCT) application using Signal Processing (SP), which optoelectronic devices are shown in detail, according to the operation in each of them. Using the following principal components: LM629N-8 digital controller, an Arduino Mega 2560 single-board, an L298N motor driver, and brushed DC motor with an encoder incremental to develop with the LP, a GUI developed in LabVIEW Platform to operation it. A DAQ USB-1208 single-board, a single-board Arduino UNO and an H-bridge are used to develop an algorithm for a closed-loop velocity control to reduce the angular error of the Scanning Aperture operation.

FUNDAMENTAL BASICS

The fundamental basics used in this chapter are presented in this section. The first one deals with Signal Processing (SP), followed by briefly introduced Laser Scanners (LS) later, Trapezoidal Velocity Profile (TVP) concept, which focuses on explaining a motion control method to determine a control variable in the motion control system, this variable is called Trapezoidal Velocity for servomotors. Subsequently, Digital Closed-Loop Control (CLC) subtopic explains a motion control system in closed-loop configuration using the embedded controller LM629N-8. Finally, the Dynamic Triangulation Method as the principle of functioning for a Laser Scanner is presented.

Signal Processing

Signal processing is the analysis, conversion and manipulation of physical signals from environments surrounding systems. These signals are primarily used to detection of condition in such systems. One of the purposes of signal processing is to acquire a physical signal and transform it, by means of electronics transductors, to an electric signal, which can be either voltage level or electric current, so it can be used to state detection and influence of this physical signals on different systems. Signals acquired by this processing methods can be analyzed mathematical, to determine information or characteristics of some physical phenomenon. According to the IEEE Signal Processing Society, signal processing is a branch of electrical engineering that models and analyzes data representations of physical events.

Physical signals present at any environment are considered analog. To use analogic signals on digital systems, they must be transformed to digital ones, this task it's commonly achieve using Analog-to-digital converters. Signal processing is applied in multiple fields, such as audio signals which are converted to electric signals, it's used on image or videos processing. The main devices for signal conversion are: signal compressors, digital signal processors and digital filters. There are also some integrated circuits that realize the processing of using electronic circuits or software. Some of the mathematical methods used for signal processing are differential equations, transforms theory, time-frequency analysis, spectral estimations, calculus and statistical theories (Madisetti, 2009).

Laser Scanners

Laser Scanners are widely used as measuring systems in all kind of operations that involves gathering of physical information of the environment using contactless methods, such as distance measuring, surface topography and other characteristics. A scanner used for spatial documentation can be found in (Haddad, 2011). As optical devices, this kind of scanners are capable of providing the same amount of information of their surroundings and at higher speeds than scanners, that utilize different instruments to measure (such as stylus instruments). Such as (Zhongdong, Peng, Xiaohui, & Changku, 2014), which presents a high dynamic range imaging laser scanner. And (Xiang, Chen, Wu, Xiao, & Zheng, 2010) were a study of a fast linear scanning for a new laser scanner is developed.

As mentioned in (Lindner, Laser Scanners, 2016), there exist several kinds of laser scanner in the industrial and research field. Each one of those, use different methods to perform measurements, but all of them have as main task, the scanning of a surface by an optical application and the reception of this information (by a sensing element) to further post-processing either using mathematical methods or directly using powerful processors.

The main objective of the different laser scanners will depend on the application that is being used. They can be used to monitor welding quality, to get a structure tomography or mobile robot navigation. As remote sensing technology a laser scanner such as Lidar (Light detection and ranging) systems can be used for multiple applications, due to its operations capacities, and with slightly changes on algorithms and elements in (Alexander, Erenskjold Moeslund, Klith Bøcher, Arge, & Svenning, 2013) a scanner its presented used to known understory light conditions in forest ecosystems. Recent laser scanner applications are extend to multiple fields. Used to determine vegetation condition in plantations. To detect road edges in mobile robots trajectory. As reception system of geo-reference points from specialized terrains. And even dental intra oral scanner to perform a high accurate 3D image of the oral cavity.

According to (Lindner, Laser Scanners, 2016) laser scanners are divided by: used measuring signals (principle applied by the receptive part of the scanner), measurement acquisition methods and system design itself. The used measuring signals of scanners include laser, fast and long range signals (long distances). As disadvantage can be mentioned, that the readings of laser signal depend on the scanning surface, due this they cannot be used underwater (Lindner, Laser Scanners, 2016). Sonar is another measuring signal. This is cost-efficient in comparison to laser. No matter environmental or surface condition (can even be used underwater) it can work properly, but has less precision for the measurement results. There is optical signals which are images obtained by different methods. Optical signals are most close to human vision. As main advantage, the entire surrounding of the system can visualize. But, due the great quantity of information that images require, systems receives more than needed for its measurements, consuming more time. Also, lack of illumination can make system unable to work correctly just as human sight.

The measurement methods include Time-of-flight (TOF), which measures the total traveling time of a laser beam, emitted and reflected to and from a scanning surface. Phasing, whose measurement principle uses the phase differences between the transmitted and received laser beam from a scanning surface. Static and dynamic triangulation (Sergiyenko, et al., 2011), which measures the system's receptive element angle of the reflected laser beam. The last measuring method its imaging, these scanners detect the intensity of the reflected laser beam.

To categorize laser scanners by design it is important to emphasize on the application for which the system is designed. If the scanner has monitoring purposes of structures such as institutional building, the design will be a stationary laser scanner. The design use tripods or special bases strategically located around the building to accomplish the main task. If the scanner is used for autonomous robot navigation applications, the design then should be a mobile laser scanner. The laser system will be in charge of collecting the surroundings information of the robot path. In mobile robots, scanners can be designed within robots structures. In (Fu, Menciassi, & Dario, 2012) the development of an indoor low-cost scanner for miniature mobile robots is realized. In hand-held laser scanners mobility and flexibly are the principal features. Hand-held scanners are the smallest of three kind of designs they are mostly used as barcode reader or hand-held distance measurement devices.

Trapezoidal Velocity Profile

"Kinematic" defines the movements of tangible objects omitting the causes that originate them (Singh, 2008). This branch of physics analyzes the trajectory generated by the object movement and study this movement as function of time. However, a trapezoidal velocity trajectory is analyzed in this section as a controlled variable used in Motion Control Systems (MCS) of electric machines.

In the last 50 years, MCS have been required for automatic measurements, using a variety of techniques to control the movement of DC motor shafts, these electric machines are used in industrial equipment, as an example CNC machines (Computer Numerical Control) and robots. Using a trapezoidal velocity profile as control variable results advantageous, due to the automated routines with high performance of industrial operations are stable under conditions of precision and high velocity using Direct Current Motors (DC motors) (Presicion Motion Controller LM629, 2013). The controlled systems by MCS are the main devices to be interpreted to obtain high-performance in consequence for the above, Maxon Motors supplies DC Motors with high-quality which are made using permanent magnets and winding rotors without iron (Maxon Academy, n.d.). Thus, the iteration (magnetic field) between the magnets of the stator and the rotor core is deleted. That iteration is a physical phenomenon called (cogging-torque), likewise, these motors offer a reduction of breakaway torque by using ball bearings, using brushes of a precious material (could be gold, silver, rhodium, palladium or platinum) inclusive. Thus, these electrical machines keeping a lineal rotation of the shaft DC motor under low velocities.

The ideal trapezoidal velocity profile is shown in Figure 1, where ω is velocity and t is time. This figure shows that mainly three time intervals have to be considered. The time interval $\Delta t = (t_a - t_0)$ uses a constant rate of change from the zero velocity at t_0 until reaching the maximum velocity defined as $\omega_{max}(t_a)$. Starting from there to the moment t_b with maximum constant velocity. Finally, the last third of the trajectory represents a decreasing slope with the same rate of change as used for the initial acceleration. Therefore, a deceleration is applied to the motor shaft, reaching the time t_φ which is the time required to obtain the desired angular position when integrating the trapezoidal profile. The trajectory can be modified in relation to the

Figure 1. Trapezoidal velocity profile

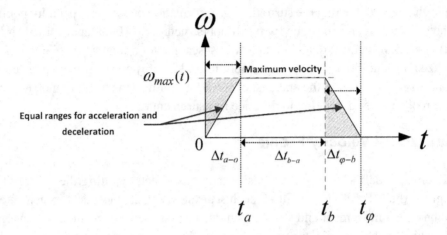

desired positioning time. A motion controller is in charge of altering the controlled variable in consequence, the LM629 controller is used to the Laser Positioner of the LSS. Thus, the controlled variable is the trapezoidal velocity profile for some applied servomotor (Buccella, 1997). Figure 2 defines how the trapezoidal trajectory can be deformed in the triangle form using high maximum velocities thus, obtaining short positioning times interval

Recognizing, that there are different types of motion profiles for electric motors due to ensuring that accelerations submitted by motors with load require a torque less than the maximum electrical torque of the motor. On the other hand, S-curve profiles are also implemented in SCMs, however these have disadvantages, such as a slower movement and implementation is difficult (Lu & Chen, 2016).

$$
\omega_o(t) = \begin{cases} \dfrac{\omega_{max}(t)}{t_a}, & 0 \le t \le t_a \\[2ex] \omega_{max}(t), & t_a \le t \le t_b \\[2ex] \omega_{max} - \dfrac{\omega_{max}}{t_\varphi - t_b}, & t_b \le t \le t_\varphi \end{cases}
\tag{1}
$$

where:

Maximum velocity: ω_{max}.

Acceleration time: Δt_{a-0}.

Maximum velocity time: Δt_{b-a}.

Deceleration time: $\Delta t_{\varphi-b}$.

Positioning time: t_{φ}.

Figure 1 shows two discontinuous blue lines. Starting with the horizontal blue line, to represent the value of the maximum constant velocity using the variable $\omega_{max}(t)$ and the vertical blue line indicates the start time to deceleration of the motor axis applied at moment t. Therefore, it is possible to observe a symmetric rectangle defined in orange color where it is remarkable to see an inverted triangle which is the area that compensates the third part of the trapezoid on its left side, in such a way that it is possible to obtain the integral of the Equation 2 to obtain the final angular position given by $\varphi_o(t)$. Thus, the displacement complete traveled by the motor shaft is the final position from the zero position until t_b and represented by:

$$\varphi_o\left(t_{\varphi}\right) = \omega_{max} \cdot t_b = \int_0^{t_{\varphi}} \omega(t)\,dt \qquad (2)$$

Digital Closed-Loop Control

Figure 3 shows a block diagram representing a Motion Control System (SCM) for the Laser Positioner (LP) of LSS, the flow of information starts with a User that interacts with a developed GUI. The GUI processes configurations of commands and data that the user requests and using a read or write port through serial communication to transfer the configurations between the computer and the Microcontroller. Subsequently, the data and commands processed in the microcontroller are written or read using a parallel communication with an 8-bit port between the microcontroller

Figure 2. Deformation of the trapezoidal velocity

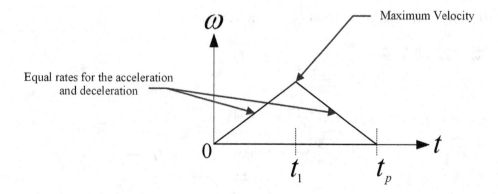

and the LM629N-8 controller, which processes data and commands with its control algorithm and supports six internal modules. A data string coming from GUI, which contains φ_r, ω_{max}, a_r, ppr and the possible gains of the PID Controller: k_p, k_i, k_d to be downloaded to the Host Interface which manages the data according to the commands. Therefore, the data φ_r, ω_{max}, a_r, ppr are sent to the Trajectory Generator. This module creates a profile of an arbitrary trapezoidal velocity, according to the entered amount of the pulses per revolution (*ppr*) converting them into accounts per revolution (*cpr*) by the rising and falling edges of the incremental signal coming from Incremental Encoder. The values assigned to the control gains are downloaded to the internal PID controller. The Summing Union block obtains the positioning error by subtracting discrete values: the *cpr* coming from Trajectory Generator, which represents the φ_r and the *cpr* measured by Quadrature Decoder block according to the quadratic signals (*ppr*) of the outputs of channels A and B of the Incremental Encoder. This provides the information of the current position of the laser beam emitted from the Laser Module and oriented to the mirror at 45 degrees to be projected to the surrounding environment in the TVS field of vision. The PID controller is fed by the position error signal, to obtain the control signal to be sent to the PWM (8-Bit) block, which is responsible for sending the control signal, which is defined as the duty cycle percentage defined by $y(t)$. The power controller is a switched mode amplifier for the signal $y(t)$ to obtain the signal of the rated armature voltage of the applied DC Motor $u_A(t)$.

Dynamic Triangulation Method

Triangulation represents a measurement method to determine angles and distances by means of trigonometric functions. Different types of triangulation are used for laser scanning systems. In static triangulation, the emitter and receiver nodes of the measuring system are fixed and motionless, which means that the capability

Figure 3. Block diagram of motion control system

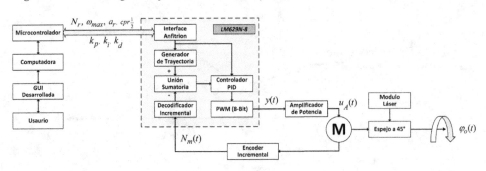

of measuring objects is limited by the size of the sensors. The lack of movement of the measurement principle elements defines the static triangulation. Dynamic triangulation utilizes two dynamic elements (nodes). An active one, that is in charge of emitting a laser beam and a passive one that will receive the reflected laser beam. Here the active element positions a laser beam towards a desired scanning object, the laser beam is reflected by the object surface (specular or diffuse) and the passive element, which is continuously rotating, receives the reflected laser beam to form a triangle for a small time period. Figure 4 shows the triangle formed between the PL, the SA and the object surface.

Using the law of sines, it is possible to calculate the distance "d" between the system and the reflection point "A". In the triangle, γ and β are known angles and with the help of a fixed and known distance of separation "a" between the active and passive nodes of the triangulation system, the distance "d" from the system to the reflection point "A" can be calculated using (Basaca-Preciado, Sergiyenko, Rodriguez-Quinonez, & Rivas-Lopez, 2012):

Figure 4. Dynamic triangulation principle top view

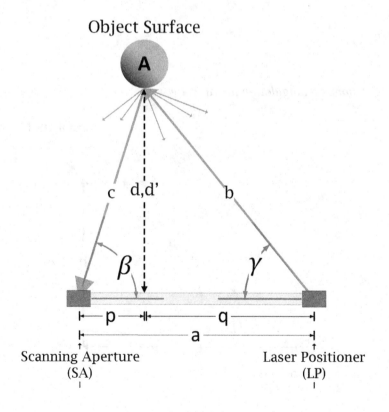

225

$$d = a\frac{(sin\beta)(sin\gamma)}{sin(\beta+\gamma)} \tag{3}$$

$$p = \frac{d}{tan\beta} = a\frac{(cos\beta)(sin\gamma)}{sin(\beta+\gamma)} \tag{4}$$

$$q = \frac{d}{tan\gamma} = a\frac{(sin\beta)(cos\gamma)}{sin(\beta+\gamma)} \tag{5}$$

Figure 5 shows the dynamic triangulation lateral view, which represents a triangle that is used to calculate the height of the object. Where η is the tilt angle of the TVS. Equation (6) is used to measure the third coordinate on the TVS field of view (FOV):

$$h = d(sin\ \eta). \tag{6}$$

In Figure 5 d'.s defined by (7), and represents a projection of the distance d on the top view.

$$d' = d(cos\eta). \tag{7}$$

Figure 5. Dynamic triangulation lateral view

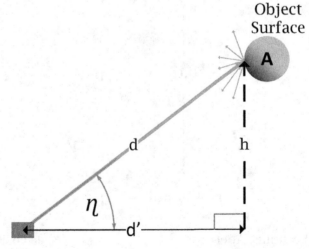

The triangulation take its name due to the dynamic characteristics of the two nodes that are in the search of the object surface. As well as the virtual triangle dynamic dimensions, due to the distance between the system and the object under observation.

LASER POSITIONER SYSTEM

The Laser Positioner System (LPS) is able to sweep a Projected Laser Beam (PLB) to the surrounding environment in a Field of Vision (FOV) of the Laser Scanning System (LSS). The principal function of the LPS is to emit a laser beam to the surfaces of some physical object under observation inside the FOV of the LSS.

Control System Operation Methodology for LP

The motion control system for the Laser Positioner (LP) is represented as block diagram in Figure 6 and consists of the following components: *User* to monitor and load data from an angular reference position $\varphi_r(t)$ arbitrary of a projected laser beam in the FOV, using a *GUI* (Graphic User Interface) executed in the Processor Central. The GUI is responsible for sending and receiving data of the angular position reference to an Arduino Mega 2560 single-board, which uses an Atmega2560 microcontroller. This single-board microcontroller is responsible for executing a Machine Code (MC) with the instructions assigned to process data, commands and physical signals that interact with the motion control system of the LP. The MC is created when compiling Source Code (SC) and developed in the Arduino IDE or the Eclipse IDE platforms. A Texas Instruments (TI) LM629N-8 *Motion Controller* designed to use in precise motion control systems; a *Power Driver* like the XY-160D module, which amplifies in terms of electrical power (voltage and current) the PWM output signal from the LM629N-8 controller to provide the nominal current and voltage to the Maxon DCX22 motor. This motor has a Maxon ENX 16 *Digital Position Sensor* to provide incremental signals of the actual angular position of the DC motor shaft which has a coupled *Mirror* with a diagonal cut at 45 degrees, the Projected Laser Beam (PLB) coming from Laser Module and the PLB is limited by Field Of Vision (FOV) in the surrounding environment of the LSS to physically determine 3D coordinates of any object under observation of a Mobile Robotic.

Figure 6. Block diagram of the laser positioner for the LSS applied for mobile robotic

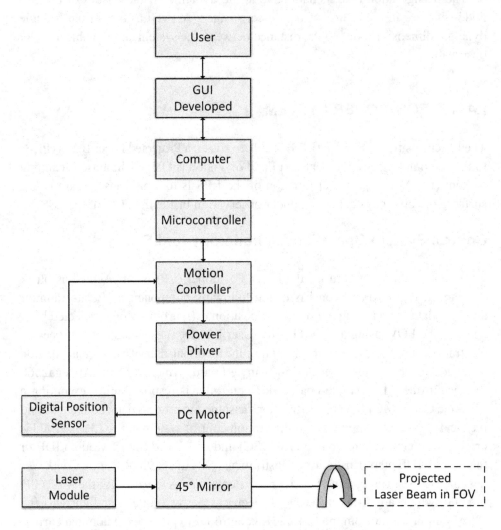

Flowchart of the Block Diagram for the Process of Laser Beam Positioning

The process of laser beam positioning is defined in this section according to the flow chart for the Laser Positioner shown in Figure 6. In relation to the section *Control System Operation Methodology,* where the components used for the LP are presented, the operation below and subsequently to detail will be briefly explained using the data flow and signals.

User and GUI

Figure 7 shows the start of the data flow, to initiate the correct functioning of the LP. The user executes the GUI developed in LabVIEW 2015 and receives visually feedback through the GUI. The GUI developed is used to parameterize the data requested by the interface and executed on the computer. The information flow is as follows: Initialize the LM629N-8 controller, then update the gains of the PID controller when downloading to the controller then, information such as: reference acceleration, maximum velocity, reference angular position, the position type: absolute or relative and incremental encoder resolution in pulses per revolution are downloaded to generate the trapezoidal velocity profile of the Maxon motor DCX22S.

Computer and Microcontroller

In Figure 8, the functional GUI block is executed by the processor central. The Atmega 2560 microcontroller is mounted on the Arduino Mega Single-board that has a USB (Universal Serial Bus) 2.0 type B female connector to make a serial communication using a USB cable of type A male to type B male, between the Arduino Mega Card and the Computer transferring and receiving data between both components.

Figure 7. User and GUI developed blocks

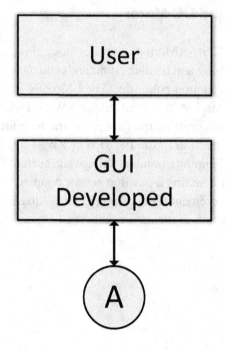

Figure 8. Computer and microcontroller blocks

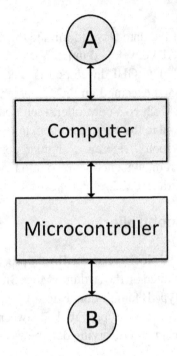

Motion Controller, Power Amplifier, Position Sensor and DC Motor

Figure 9 shows the LM629N-8 Motion controller block, this controller uses a parallel communication for writing and reading commands and data processed between the Arduino Mega and the motion controller. The LM629N-8 is an embedded circuit with the following control variable: an output PWM (Pulse Width Modulation) signal, two pins in configuration; the first pin is the magnitude of the PWM and the second pin is the direction of the PWM. The PWM signal is amplified by the XY-160D controller, obtaining a nominal voltage and current for the DCX22 Maxon actuator. This actuator contains a position sensor coupled to the motor shaft, the position sensor is an Incremental Encoder (IE) for feedback to the LM629N-8 the data of the current position of the motor shaft.

Figure 9. Motion controller, power driver, incremental encoder, and DC motor blocks

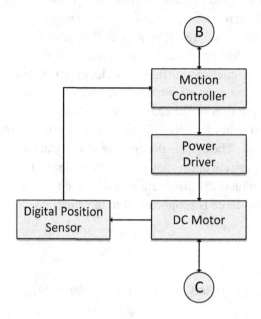

Laser Module, 45° Mirror and Projected Laser Beam

In figure 10, shows a block that represents a mirror cuts to 45 degrees which is coupled to the DC motor Maxon DCX22S shaft, in the mirror a laser beam is projected towards the surrounding environment that coming from the Laser Module.

MODELING AND SIMULATION OF A POSITIONER LASER

This section presents an analysis of the Positioner Laser using modeling and simulation. The model of the PL is developed and proposed in MATLAB – Simulink. The theoretical DC Motor model was derived, using the commonly known scheme

Figure 10. Laser module, 45° mirror and projected LB in FOV

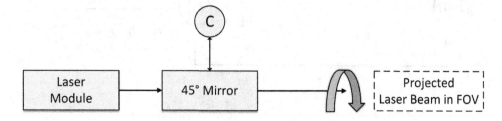

that represents the electrical part and mechanical part of a DC motor. Being: Supply voltage for the motor armature $u_A(t)$ for DC motor energization; Electric resistance R. Electric inductance L. As a result there is an Electromotive Force (FEM) ε which represents the induced voltage generated by the rotation of the motor shaft, when the rotor collectors have no contact with the active electric current brushes. The following concepts are used: J moment of rotational inertia of the motor, which represents a resistance to the rotation of the axis when applying an angular acceleration; B represents the coefficient of viscous friction, $M(t)$ represents the torque produced by the DC motor shaft. The $\omega_o(t)$ is the angular velocity and finally $\varphi_o(t)$ represents the current angular position of the motor. DC motor. Below is a mathematical model of the motor that interacts the electrical part with the mechanical part.

The electromotive force is proportional to the angular velocity

$$\varepsilon(t) = K_u \cdot \omega_o(t). \tag{8}$$

The velocity is the derivative of the final angular position

$$\omega(t) = \frac{d\varphi_o(t)}{dt}. \tag{9}$$

Using the conservation of energy by Kirchhoff's first law

$$u_A(t) - \varepsilon(t) = R \cdot I(t) + L \frac{dI(t)}{dt}. \tag{10}$$

Figure 11. DC motor model

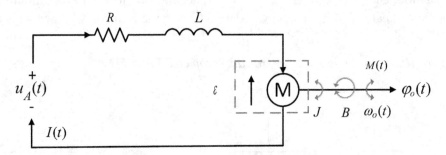

The equation (8) is replaced in the equation (10) to obtain the following equation (11)

$$u_A(t) - K_u \cdot \frac{d\varphi_o(t)}{dt} = R \cdot I(t) + L \frac{dI(t)}{dt}. \tag{11}$$

An electromagnetic moment is generated when turning the motor shaft by the magnetic field

$$M(t) = K_M \cdot I(t). \tag{12}$$

An inertial moment exists by the dynamic friction

$$M(t) - \beta \frac{d\varphi_o(t)}{dt} = J \frac{d^2\varphi_o(t)}{dt^2}. \tag{13}$$

Thus, the electromagnetic moment is the sum of the inertial moment and the viscosity coefficient

$$M(t) = \beta \frac{d\varphi_o(t)}{dt} + J \frac{d^2\varphi_o(t)}{dt^2}. \tag{14}$$

The equation (12) is replaced in the following equation (15) by the equation (14)

$$K_M \cdot I(t) = \beta \frac{d\varphi_o(t)}{dt} + J \frac{d^2\varphi_o(t)}{dt^2}. \tag{15}$$

The electrical current $I(t)$ is replaced in the equation (10)

$$I(t) = \frac{J \frac{d^2\varphi_o(t)}{dt^2} + \beta \frac{d\varphi_o(t)}{dt}}{K_M}. \tag{16}$$

$$\frac{dI(t)}{dt} = \frac{J\dfrac{d^3\varphi_o(t)}{dt^2} + \beta\dfrac{d^2\varphi_o(t)}{dt^2}}{K_M}. \tag{17}$$

Of the equations (16) and (17) it obtains the equation (18) to define the DC Motor model

$$u_A(t) - K_u \cdot \frac{d\varphi_o(t)}{dt} = R \cdot \frac{J\dfrac{d^2\varphi_o(t)}{dt^2} + \beta\dfrac{d\varphi_o(t)}{dt}}{K_M} + L \cdot \frac{J\dfrac{d^3\varphi_o(t)}{dt^2} + \beta\dfrac{d^2\varphi_o(t)}{dt^2}}{K_M}. \tag{18}$$

Modeling Using Simulink of MATLAB

This section presents a model of the Maxon motor DCX22S is simulated according to Figure 12. The used principal parameters of the Maxon motor DCX22S are presented in Table 1. This model supports an approach about implementing a motion control system using a profile generator to obtain a trapezoidal velocity as a reference control variable $\omega_r(t)$ of the controlled system which is the Maxon motor DCX22S. Furthermore, Figure 16 shows a comparison between a proportional controller and a trapezoidal velocity controller, determining, an advantage when using a trapezoidal velocity as a control variable, obtain a high grade of precision and a stable behavior of the motor DCX22S

Simulation Using MATLAB Simulink

For the simulation of the Motion Control System (SCM) model proposed for this chapter, different block diagrams were developed, these diagrams were unified in a complete model to represent the SCM. Starting with the block diagram of Figure 12 to represent the Motor Maxon DCX22S according to the motor parameters in Table 1. The function $u_A(t)$ which enters to summing union to obtain the error signal, an induced voltage is representing by ε to be entering to summing union which is multiplied by a constant kb previously, creating a closed-loop configuration. The electric part of the DC motor is constituted by L and R opposing to the flow of $I(t)$ producing a torque $M(t)$ on the shaft multiplied by KM and $M(t)$ is affected by a block that represents the mechanical part of the DC motor where, there is a moment of inertia opposing to the rotation, the output signal of the mechanical part block is the angular displacement of the motor shaft where K is the conversion of radians to degrees thus, the angular velocity is presented by $\omega_o(t)$.

Figure 12. Diagram block for DC motor Maxon DCX22S model

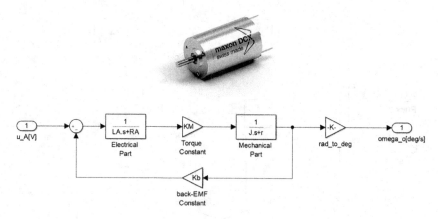

Table 1. Parameters of the DC motor Maxon DCX22S

Description	Parameter	Unit
Armature voltage		12
Nominal Velocity		6710
Electrical resistance		2.12
Electrical Inductance		0.130
Rotational Inertia		5.12
Motor Torque		14
Motor Torque Constant		13.8
Electrical Current		1.05

The LM629N-8 controller uses a *Trajectory Generator* among its internal modules. This is shown in Figure 13. To obtain the trapezoidal velocity, two pulse train blocks Pulse_1 and Pulse_2 are used, which produces two signals with the same amplitude, one of them is contrary to the other and the pulse train Pulse_2 has a time delay necessary due to both signals pass in synchronization by an adder block to form a continuous signal of acceleration and deceleration in sequence, observed in Figure 13 in the first red graph. Then this continuous signal goes through a block Integrator_1, which integrates the acceleration and deceleration to obtain the trapezoidal profile of the velocity, observed in the second orange graph. Between these ramps a signal with constant amplitude is shown representing the maximum constant velocity $\omega_{max}(t)$. This velocity passes through another Integrator_2 block by integrating the area below the trapezoidal velocity profile to obtain the trajectory of the reference position of the applied actuator shaft $\varphi_r(t)$ observed in the third graph.

Figure 13. Trajectory generator model and graphics

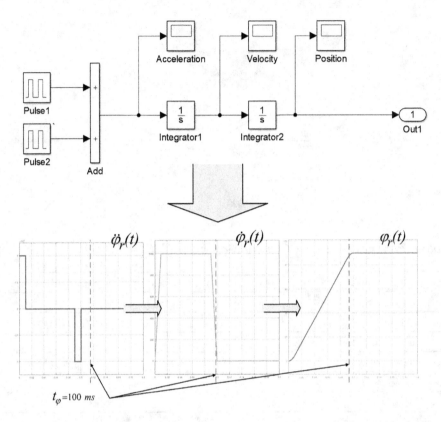

The Motion Control System (SCM) is represented with the block diagram in Figure 14. Where the initial block represents the Profile Generator of the trapezoidal velocity signal $\omega_o(t)$ obtaining the reference position $\varphi_r(t)$ that is the profile generator signal enters to summing union that produces the positioning angular error signal $\varphi_e(t)$. This signal enters through a configurable controller PID obtaining, the control signal $y(t)$ to drive the DCX22S block, which represents the controlled system. This is a Maxon DC motor DCX22S executes a rotation under its shaft. The signal block DCX22S is the final angular position $\varphi_o(t)$ that enters to block Encoder to convert this analogic signal to a digital signal to be feedback it to the summing union, obtaining a closed-loop configuration of the SCM. The Graph in Figure 14 is notable for a minimal time delay for obtaining $\varphi_o(t)$ due to the friction effects suffered and moment of inertia by the Maxon DCX22S motor (Maxon Academy, n.d.).

Figure 14. Motion control system using a profile generator

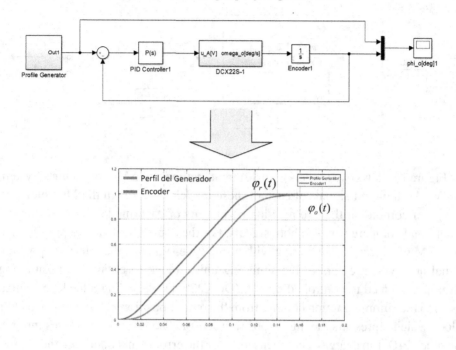

Comparison of the Output Signal φ₀(*t*) for Different Models

The Laser Positioner projects a green laser inside the LSS field-of-view, seen in Figure 4 using a MCS implemented previously which is seen in Figure 15 (Lindner, et al., Issues of exact laser ray positioning using DC motors for vision-based target detection, 2016). Fig. 15, there is a user block represents to a user that enters the value of reference angular position $\varphi_m(k)$ and also, the value of the gain of the proportional action for the control system, using a GUI (Graphical User Interface). The digital closed-loop control is implemented by an Arduino Uno, this single-board microcontroller executes an implemented proportional control algorithm. The Algorithm contains a summing union (the circular block) where the reference angular position $\varphi_r(t)$ is entered plus the negative feedback signal $\varphi_m(k)$ of the digital sensor position, which provides a digital signal at a time instant using an instant sample (k). The variable $\varphi_m(k)$ represents the digital information of the current angular position of the motor shaft $\varphi_o(t)$. From the summing union, there is an output error signal $\varphi_m(k)$ which feeds the position controller to obtain the control variable $y(k)$ as percentage of the duty cycle of the PWM signal. A power controller is used to amplify the duty cycle, obtaining a nominal armature voltage for the DC motor (D / A converter) using an external power supply and finally obtaining the desired angular position.

Figure 15. Block diagram for a MCS using a proportional controller

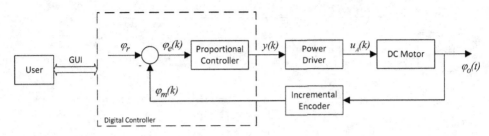

Figure 16 shows two models with different methods for a Motion Control System (SCM), both output signals of each of these models go through the Mux block, in charge of combining the two positioning outputs of each model to visualize both graphs and compare between both signals. For the control system 1, represented to the SCM of Figure 15, beginning with a step type signal as the reference position signal $\varphi_r(t)$ which enters to the summing union block to obtain the positioning error $\varphi_e(t)$ to feed to controlled system (DCX22S-1). Providing a feedback signal $\varphi_m(k)$ to the summing union coming from the output position $\varphi_o(t)$ y the Encoder-1 Block which represents the digital position sensor. The signal $\varphi_e(t)$ passes through the block PID-1 in charge of compensating for the error signal obtaining the control variable $y(t)$ that is fed to the block DCX22S which represents the DC motor Maxon DCX22S this produces a moment of force on the axis of power transmission to cause angular displacement and obtaining the signal of the output position $\varphi_o(t)$. The flow of signals passing through the second system was explained using Figure 14.

Figure 16. Comparison between two control systems using $\varphi_r=90°$

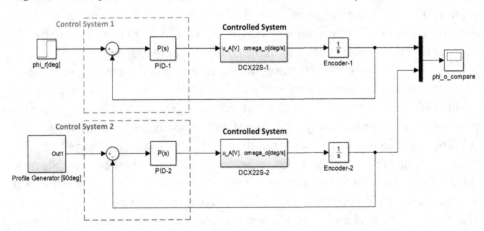

   Figure 17 shows the comparison between two control systems. The result of the comparison is shown by Simulink block "phi_o_compare", which visualizes the results (as shown on the Fig.14), observing and comparing two different angular positions of signal $\varphi_o(t)$ of Figure 16. The signal of blue color corresponds to the control system 1 which uses as input a unit step as the reference position signal $\varphi_r(t)=$ 90 degrees. However, the signal $\varphi_o(t)$ presents an overshoot in the transient state, and quickly compensated for the error to position itself at 93.13 degrees in the Figure 17 into the simulation. The orange signal represents the output signal $\varphi_o(t)$ of the second one control system which uses a trapezoidal velocity profile as the control variable applied, $\varphi_o(t)$ is not the overshoots in transient state and remains stable in the stationary zone, allowing a smooth movement without considerable load on the applied motor shaft obtaining a precise response at $\varphi_o(t)=9.99$ degrees. Therefore, there is an evidence when using a trapezoidal velocity profile as control variable to the DC motor as the controlled system. This evidence is advantage which consists in keeping track of the final angular position using the incremental signals by the encoder, avoid overshoots of the signal $\varphi_o(t)$ with high precision demonstrating in Figure 17.

Figure 17. Comparison between the graphics of $\varphi_o(t)$ using different motion controls

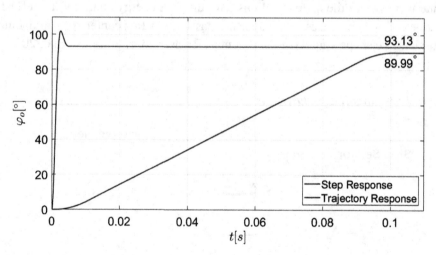

SCANNING APERTURE

Parts and Principles

Scanning Aperture is the name given to the passive element of the Laser Scanning System. Its main purpose is to receive the reflected laser beam from the object surface. Extensive information about this Aperture and the TVS can be found in (Rivas-Lopez, Sergiyenko, & Tyrsa, 2008), (Basaca-Preciado, Sergiyenko, Rodriguez-Quinonez, & Rivas-Lopez, 2012), (Sergiyenko, et al., 2009) and (Lindner, et al., 2016).

The Scanning Aperture is depicted in Figure 18 and consists of a brushed DC motor, a 45° cut mirror mounted on the shaft of the motor. This mirror has a base provided with a notch that is sensed by an opto-switch called "Zero Sensor" fixed in optical contact with the rotating base and provides a signal every time the motor completes a full rotation. Furthermore, the SA contains two biconvex lenses called "Objectives" that will focus the energy captured by the aperture and a photo receiver (PR sensor) called "Stop Sensor" to transform the reflected laser beam into an electric signal.

The main photoelectric signal on this process is the Stop Sensor signal provided by the phototransistor at the top of the SA. this signal is generated as the SA rotates and due to the reflection characteristics of the laser beam, the voltage level of the signal increases as the aperture aligns with the spot energy center of the reflected laser beam, and decreases as the aperture passes over this center spot producing a Gaussian-like shape signal (Flores-Fuentes, Rivas-Lopez, & Sergiyenko, 2014).

Figure 18. Scanning aperture (SA)

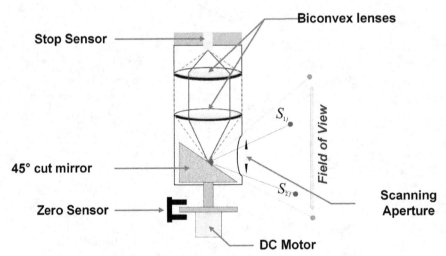

Figure 19 shows graphically how the signal generates through rotation time of the SA.

After an object has been detected, the system performs calculations to determine the B_{ij} (where i and j are the horizontal and vertical points in the full 3D scan) angle of the triangle formed with the reflected laser beam. Now, with B_{ij} calculated and with a known value of γ it is possible to get the 3D coordinates of the object under observation.

Two variables are initialized to perform angle calculations, $N_{2\pi}$ and N_A. Figure 3 shows all photoelectric signals used by the SA. $N_{2\pi}$ is the number of count pulses from the reference standard frequency (f_p at a complete rotation of the mirror. N_A is the number of count pulses from the Zero Sensor signal to the Stop Sensor that gives

Figure 19. Light to voltage conversion

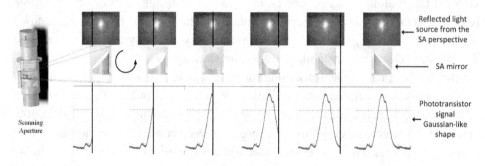

Figure 20. Photoelectric signals and angle detection

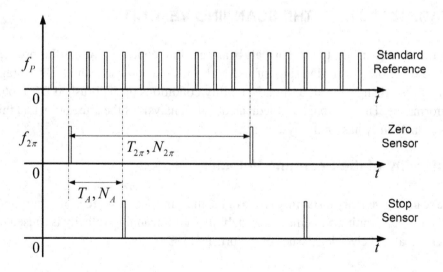

a signal when the reflected laser beam has been detected. If the laser beam hits an object on the TVS field of view (FOV) B_{ij} will be the angle at which the reflected laser beam was detected. This angle is calculated using (8).

$$B_{ij} = 360 \frac{N_A}{N_{2\pi}} . \tag{19}$$

Uncertainty Sources

The Scanning Aperture has several uncertainty sources due to its operating principles, such as energy saturation of the photo-receiver by daylight operation, the striking distance from the system to the object under observation, problems with the reference frequency, obstructions on the SA mirror and lenses, poor velocity stability of the brushed DC motor and more. Each one of these uncertainty sources has multiple reasons and factors that may affect its influence on the system. This chapter focuses on one of the most important sources of uncertainty on the system measurements: The brushed DC motor instability of rotation velocity.

Due to the measurement principle the calculation of the B_{ij} angle has a direct relation with the scanning velocity, due to $N_{2\pi}$ are count pulses on a full rotation of the motor, therefore if the motor presented changes on the rotation velocity, the period of the detected signal will be different between consecutive rotations, as well as the count pulses. Thus for several detection periods with the reflected laser beam on the same position the final calculated angle can present differences.

STABILIZATION OF THE SCANNING VELOCITY

Friction losses inside the motor ball bearings, low contact between the brushes due to deterioration during time of use and poor maintenance, suddenly voltage and current changes result in high instability conditions of the brushed DC motor performance. This subchapter is dedicated to the analysis of the affectation that this motor instability has on the system signals.

Instability of the Scanning Velocity

The scanning velocity instability is one of the most important sources of uncertainty in the Laser Scanning System. This instability of the scanning velocity is caused by mechanical and electrical issues of the brushed DC motor.

Experimentation was made using the following algorithm: Manipulate the LP to position at the desired constant angle. Locate a surface or object to detect at a known distance in which the laser beam will be reflected. Drive the SA with a theoretically constant velocity for an established time of 1 minute, in which the SA will be rotating. Therefore the reflected laser beam energy will be reaching the Stop sensor and the Gaussian-like signal will be generated several times. By repeating these tests and changing the environmental conditions, as well as the rotation velocity of the brushed DC motor it will be possible to analyze the system's photoelectric signals and quantify the influence of the motor instability. Also since a wide range of data will be generated it can be used to further comparative statistical analysis against data collected after possible stabilization of the motor. The analysis of the system signals and the influence of the motor instability on them, is explained on the following paragraph and figures.

Locate the LP to known desired angles and drive the SA for 1 minute at different scanning velocities, first the SA brushed DC motor was driven at the lowest possible speed, the LP was located at 90° and the signal from the PR or Stop Sensor were measured using an oscilloscope.

For the following analysis two concepts must be defined. The first one is "detection point" which is the center spot energy of the reflected laser beam. Hence the point where the laser beam hits the object under observation. The second one is "detected period" which is the time between starts of consecutive Gaussian-like shapes signals. From Figure 21 to 23 it can be seen that for the same detection point the signal has

Figure 21. SA at lowest possible velocity, first detection period of the PR

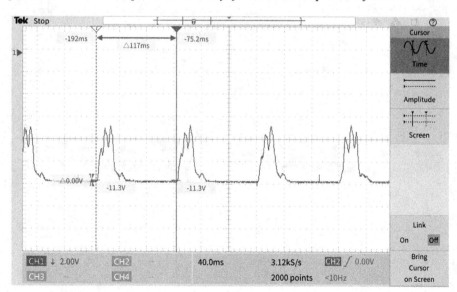

Figure 22. SA at lowest possible velocity, second detection period of the PR

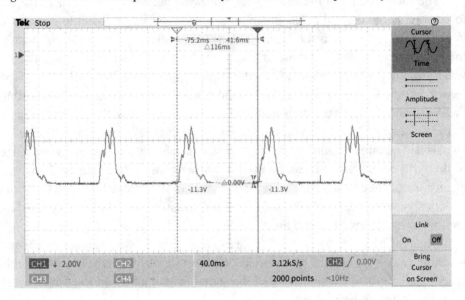

Figure 23. SA at lowest possible velocity, third detection period of the PR

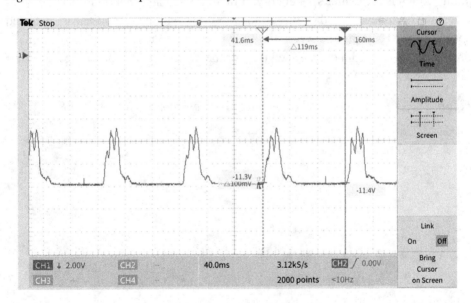

differences between consecutive periods. From Figure 21, the first detected period is 117 ms (which is the time the motor needed to complete a rotation). Figure 22 shows the next period, with a period of 116 ms . Hence, this time the motor performs the full rotation faster than the previous one. Last Figure 23 shows the third consecutive detection period of the signal now with 119 ms of time interval. Due to reasons mentioned before in the chapter, the motor needed more time to complete a full rotation. Ideally all detected periods should have the same time interval. It can be seen from Figure 21 to 23 that none of the 3 consecutive detected periods have the same time interval. This, due to the instability of rotation of the brushed DC motor of the aperture, since the motor does not have a stable rotation velocity it can complete a full rotation in less or more time, this will produce that the detected signal of the PR will be shifting on the time scale for the same detection point and produce an error on the detection angle of the system.

To demonstrate that this phenomenon is presented for any kind of environmental and performance conditions, the same experimentation was made with changing some system operation conditions, such as velocity of rotation, daytime operation and object's under observation position.

Therefore Figure 24 shows the Stop sensor signal isolated from the daylight at the lowest possible velocity. As it can be seen, the form of the detected signal has changed, due to now the reflected laser beam is the only light source that reaches the SA and hence there is no optical noise that affects the signal shape. Even though the signal shape has changed, the differences of the signal periods still have a presence on this operation mode.

Figure 25 represents the signal of detection at an augmented velocity and isolation from daylight. Since the brushed DC motor is rotating faster than the ones on Figures 21 to 23 the signal periods have been reduced in time. In theory, this velocity change will generate fewer friction problems and the motor will have more stability of rotation. However, it can be seen that the detected periods still have differences between them due to motor instability.

Closed-Loop Control of the Brushed DC Motor

The brushed DC motor of the Scanning Aperture is being driven in open-loop configuration. Hence if a disturbance of rotation velocity occurs it cannot be reduced, avoid or processed by the system. This disturbances generate error on the coordinate detection of the system. Thus, a closed-loop velocity control is proposed to stabilize the motor velocity. The objective of the control implementation is to decrease the effect of the instability in the 3D coordinate.

To perform a closed-loop control, a sensor is needed, which provides feedback from the plant about the actual state. Many control applications use high-cost elements as

Figure 24. Signal periods at lowest possible velocity isolated from the daylight

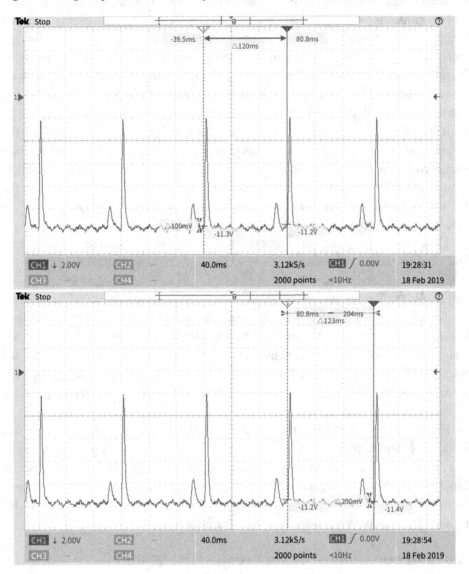

sensors and actuators of the controlled plants, such as the ones implemented in the industry and that there are many sensors on the market, which can be implemented in the system, such as encoders, tachometers, and others. The feedback of the closed-loop control, will be done with the Zero Sensor. The sensor provides a signal every time the motor completes a full rotation, this signal will be acquired using a Measurement Computing DAQ USB-1208LS in its counter input. The velocity of the brushed DC motor can be known, measuring the time between consecutive

Figure 25. Stop sensor signal at augmented velocity

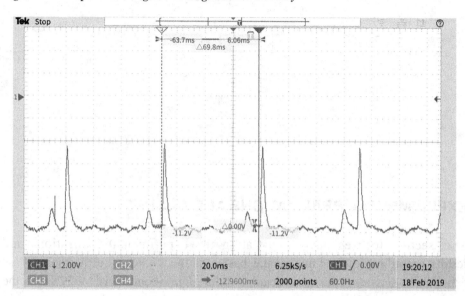

pulses. This velocity will feedback to the controller, to compare it with the desired velocity. After comparison has been made control output will be generated for the brushed DC motor.

The signals acquired by the DAQ are processed using LabVIEW, where a programmed VI with the PID control algorithm determine the required control output to the power stage.

To drive the brushed DC motor, an Arduino UNO was programmed to provide a PWM (Pulse Width Modulation). A PWM is a technique in which the duty cycle and frequency of a signal is modified to control the energy quantity supplied to a load. Duty cycle of a signal it is the on\off time of said signal in a time period. The duty cycle of the controller it's regulated by the control output of the PID algorithm. The control output will be in function of the error between the real and reference velocity of the motor. The PWM with its duty cycle is supplied to an H-bridge utilized as the power stage of the circuit, since the Arduino Outputs doesn't have the necessary current to drive the brushed DC motor of the Scanning Aperture.

Figure 26 shows the block diagram of the closed-loop control system, LabVIEW VI interface provides the desired velocity for the PID function of the program.

Figure 26. Block diagram of the closed-loop

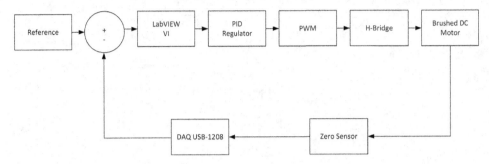

EXPERIMENTAL REALIZATIONS AND RESULTS

This section focuses on demonstrating the experimental realizations and results according to the fundamental basics of this chapter. Furthermore, these experimentations are explained in detail, using the digital controller LM629N-8 for the implementation of the Laser Positioner System in closed-loop and the implementation of a velocity controller in a closed-loop configuration for the Scanning Aperture. Finally, in consequence, the results of the experimental realizations are explained.

Experimentation of Laser Positioner System Implementation

This section presents the works of the implementation through the methodology of the section Laser Positioner System. The evidences of the principal experimental result. The execution of the Laser Positioner from a GUI; demonstrates how the experimentation was done to move the laser beam using the TVS No. 3. Additionally, the experimental conditions are presented using the Tables 2 to 9 showing, a high precision and short positioning times.

The integration was made using the data sheet of the LM629 Controller (Presicion Motion Controller LM629, 2013). Figure 27 shows, a) the first test connections for the LM629N-8 controller, the blue wires enable the communication in parallel of the "Host Interface Module" to 8 bits. And b) shows the first integration of experimentation which presents the Arduino Mega 2560 Single-board, a Maxon DC motor RE-max 29 (Maxon Academy, n.d.) and a L298 power driver module.

Figure 28 shows the scope of the control system proposed in the section "Laser Positioner System", because of the control system has two modes of operation: position mode and velocity mode, the motion control system proposed in this chapter is functional for turning the TVS system No. 3 using its three main actuators which are used for the Laser Positioner and the Scanning Aperture.

Figure 27. a) Enabled I/O port of LM629N-8; b) The first experimental setup

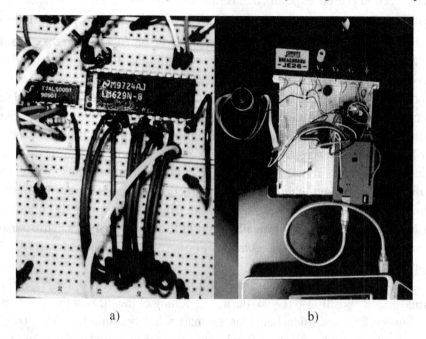

a)   b)

Figure 28. Three motion controller systems for the TVS No. 3

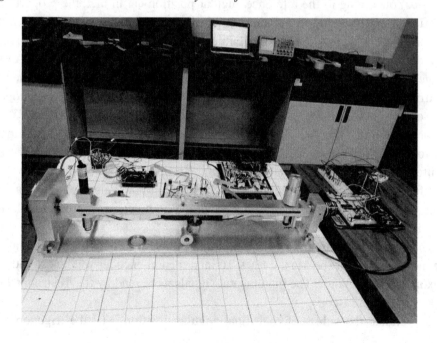

This section presets experimental results measuring, the average position error relative angular $\overline{\varphi}_e'$. the average value of the positioning time $\overline{t}_\varphi'$.as the minimum time in ms to move the Maxon DCX22S motor (Maxon Academy, n.d.), according to different angular positions $\varphi_r(t)$=(1°, 5°, 30°, 90°, and 360 °). Consequently, the experimental results present a stable control system and with an optimal performance for the PL of the TVS No. 3 using the LM629N-8 controller. The experimental setup was used to have the operational control system for the Laser Positioner (PL), it was possible to know its scope and limits under the experimentation made. A Graphical User Interface was developed on LabVIEW platform using several modifications to improve the performance of the same. Obtaining more fails in the serial communication by having delays of the bidirectional transfer of data between the computer and the Arduino 2560 single-board. The tuning of the PID controller of the LM629N- 8 was made empirically, in order to avoid overshoots in the transient and stationary zone of the final angular position $\varphi_o(t\rightarrow\infty)$ using the Maxon DCX22S motor shaft.

Figure 29 shows six different images from A) to F) are some of the responses of the final angular position $\varphi_o(t\rightarrow\infty)$ during the tuning of the PID controller. In image A), using φ_r= 2 ° a sinusoidal behavior is remarkable, obtaining $1.93 ° \leq \varphi_o(t) \leq 2.2$ °. Also, for the following images: B) to F), the tuning was started with a proportional coefficient ≤ 100 units, observing the motor behavior and the final angular position $\varphi_o(t\rightarrow\infty)$ oncoming to the reference angular position φ_r. In fact, the proportional coefficient was used up to 1,500 units as a limit. Therefore, by using the controller proportional coefficient, it was not possible to approximate the reference value from φ_r under short positioning times $t_\varphi \leq 500$ ms, the oscillatory behavior of the A to C images were eliminated by including the controller integrative coefficient, the PID controller effects were changed seen in the image D) to F), where the stationary zone of the signal $\varphi_o(t\rightarrow\infty)$ is stable. However, the behavior of the Maxon DCX22S motor was similar and was not identical under the same experimental conditions due to the friction effects and physical characteristics of the DCX22S motor were omitted due to the complex analysis within the proposed work of this chapter.

After the PID controller tuning was achieved, with the values proposed at two levels in Table 2, showing results of $\varphi_o(t)$ with a high degree of precision, during the empirical tests in the Tables 3 to 8.

To validate the response of the positioning time t_φ the incremental signals of channels A and B of the Maxon ENX 16 incremental encoder were used with the Maxon DCX22S motor. The incremental signals were read by the Tektronix TBS 2000 Series oscilloscope.

To validate the values of each final angular position $\varphi_o(t)$ the trigonometric function $\varphi_o = arctan\left(\dfrac{a}{b}\right)$ was used, being as the opposite side and b as the adjacent

Figure 29. Experimental results using the laser positioner

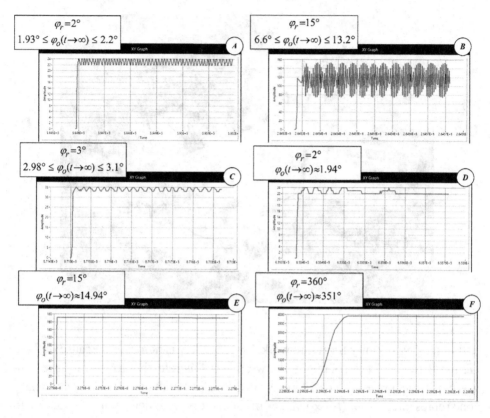

Table 2. Experimental factors

Factors of the PID Controller	Level 2	Level 1
Proportional coefficient	1,000, 1,050,1,100, 1,200,1,300, 1,500	1,000, 1200
Integrative coefficient	10, 15, 100, 200, 250	5, 10, 50, 100

side, working according to the experimental setup of Figure 30 which is an evidence, where there are imaginary lines to obtain φ_o where, $a=2$ m and b the value in meters displaced from a point of light reflected as the zero position $\varphi_o =0$. On the surface of the wall due to the laser beam emitted by the PL, until the move is stopped when the point of reflected light reaches the final angular position $\varphi_o(t)$.

The experimental factors k_p, k_i, a_r, ω_{max}, N_m, φ_o, φ_e, t_φ presented in the Tables 3 to 8, show the behavior of the Maxon DCX22S motor shaft under the experimental conditions. However, in quantitative terms it is possible to reach those values if the tuning of the PID controller is improved for high velocities and accelerations

Figure 30. Experimental setup

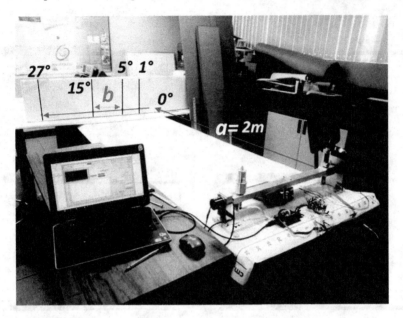

without losing stability for future work. The accuracy is notorious because of there are oscillations of ∓ 1 count and the word "Osc" is referring to the displacement was not stable, presenting oscillations during the test in the transient and stationary zone represented by φ_o ($t\to\infty$) values of the factors were determined according to a high degree of accuracy.

According to the above tables using 16 tests for each angular position described it was possible to obtain the values for $\overline{\varphi}_e'$, $\overline{t}_\varphi'$, and $\overline{\varphi}_o'$ defined, using a minimum resolution like $\rho_{min}=0.088°$. Next, Table 9 is presented to describe the values obtained for the chapter proposal using the Laser Positioner. Taking as a reduction of averaging positioning error less than 12 percent and averaging positioning time less than 500 ms.

Experimentation of Control Implementation in Scanning Aperture

In this subchapter the experimentation with closed-loop control of the SA is presented. By reproducing previous experimental algorithm and algorithm just with the addition of the closed-loop control. All this in order to compare performances and results between the two operation modes. The experimental results are shown in comparative diagrams and tables below.

Table 3. Experimental results using $\varphi_r=1°$ and $N_r=11$

Test Number								
1.	1000	100	15	30	10	0.88°	12%	15 ms
2.	1000	100	15	40	10	0.88°	12%	12 ms
3.	1000	100	10	30	10	0.88°	12%	22.4 ms
4.	1000	100	10	40	10	0.88°	12%	22 ms
5.	1000	200	15	30	Osc	-	-	-
6.	1000	200	15	40	10	0.88°	12%	11.2 ms
7.	1000	200	10	30	11	0.96°	4%	827 ms
8.	1200	200	10	40	Osc	-	-	-
9.	1200	100	15	30	10	-	-	15.2 ms
10.	1200	100	15	40	Osc	-	-	-
11.	1200	100	10	30	10	0.88°	12%	22 ms
12.	1200	100	10	40	Osc	-	-	-
13.	1200	200	15	30	10	0.88°	12%	11.9 ms
14.	1200	200	15	40	10	0.88°	12%	12.3 ms
15.	1200	200	10	30	Osc	-	-	-
16.	1200	200	10	40	11	0.96°	4%	178 ms

Figure 31. Test No. 7 for $\varphi_r=1°$ with oscillations of ± 1 cuenta

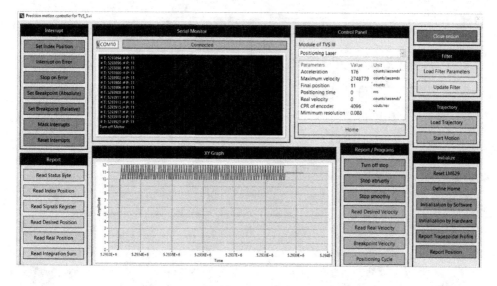

Table 5. Experimental results using $\varphi_r=5°$ and $N_r=57$

Test Number								
1.	1200	10	15	30	57	5°	0%	438 ms
2.	1200	10	15	40	56	4.92°	1.6%	49.6 ms
3.	1200	10	10	30	57	5°	0%	95.2 ms
4.	1200	10	10	40	Osc	-	-	-
5.	1200	5	15	30	57	5°	0%	62.4 ms
6.	1200	5	15	40	56	4.92°	1.6%	45.6 ms
7.	1200	5	10	30	57	5°	0%	70 ms
8.	1200	5	10	40	57	5°	0%	62.1 ms
9.	1300	10	15	30	Osc	-	-	-
10.	1300	10	15	40	57	5°	0%	60 ms
11.	1300	10	10	30	Osc	-	-	-
12.	1300	10	10	40	Osc	-	-	-
13.	1300	5	15	30	Osc	-	-	-
14.	1300	5	15	40	Osc	-	-	-
15.	1300	5	10	30	Osc	-	-	-
16.	1300	5	10	40	Osc	-	-	-

Figure 32. Test No. 1 for $\varphi_r \approx 1°$. Obtaining $N_m=10$ $t_\varphi=15ms$.

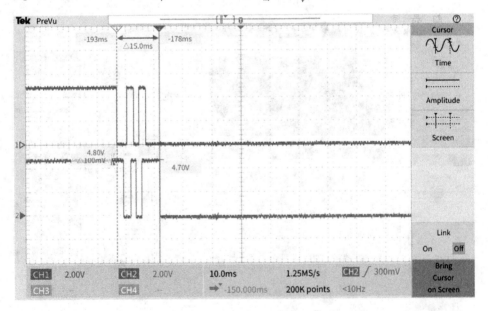

Figure 33. Test No. 1 for $\varphi_r=5°$ without oscillations

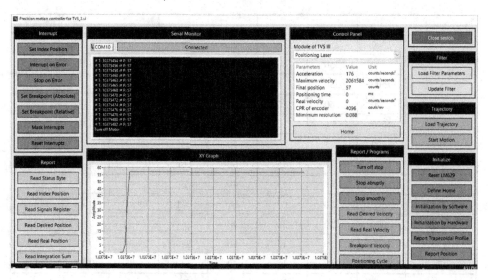

Figure 34. Test No. 6 for $\varphi_r \approx 5°$. Obtaining $N_m=56$ $t_\varphi=45.6ms$.

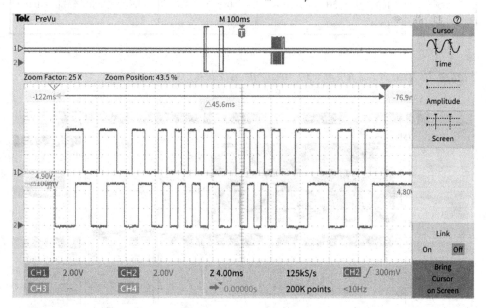

Table 6. Experimental results using $\varphi_r=30°$.and $N_r=341$

Test Number								
1.	1500	250	30	60	Osc	-	-	-
2.	1500	250	30	40	341	29.97°	0.1%	111 ms
3.	1500	250	15	60	Osc	-	-	-
4.	1500	250	15	40	Osc	-	-	-
5.	1500	50	30	60	341	29.97°	0.1%	115 ms
6.	1500	50	30	40	341	29.97°	0.1%	125 ms
7.	1500	50	15	60	Osc	-	-	-
8.	1500	50	15	40	Osc	-	-	-
9.	1000	250	30	60	341	29.97°	0.1%	132 ms
10.	1000	250	30	40	341	29.97	0.1%	173 ms
11.	1000	250	15	60	340	29.88°	0.4%	134 ms
12.	1000	250	15	40	340	29.88°	0.4%	135 ms
13.	1000	50	30	60	Osc	-	-	-
14.	1000	50	30	40	Osc	-	-	-
15.	1000	50	15	60	341	29.97°	0.1%	570 ms
16.	1000	50	15	40	Osc	-	-	-

Figure 35. Test No. 2 for $\varphi_r=30°$ without oscillations

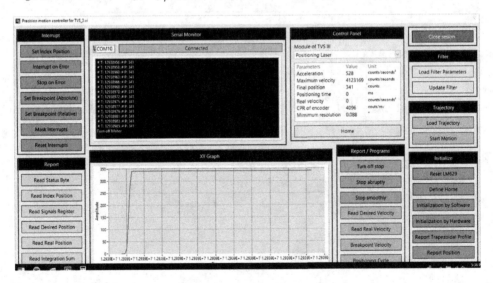

Figure 36. Test No. 5 for $\varphi_r \approx 30°$. Obtaining $N_m = 341$ $t_\varphi = 115ms$.

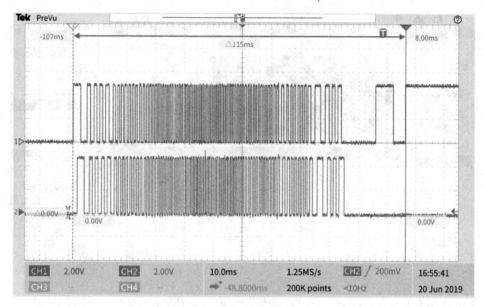

Table 7. Experimental results using $\varphi_r = 90°$.and $N_r = 1024$

Test Number								
1.	1100	250	30	60	Osc	-	-	-
2.	1100	250	30	40	Osc	-	-	-
3.	1100	250	15	60	Osc	-	-	-
4.	1100	250	15	40	1024	90°	0%	464 ms
5.	1100	100	30	60	1024	90°	0%	241 ms
6.	1100	100	30	40	1024	90°	0%	1s
7.	1100	100	15	60	1023	89.91°	0.1%	243 ms
8.	1100	100	15	40	1024	90°	0%	2.17 s
9.	1000	250	30	60	1024	90°	0%	547 ms
10.	1000	250	30	40	1023	90°	0%	170 ms
11.	1000	250	15	60	Osc	-	-	-
12.	1000	250	15	40	1023	89.91°	0.1%	244 ms
13.	1000	100	30	60	1024	90°	0%	360 ms
14.	1000	100	30	40	Osc	-		-
15.	1000	100	15	60	1023	89.91°	0.1%	248 ms
16.	1000	100	15	40	Osc	-	-	-

Figure 37. Test No. 4 for $\varphi_r = 90°$ without oscillations in stably zone

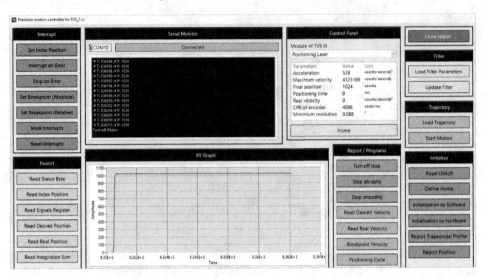

Figure 38. Test No. 12 for $\varphi_r \approx 90°$. Obtaining $N_m = 1023$ $t_\varphi = 244ms$.

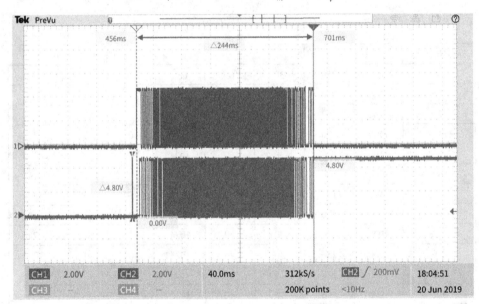

Table 8. Experimental results using $\varphi_r=360°$ and $N_r=4096$

Test Number								
1.	1200	15	30	60	4096	360°	0%	371 ms
2.	1200	15	30	40	4096	360°	0%	387 ms
3.	1200	15	15	60	4096	360°	0%	547 ms
4.	1200	15	15	40	Osc	-	-	-
5.	1200	10	30	60	Osc	-	-	-
6.	1200	10	30	40	4096	360°	0%	2s
7.	1200	10	15	60	Osc	-	-	-
8.	1200	10	15	40	4096	360°	0%	575 ms
9.	1000	15	30	60	Osc	-	-	-
10.	1000	15	30	40	4096	360°	0%	367 ms
11.	1000	15	15	60	4096	360°	0%	703 ms
12.	1000	15	15	40	Osc	-	-	-
13.	1000	10	30	60	4096	360°	0%	360 ms
14.	1000	10	30	40	4095	359.91°	0.02%	355 ms
15.	1000	10	15	60	Osc	-	-	-
16.	1000	10	15	40	4096	360°	0%	508 ms

Figure 39. Test No. 16 for $\varphi_r=360°$ with oscillations in stationary zone

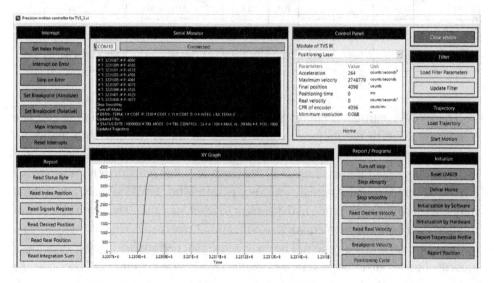

Figure 40. Test No. 1 for $\varphi_r \approx 360°$. Obtaining $N_m = 4096$ $t_\varphi = 371ms$.

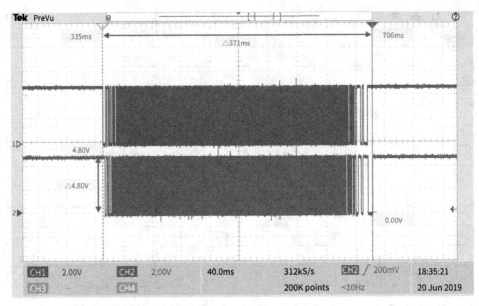

Table 9. Experimental results of the factors: $\overline{\varphi}_e'$, $\overline{t}_\varphi'$, and $\overline{\varphi}_o'$

Maxon Motor DCX22S	1°	2°	3°	5°	15°	30°	45°	90°	180°	360°	Unit
	10.4	3.30	1.11	0.4	0.14	0.17	0.01	0.03	0	0.02	%
	104	42	154	110	227	186	183	314	365	463	*ms*
$\overline{\varphi}_o'$	0.89	1.93	2.96	4.98	14.97	29.94	44.99	89.97	179.98	359.99	*deg*

The PWM controls the energy supplied to the motor using the duty cycle. The frequency at which the duty cycle is working has direct relation with the motor performance due to its electromechanical characteristics. Due to inductance of the motor the current of the windings will not reach its value immediately. When using PWM the same situation repeats on every period. Thus PWM signals a very high frequencies will give less time to the current to reach its value. Therefore velocity of rotation used in this experimentation was increased, due to mechanical characteristic of the controller and the system. The analyzed signal periods are smaller than those presented in the uncertainty sources subchapter. Previous captures of oscilloscope screen showed periods with a time duration around 120 ms, for this experimentation, detected periods length are around 77 ms.

At the experimentation time the lowest possible velocity for the controller was 12 rounds per second. Theoretically with the closed-loop control, detected periods of the signal will be more stable. Therefore difference between them should be lower.

Figure 27 shows comparison between the non-controlled and controlled performance of the SA. By analysis of the Zero sensor signal and the difference of consecutive periods with the closed-loop and open loop operation of the system. It can be seen that signal with the control method proposed, present differences between consecutive periods of 0.1 ms. Two screen captures on the left side of the Figure show this. The right side shows the signal performance at open-loop configuration. In this case differences could reach 1.1ms between consecutive periods. This is about 10 times higher than differences with the closed-loop control.

CONCLUSION

It can be conclude that the two subsystems presented in the chapter are part of a laser scanner or TVS with high accuracy, working velocity and resolution. This laser scanner has a wide range of application due to its simplicity and versatility for a Laser Scanning System applied for Mobile Robotic. The TVS stills has improvement areas.

Figure 41. Comparison, signal periods with SA closed-loop control against non-controlled

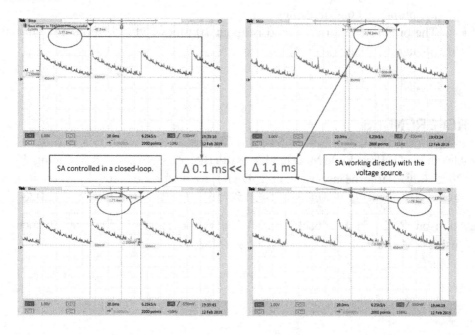

From the Laser Positioner work it can be conclude the following:

- It was possible to measure and obtain a reduction to the positioning error of the Laser Positioner (PL) of the Technical Vision System (TVS) prototype No.3, using an arbitrary trapezoidal velocity profile and generated by the digital controller LM629N -8. The actuator of the PL was the Maxon DCX22S engine during all the experimental tests. The results of the error reduction were obtained by experimental setup.
- The proposed theoretical method defines the angular positioning trajectory with an accuracy of ± 1 counts as the minimum resolution of the encoder used, without presenting instability. This theoretical method was implemented, using the digital controller LM629N-8 in closed-loop, adjusting parameters that constitute the profile of the reference trapezoidal velocity profile.
- Modeling and simulation of the ideal operation for the digital controller LM629N-8 using Matlab-Simulink according to the theoretical method. Obtaining the variable of the final angular position of the motor shaft with error less than or equal to 12.0%.

From the Scanning Aperture work it can be conclude that:

- The SA is a useful tool as receptive part of a system.
- The SA instability of rotation has a direct relation to angle detection errors.
- The instability of rotation is satisfactory reduced with the implementation of the proposed control method.
- The proposed control method is about 10 times better than the non-control operation of the SA.

REFERENCES

Alexander, C., Erenskjold Moeslund, J., Klith Bøcher, P., Arge, L., & Svenning, J.-C. (2013, July). Airborne laser scanner (LiDAR) proxies for understory light conditions. *Remote Sensing of Environment, 134*, 152–161. doi:10.1016/j.rse.2013.02.028

Atkinson, P. (2012). Feedback Control Theory for Engineers. Springer Science & Business Media. doi:10.1007/978-1-4684-7453-4

Basaca-Preciado, L., Sergiyenko, O., Rodriguez-Quinonez, J., & Rivas-Lopez, M. (2012). *Optoelectronic 3D Laser Scanning Technical Vision System based on Dynamic Triangulation. In Photonics Conference (IPC) in Burlingame California* (pp. 648–649). IEEE. Retrieved from http://ieeexplore.ieee.org/xpl/articleDetails. jsp?arnumber=6358788

Bass, M., DeCusatis, C., Enoch, J., Lakshminarayanan, V., Li, G., MacDonald, C., & Stryland, E. (2009). *Handbook of Optics* (D. M. Bass, Ed.; 3rd ed.; Vol. 1). McGraw-Hill Professional.

Buccella, T. (1997). *Servo Control of a DC Brush Motor (AN532)*. Microchip.

Flores-Fuentes, W., Rivas-Lopez, M., Sergiyenko, O., Rodriguez-Quinonez, J. C., Hernandez-Balbuena, D., & Rivera-Castillo, J. (2014, March 6). Energy Center Detection in Light Scanning Sensors for Structural Health Monitoring Accuracy Enhancement. *IEEE Sensors Journal, 17*(7), 2355–2361. doi:10.1109/ JSEN.2014.2310224

Fu, G., Menciassi, A., & Dario, P. (2012, October). Development of a low-cost active 3D triangulation laser scanner for indoor navigation of miniature mobile robots. *Robotics and Autonomous Systems, 60*(10), 1317–1326. doi:10.1016/j. robot.2012.06.002

Haddad, N. (2011, June). From ground surveying to 3D laser scanner: A review of techniques used for spatial documentation of historic sites. *Journal of King Saud University - Engineering and Science, 23*(2), 109–118. doi:10.1016/j. jksues.2011.03.001

Lindner, L. (2016). Laser Scanners. In O. Sergiyenko, J. Rodriguez-Quinonez, O. Sergiyenko, & J. Rodriguez-Quinonez (Eds.), Developing and Applying Optoelectronics in Machine Vision (p. 38). Hershey, PA: IGI Global. doi:10.4018/978-1-5225-0632-4.ch004

Lindner, L., Sergiyenko, O., Rivas-Lopez, M., Ivanov, M., Rodriguez-Quinonez, J., Hernandez-Balbuena, D., . . . Mercorelli, P. (2017). Machine vision system errors for unmanned aerial vehicle navigation. In *Industrial Electronics (ISIE), 2017 IEEE 26th International Symposium on* (pp. 1615-1620). Edinburgh, UK: IEEE. doi:10.1109/ISIE.2017.8001488

Lindner, L., Sergiyenko, O., Rivas-Lopez, M., Rodriguez-Quinonez, J., Hernandez-Balbuena, D., Flores-Fuentes, W., . . . Kartashov, V. (2016). Issues of exact laser ray positioning using DC motors for vision-based target detection. In *2016 IEEE 25th International Symposium on Industrial Electronics (ISIE)* (pp. 929-934). Santa Clara, CA: IEEE.

Lindner, L., Sergiyenko, O., Rodriguez-Quinonez, J., Rivas-Lopez, M., Hernandez-Balbuena, D., Flores-Fuentes, W., ... Tyrsa, V. (2016). Mobile robot vision system using continuous laser scanning for industrial application. *The Industrial Robot*, *43*(4), 360–369. doi:10.1108/IR-01-2016-0048

Lu, T.-C., & Chen, S.-L. (2016). Genetic algorithm-based S-curve acceleration and deceleration for five-axis machine tools. *International Journal of Advanced Manufacturing Technology*, *87*(1-4), 219–232. doi:10.100700170-016-8464-0

Madisetti, V. (2009). *Digital signal processing fundamentals*. CRC Press. doi:10.1201/9781420046076

Maxon Academy. (n.d.). Retrieved November 09, 2016, from www.maxonmotor.com

Presicion Motion Controller LM629. (2013, May 6). Retrieved February 27, 2019, from http://www.ti.com/product/LM629?jktype=recommendedresults

Rivas-Lopez, M., Sergiyenko, O., & Tyrsa, V. (2008). *Machine Vision: Approaches and Limitations*. InTech. Retrieved from http://cdn.intechopen.com/pdfs/5210/InTech-Machine_vision_approaches_and_limitations.pdf

Sergiyenko, O., Hernandez, W., Tyrsa, V., Devia Cruz, L., Starostenko, O., & Pena-Cabrera, M. (2009, May 21). Remote Sensor for Spatial Measurements by Using Optical Scanning. *Sensor*, 5477-5492. doi:10.339090705477

Sergiyenko, O., Tyrsa, V., Basaca-Preciado, L., Rodriguez-Quinonez, J., Hernandez, W., Nieto-Hipolito, J., . . . Starostenko, O. (2011). Electromechanical 3D Optoelectronic Scanners: Resolution Constraints and Possible Ways of Improvement. In Optoelectronic Devices and Properties. InTech. doi:10.5772/14263

Singh, S. K. (2008). *Kinematics Fundamentals*. Houston, TX: Connexions.

Wescott, T. (2006). *Applied Control Theory for Embedded Systems*. Newnes.

Xiang, S., Chen, S., Wu, X., Xiao, D., & Zheng, X. (2010, February). Study on fast linear scanning for a new laser scanner. *Optics & Laser Technology*, *42*(1), 42–46. doi:10.1016/j.optlastec.2009.04.019

Zhongdong, Y., Peng, W., Xiaohui, L., & Changku, S. (2014, March). 3D laser scanner system using high dynamic range imaging. *Optics and Lasers in Engineering*, *54*, 31–41. doi:10.1016/j.optlaseng.2013.09.003

KEY TERMS AND DEFINITIONS

Active Node (of the System): The active node of the system is the one in charge of emitting a signal, in the TVS case the active node y emitting a laser beam, that works as primordial element for the triangulation.

Diffuse Reflection: Phenomenon when light waves reflect from a surface in a scattered way in several angles rather than just one (specular).

Laser Positioner: This is a capable optoelectronic device to move a laser beam in a precise form, using different methods, these Laser Positioner are used in Laser scanning systems as a principal active element of them.

Microcontroller: This is a digital embedded electronic microcircuit generally is packaging, it contents various modules to process basic or complex tasks due to has a central process unit, memories, inputs/outputs ports, and like minimum a system bus, to run developed algorithms to give some specific solution. Currently, this device has a great relationship between performance and low-cost.

Mixed Reflection: A reflection that is both diffuse and specular, this is the kind of reflection that most of the material present.

Motion Control System: This is integrated by a group of different components, these components are of different disciplines such as mechanical, electronic, control, and programming, to obtain a control variable, which is elemental in mechatronics discipline due to these systems are responsible to process control signals to move joints and mechanical systems.

Passive Node (of the System): This node is considered passive due to always be waiting for a signal to be received, this subsystem of the TVS works as the feedback for the entire triangulation.

Power Driver: This electronic circuit which represents a switch-mode power amplifier for driving DC Motors. It is essential to transmit power (voltage nominal and electrical current) for the applied actuator in a motion control system.

Specular Reflection: Is the mirror like reflection, also known as regular, considered this way when any kind of wave (in this case light), reflect as a mirror from a surface.

Trapezoidal Velocity Profile: This is a defined concept as an internal control variable in some precise motion controllers, where the acceleration and deceleration have equal rates, the trapezoidal velocity is integrated mathematically to obtain the final angular position.

Chapter 7
ZMP–Based Trajectory Generation for Bipedal Robots Using Quadratic Programming

Sergei Savin

(iD) https://orcid.org/0000-0001-7954-3144

Innopolis University, Russia

ABSTRACT

In this chapter, the problem of trajectory generation for bipedal walking robots is considered. A number of modern techniques are discussed, and their limitations are shown. The chapter focuses on zero-moment point methods for trajectory generation, where the desired trajectory of that point can be used to allow the robot to keep vertical stability if followed, and presents an instrument to calculate the desired trajectory for the center of mass for the robot. The chapter presents an algorithm based on quadratic programming, with an introduction of a slack variable to make the problem feasible and a change of variables to improve the numeric properties of the resulting optimization problem. Modern optimization tools allow one to solve such problems in real time, making it a viable solution for trajectory planning for the walking robots. The chapter shows a few results from the numerical simulation made for the algorithm, demonstrating its properties.

INTRODUCTION

Walking robots are one of the focuses of modern robotics, and had remained so for the last decades. Even though there had been a significant amount of research and significant achievements in the field, it still remains one of the main challenges for robotics research.

DOI: 10.4018/978-1-5225-9924-1.ch007

Stable walking on horizontal surfaces had been demonstrated a variety of times. One of the best known examples are ASIMO robots produced by Honda (Chestnutt et al., 2005; Chestnutt et al., 2007; Sakagami et al., 2002). That robot was able to walk stably, climb stairs and was one of the first anthropomorphic bipedal robots. Early ASIMO models used energy inefficient gaits and were not demonstrated being able to walk on rough terrain.

It should be noted that although at this stage of humanoid robot development, the best implemented models were slow and energy-inefficient, there were parallel research such as Raibert's hoppers and runners (Thompson & Raibert, 1990, Pratt 2000) and passive dynamic walkers (McGeer, 1990a, 1990b). Those robots exhibited not only stable but energy efficient gaits and provided ideas for the following generations of walking robots of different types.

A number of highly dynamic quadruped robots had been developed, including BigDog by Boston Dynamics (Raibert et al., 2008; Playter et al., 2006). Those robots provided a demonstration of capabilities of full-scale controllable walking robots, including demonstrations of walking on ice, grassland, forest undergrowth, climbing stairs and withstanding kicks from humans. Those early quadrupeds inspired and were followed by newer models such as ANYmal by ETH Zurich (Hutter et al., 2016 Hutter et al., 2017), Spot and Spot Mini by Boston Dynamics and others. Unlike bipedal robots, quadrupeds have lesser constraints on their admissible regimes of walking and have better possibilities to remain vertically stable. This allows them to demonstrate more dynamics motion and serve as pioneers in solving many of the tasks that are common for walking robots.

The progress in the walking robotics was highly improved by DRC (DARPA Robotics Challenge) and its co-events (such as DARPA virtual challenge). That event assembled a number of different walking robots and allowed the teams to test them against a set of challenges, including egress from a car, operating a valve, walking over rough terrain and others. This allowed not only to assess the current state of the walking robotics (Atkeson et al., 2015, 2016), but also to point out the best control strategies (Long, 2017; Feng et al., 2015; Karumanchi et al., 2017; Tsagarakis et al., 2017) and to discuss the future research directions needed for the further improvements in the field as became evident during the event. Most of the teams that took part in the project provided extensive discussions of their results (Yi et al., 2015; Johnson et al., 2015, 2017; Radford at al., 2015), which allowed to easily assess different methodologies and compare further projects to those approaches using them as state of the art indicators.

Since then, a lot of attention has been given to the motion of bipedal robots through uneven terrain. This is a challenging task with a number of independent components. For example, it can be divided into the tasks of 1) moving over unknown terrain where the robot cannot plan its footsteps in advance, 2) moving over partial footholds, 3) recovering balance after making a step with supporting surface properties different from what was expected, 4) moving in regimes other than walking, such as running or jumping and others. There are some theoretical and applied works that provide solutions for this problem (Zheng et al., 2010; Kanoulas et al., 2018; Deits & Tedrake, 2014; Focchi et al., 2018; Short & Bandyopadhyay, 2018; Mastalli et al., 2018).

Center of Mass Trajectory Generation and Zero Moment Point

The problem of center of mass trajectory generation had remained central to the bipedal robotics. This is related to the fact that incorrect motion of the center of mass can lead the robot to lose its vertical balance and fall. One of the most successful approaches for detection of the potential loss of balance is ZMP (zero moment point) criterion (Kajita et al., 2003; Vukobratović & Borovac, 2004). That criterion poses constraints on the position and dynamics of the center of mass, treating it as a point mass and taking into account the current shape of the support polygon (the convex hull of all support points on the robot's feet).

Zero moment point criterion can also be treated as a tool for the center of mass (CoM) trajectory generation. A number of methods had been developed that use zero moment point criterion to generate trajectories, which can be feasibly executed by the robot, without it falling over. One of the direct approaches is to treat ZMP criterion as a differential equation that can be solved for the given boundary conditions. In (Panovko et al., 2016) this approach was used to generate a stable center of mass trajectory for the case of robot verticalization (sit-to-stand motion). An alternative approach is to reformulate the task as a control problem, as it was done in (Kuindersma et al., 2014). It has been shown that ZMP criterion can be used to generate a linear quadratic regulator, which in turn allows to produce an optimal center of mass trajectory.

We should note that traditional ZMP pipeline requires to first having a set of footholds with information about when they are obtained and when they are lost, and then making a ZMP trajectory using those footholds. There is an alternative to this method, allowing to first generate an arbitrary ZMP trajectory and then place footholds along it, which in some cases might be faster (Jatsun et al., 2018). This approach has a number of challenges, such as timing the footholds, generating ZMP trajectories for arbitrary goals, avoiding obstacles and so on.

In this paper, we propose solving ZMP-CoM trajectory generation with the use of convex optimization. Building upon the results shown in (Kuindersma et al., 2014), we can formulate the trajectory generation problem as a quadratic program, which would allow the use of highly efficient convex optimization tools to find a numeric solution. These tools, including modern implementations of the interior point methods, allow solving quadratic programs in the real time, using on-board computers of walking robots (Boyd & Vandenberghe, 2004). This is further facilitated by the availability of software packages for convex optimization, such as Gurobi, MOSEK, CVX, CVXGEN and others, which are being actively used in robotics and control applications (Mattingley & Boyd, 2012; Hanger et al., 2011).

There are two alternative ways of using quadratic programming in ZMP- CoM trajectory generation. One is to treat is as a receding horizon model predictive control problem, iteratively generating new trajectories for a certain time window, relative to the current time. Another approach is to generate the entire trajectory at once, producing a single big problem instead of a multitude of smaller ones. In the chapter, the relative advantages of these two approaches will be discussed and analyzed.

The chapter aims at presenting a comprehensive view on the modern methods of center of mass trajectory generation, with the focus on the ZMP-based methods, and at presenting new research results. The limitations of the zero moment point methods will also be presented, and some of the existing alternatives, such as Contact Wrench Cone stability criterion (Hirukawa et al., 2006, 2007) will be discussed. The results in the chapter will supported by simulation and numeric experiments, with discussion of their generality and applicability to the real world scenarios.

FOOTSTEP PLANNING

As mentioned before, traditional approach to ZMP-based center of mass trajectory generation requires a footstep plan (a set of footholds with information about when they should be obtained and lost). The problem of generating a footstep plan is highly non-trivial, as it directly inherent discrete and non-convex nature of the motion planning problem for legged systems. However, when the motion takes place over a flat convex obstacle-free region this problem becomes much simpler, even though it still requires the reasoning about the kinematics of the robot in order to find admissible footholds (Kuindersma et al., 2016).

When obstacles are present, footstep planning becomes more challenging. There are several popular families of methods for solving this problem. One is to find a decomposition of the original non-convex free space into convex obstacle-free regions. In (Deits & Tedrake, 2015a) this is done with an iterative algorithm based on semi-definite programming. This approach fares well against a large number of convex obstacles, and as the authors argue, any non-convex obstacle can be easily decomposed into a set of convex ones. This algorithms is not restricted to walking robots and can be used for three dimensional motions as well, for example for quadrotors (Deits & Tedrake, 2015b). However, this approach still assumes that the input data contains information about obstacles as entities. An alternative solution is given in (Savin, 2017), where stereographic projection-based method is introduced. This method operates directly on point clouds, which makes it simpler to implement for systems where LIDAR data is available. That method was also shown to be usable for quadrotors and in-pipe robots. Another alternative method was presented in (Jatsun et al., 2017).

In order to decide which regions are suitable to use for footholds, a deep learning-based method can be used, as proposed in (Savin & Ivakhnenko, 2019). The proposed method requires collection of data based on simple simulation procedure. The aim of the procedure is to determine if a region is suitable for foothold placement, using numeric optimization. This is a fast method and it allows to rapidly generate a large training dataset. Using standard convolutional layers in structures typical for image classification tasks, it is possible to design and train an agent that would be capable of telling if the region is suitable or if it should be discarded.

An alternative family of methods is based on random search and graph search. For example, a footstep plan can be generated along a feasible path, found by RRT algorithm (Rapidly exploring Random Trees). The RRT algorithm was studied in (LaValle, 1998; LaValle & Kuffner, 2000), and the implementation for footstep planning can be seen, for example, in (Xia et al., 2011).

When the terrain is not horizontal and has changing slope, traditional footstep planners may no longer work adequately. One of the reasons is that the frictions cone constrains can be violated if the friction coefficient isn't sufficiently large and the slope isn't sufficiently shallow. On the other hand, as shown in paper (Savin et al., 2019a), based on the orientation of the robot feet, the stability region (an analog of support polygon in the classical ZMP formulation) can actually increase. This means that treating uneven terrain the same way as flat terrain may not only make the footstep plan unstable (in the sense of vertical stability) but also make it unnecessarily conservative.

One of the ways developed for walking over uneven terrain is contact wrench cones, an instrument and methodology which generalizes ZMP (Hirukawa et al., 2006, 2007). However, they are associated with additional computational challenges and can be seen as a movement towards full model motion planning from a simplifies model motion planning, represented by ZMP. It is not yet clear if the contact wrench cones will completely supersede ZMP, will co-exist with it as a special tool for rough terrain or with both methodologies will be superseded by full model motion planning.

The way to place footholds on the terrain is in general a mixed-integer nonlinear optimization problem, as it requires to decide how many steps to take and where to place them on a generally non convex free space. If the free space was tiled with convex obstacle-free regions, then the problem becomes a mixed integer convex optimization, as long as we ignore feet orientation. Namely, we can formulate a problem of deciding at which convex obstacle-free region the foot needs to be places and where on that region it needs to be placed. There are ways to add variable number of footholds into that formulation. Mixed integer problems are solved with branch and bound methods, which rely on a relaxation of integer constraints. Branch and bound methods can be computationally expensive, but there are well-developed solvers for these problems, such as Gurobi.

If feet rotation is necessary to include in the footstep plan formulation, there are several methods of doing it. One is to approximate rotation are a piece-wise linear function, for example by approximating sines and cosines as piecewise linear functions, or by approximating unit circle as a piecewise linear function. It can then by treated either as a mixed integer program with linear approximation or as a convex program with constraint relaxation (Savin et al., 2019b).

ZMP Trajectory Planning as an Optimization Problem

By definition, zero moment point **p** is a point on the supporting surface for which the net torque of reaction forces is balanced. The condition of vertical stability given by Vukobratovich states that zero moment point should always lie inside support polygon of the robot in order for the robot to be able to remain vertically stable. The support polygon is a convex hull of all contact points between the robot and the supporting surface. If the robot moves on flat feet over flat horizontal terrain, then when it stands on one foot, the sole of that foot is the support polygon. If the robot stands on two polygonal feet, then the convex hull of the corners of those feet is the support polygon.

In order to use ZMP to generate trajectory of the center of mass, one requires a simplified model of the center of mass dynamics as a function of zero moment point **p**. One of such models is called linear inverted pendulum, and is given as follows:

$$\frac{z_C}{g} \ddot{\mathbf{r}}_C - \mathbf{r}_C = -\mathbf{p} , \tag{1}$$

where $\mathbf{r}_C = [x_C \, y_C]^T$ is the position of the center of mass of the robot on the horizontal plane, and g is the gravitational acceleration. This model assumes that the vertical acceleration of the center of mass remains zero at all times and that the height of the center of mass remains constant.

Let us represent the trajectory of the center of mass as a discrete sequence of points $\mathbf{r}_C[k]$, where $k=1,...,N$, and N is the number of points on that trajectory. The time between consecutive points is Δt seconds. Using finite differences to represent derivatives we can require that on that trajectory there was a discrete set of acceleration points given as follows:

$$\Delta t^2 \cdot \ddot{\mathbf{r}}_C[k] = \mathbf{r}_C[k+1] - 2\mathbf{r}_C[k] + \mathbf{r}_C[k-1] . \tag{2}$$

With this we can introduce a cost function on the previously introduced accelerations of the center of mass:

$$J_a = \sum_{k=2}^{N-1} \|\ddot{\mathbf{r}}_C[k]\|_2^2 . \tag{3}$$

The idea behind this cons function is that the trajectory of the center of mass should have minimal accelerations if possible. With this, we can formulate a quadratic program for trajectory planning:

$$\begin{aligned}
&\underset{\mathbf{r}_C,\ddot{\mathbf{r}}_C}{\text{minimize:}} \quad \sum_{k=2}^{N-1} \|\ddot{\mathbf{r}}_C[k]\|_2^2 , \\
&\text{subject to:} \\
&\begin{cases}
(z_C/g) \cdot \ddot{\mathbf{r}}_C[k] - \mathbf{r}_C[k] = \mathbf{p}^*[k], \\
\Delta t^2 \cdot \ddot{\mathbf{r}}_C[k] = \mathbf{r}_C[k+1] - 2\mathbf{r}_C[k] + \mathbf{r}_C[k-1] \\
\mathbf{r}_C[1] = \mathbf{r}_{C0}, \\
k \in 2,...,N-1,
\end{cases}
\end{aligned} \tag{4}$$

where $\mathbf{p}^*[k]$ are discrete points of the desired ZMP trajectory. There is an additional constraint in that program: $\mathbf{r}_C[1] = \mathbf{r}_{C0}$. It is needed to specify the beginning of the center of mass trajectory.

From the formal standpoint, this program does exactly what is needed. However in practice, it can have no solutions. This might be a result of numerical errors or the particular initial conditions. It is also important to realize that here, the desired ZMP trajectory is given by a set of points $\mathbf{p}^*[k]$. However, ZMP principle only requires that the points of this trajectory lie inside support polygon. Our requirement that those points follow a predetermined trajectory is superfluous and overly conservative, which is the reason for bad numeric properties of the resulting quadratic program.

In order to fix this problem, we introduce a slack variable. Slack variables are used in numeric optimization when the exact constraint is too tight and interferes with the numeric procedures, while not being strictly necessary for obtaining the desired result. The slack variable $\mathbf{s}[k]$ introduced here is also a sequence associated with the trajectory. It allows us to rewrite the ZMP dynamics constraint as follows:

$$(z_C/g) \cdot \ddot{\mathbf{r}}_C[k] - \mathbf{r}_C[k] = \mathbf{p}^*[k] + \mathbf{s}[k] . \tag{5}$$

This shows physical interpretation of the slack variable: it is an error in the ZMP trajectory, a difference between the desired ZMP trajectory and the one that the planned will provide. In order to keep the slack to a minimum we introduce a slack (ZMP error) cost function:

$$J_s = \sum_{k=1}^{N} \left\| \mathbf{s}[k] \right\|_2^2 . \tag{8}$$

With this, the quadratic program for trajectory planning can be rewritten as follows:

$$\underset{\mathbf{r}_C, \ddot{\mathbf{r}}_C, \mathbf{s}}{\text{minimize:}} \quad w_a \sum_{k=2}^{N-1} \left\| \ddot{\mathbf{r}}_C[k] \right\|_2^2 + w_s \sum_{k=1}^{N} \left\| \mathbf{s}[k] \right\|_2^2 ,$$

subject to: $\qquad\qquad\qquad\qquad ,$ $\qquad\qquad$ (9)

$$\begin{cases} (z_C/g) \cdot \ddot{\mathbf{r}}_C[k] - \mathbf{r}_C[k] = \mathbf{p}^*[k] + \mathbf{s}[k] \\ \Delta t^2 \cdot \ddot{\mathbf{r}}_C[k] = \mathbf{r}_C[k+1] - 2\mathbf{r}_C[k] + \mathbf{r}_C[k-1], \\ \mathbf{r}_C[1] = \mathbf{r}_{C0}, \\ k \in 2, ..., N-1, \end{cases}$$

where w_a and w_s are cost function weights.

Let us observe that the solution of this program naturally lies far away from the origin. This makes the problem numerically harder than it needs to be. Let us introduce a change of variables that would serve to improve the properties of the problem (9):

$$\mathbf{r}_C = \mathbf{r}_C^* + \Delta \mathbf{r}_C , \tag{10}$$

where $\mathbf{r}_C^*$ is a nominal trajectory of the center of mass, which we will provide to the solver, and $\Delta \mathbf{r}_C$ are the corrections to that nominal trajectory that the solver will be making. Using desired ZMP trajectory as nominal center of mass trajectory (which is motivated by the fact that the two become identical as the motion speed of the robot tends to zero), we can use it to calculate same nominal trajectories and corrections. for the accelerations:

$$\ddot{\mathbf{r}}_C = \ddot{\mathbf{r}}_C^* + \Delta \ddot{\mathbf{r}}_C . \tag{11}$$

Then the quadratic program for trajectory planning takes the following form:

$$\underset{r_C,\dot{r}_C,s}{\text{minimize:}} \quad w_a \sum_{k=2}^{N-1}\left\|\Delta\ddot{r}_C[k]\right\|_2^2 + w_s \sum_{k=1}^{N}\left\|s[k]\right\|_2^2,$$

subject to:

$$\begin{cases} (z_C/g)\Delta\ddot{r}_C[k] - \Delta r_C[k] - s[k] = -(z_C/g)\cdot\ddot{r}_C^*[k] + r_C^*[k] + p^*[k] \\ \Delta t^2 \cdot \Delta\ddot{r}_C[k] - \Delta r_C[k+1] - 2\Delta r_C[k] - \Delta r_C[k-1] = r_C^*[k+1] - 2r_C^*[k] + r_C^*[k-1] - \Delta t^2 \cdot \ddot{r}_C^*[k] \\ \Delta r_C[1] = r_{C0} - r_C^*[1], \\ k \in 2,...,N-1, \end{cases}$$

$$(12)$$

This problem can have a solution close to the origin, which makes it numerically better than the previous ones.

SIMULATION RESULTS

In order to demonstrate how the proposed methods work, let us consider several simulations of the center of mass trajectory generation. In the first experiment, we use weights $w_a=1$ and $w_s=10$, with step length of 0.2 m, step width of 0.3 m, over four footholds. The time for each center of mass transfer between centers of the footholds is 3 second. The center of mass height is 1 m. Figure 1 shows the simulation results.

We can observe that the desired and actual ZMP trajectories are almost coinciding, although some error is still seen. This is a property related to the choice of weight in the optimization problem described above. We can also observe that the center of mass trajectory follows the ZMP trajectory, although loosely.

Figure 1. Simulation results (center of mass and zero moment point trajectories) for the first experiment

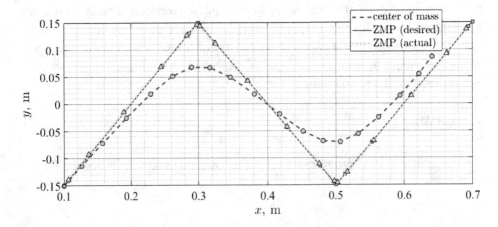

We can demonstrate that for faster movements the similarities between the center of mass trajectory and zero moment point trajectory decrease. Figure 2 shows results of the second experiment where the only difference from the first experiment lies in the change of the duration of transfer of the ZMP between centers of the feet. For the second experiment this time was diminished to 1.8 seconds.

While desired and actual trajectories of the zero moment point remained the same, the shape of the center of mass trajectory changed significantly. The end of the trajectory coiled, which is also a typical result for trajectories with faster movements.

We can also demonstrate that changing the weights in the optimization problem will change the error in following desired ZMP trajectory. Third experiment is done for the same parameters as the first one, except we changed weights to $w_a=5$ and $w_s=5$. Figure 3 shows the results.

We can observe much larger errors in ZPM trajectories. However, the errors should still be acceptable for most anthropomorphic feet designs.

Control System Design With the Optimization-Based Planner

Proposed ZMP-CoM trajectory planner is a part of the overall control system of a bipedal robot. In this section we discuss the role of the planner and the overall structure of the control system.

There are a number of parts to the control system design we present here. First is **obstacle-free regions identification module**. This module can be implemented based on the algorithms given in (Deits & Tedrake, 2015a; Savin, 2017), discussed

Figure 2. Simulation results (center of mass and zero moment point trajectories) for the second experiment with faster motion of the zero moment point

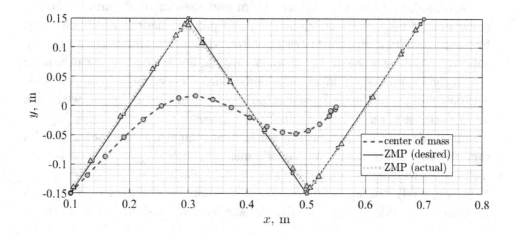

Figure 3. Simulation results (center of mass and zero moment point trajectories) for the third experiment with equal weights

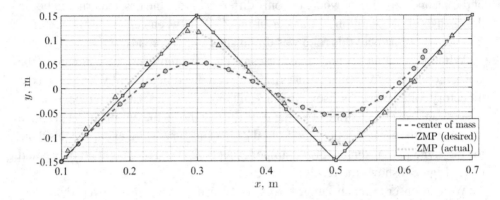

previously. Its result is a set of obstacle-free regions. The next module's task is recognition of the obstacle-free regions fit for hoot placement. This is the **obstacle-free regions classifier module**. This can be achieved by the use of a neural network-based classificatory, as was done in (Savin & Ivakhnenko, 2019). As the result of its work, only regions of admissible geometry will be considered by the following algorithms.

Next is the **footstep planner** module. This module can be implemented using big-M formulation and mixed integer quadratic programming, as was done in (Kuindersma et al, 2016). Resulting algorithm has varied running time, which is not ideal for real-time operation. During DRC 2015 and in the following experimental studies, robots usually stand still during the footstep sequence planning, which might reflect the imperfection of the current practices. The output of the planner is a footstep sequence, places in the obstacle-free regions on the supporting surface.

Next module is the **zero moment point and the center of mass trajectory generator**. It takes the footstep sequence to produce desired ZMP trajectory, formulates quadratic program (12) using solver API (such as Gurobi C++ API) and solves the resulting problem. The formulation of the quadratic problem can be done separately from solving it, in an offline procedure, for speeding up computations. The output of the module is a discrete trajectory for the center of mass, which can be approximated or used in an interpolation procedure.

Obtained center of mass trajectory can be used as input to an **inverse kinematics module**. That can be done with numeric methods, such as gradient decent-based methods, or with optimization (Gupta et al., 2018; Savin et al., 2018; Meredith & Maddock, 2004). The topic of inverse kinematics remains an active area of research for multilink mobile robots, since the problem itself is non-convex and in general requires an iterative procedure to be solved. The output of the inverse kinematics module is the desired trajectory in the joint space of the robot.

In order to make the robot follow the desired trajectory in the joint space (and by extension – follow the desired center of mass and zero moment point trajectories), a **feedback controller** is used. There are several ways to design a feedback controller for a bipedal robot, which is usually modelled as a system with explicit mechanical constraints. The controller can be a modification of a linear quadratic regulator as was done in (Mason et al., 2016; Savin et al., 2017), or as a quadratic program, as was done in (Kuindersma et al, 2016; Savin & Vorochaeva, 2017).

The sensory information is used to estimate the state of the robot and produce feedback for the controller. This is done in a state estimator module. The module can be based on a Kalman filter or some other state estimation technique taking into account the dynamics of the robot and allowing for asynchronous sensor updates.

Alternative CoM Trajectory Planning Formulations

We can observe that the proposed previously method allows to plan the entire center of mass trajectory and to correct zero moment point trajectory in one numeric optimization. There is an alternative approach. It consists of solving problems alike to (9) or (12) for a moving horizon planning. This can be seen as analogous to model predictive control (MPC). The benefits of this approach is that each quadratic program solved for this method can be smaller, as there is no need to plan ZMP trajectory far ahead. However, it still needs to be long enough to be able to take in the geometric properties of the motion, such as step lengths and single and double support phases of walking, which play significant role in the resulting shape of the center of mass trajectory.

On the down side, this approach requires continuously solving optimization problems as the robot moves, which can create large computational burden on the on-board computer. Additionally, with this approach one needs to reason about the update frequency of the ZMP-CoM planner.

Another approach is to avoid treating trajectory planning as a discrete optimization problem and instead work with it as with a continuous control problem, as was done in (Kuindersma et al., 2014). In that paper, CoM trajectory planning was solved by designing a controller by solving algebraic Riccati equation. This allows to rapidly generate CoM trajectory. However, this approach does not allow as much customization and control over the properties of the resulting trajectory as planning with quadratic programming does.

FUTURE RESEARCH DIRECTIONS

We should note that although ZMP-based trajectory generation methods are already well studied and although there are well known limitations to their practical use, they remain a baseline for motion planning tasks in humanoid robotics. There are several research directions in the ZMP methods, one of them is the seamless generalization of the method for uneven terrain, surfaces with low friction, surfaces with discrete elevations.

There are a number of promising directions regarding the use of neural networks in the ZMP-CoM trajectory planning pipeline, such as the use of neural networks are reaction predictors, use of neural network-based free space regions classifications and others. Those might constitute the directions for the future research as well.

CONCLUSION

In this chapter, the problem of zero moment point-based trajectory generation was studied. A survey of the current methods was presented, with a discussion of the existing trends in walking robotics. A more detailed discussion of ZMP-based state of the art methods for center of mass trajectory generation methods, they limitations and application was given. Additionally, footstep planning methods were touched upon.

A quadratic programming-based method for simultaneous ZMP-CoM trajectory planning was given, its numerical properties were discussed. Numeric simulation was used to demonstrate some particular properties of the method.

ACKNOWLEDGMENT

The work is performed with presidential grant MK-1537.2019.8.

REFERENCES

Atkeson, C. G., Babu, B. P. W., Banerjee, N., Berenson, D., Bove, C. P., Cui, X., . . . Gennert, M. (2015, November). No falls, no resets: Reliable humanoid behavior in the DARPA robotics challenge. In *2015 IEEE-RAS 15th International Conference on Humanoid Robots (Humanoids)* (pp. 623-630). IEEE.

Atkeson, C. G., Babu, B. P. W., Banerjee, N., Berenson, D., Bove, C. P., Cui, X., ... Gennert, M. (2016). What happened at the darpa robotics challenge, and why. *Journal of Field Robotics, 1*.

Boyd, S., & Vandenberghe, L. (2004). *Convex optimization*. Cambridge University Press. doi:10.1017/CBO9780511804441

Chestnutt, J., Lau, M., Cheung, G., Kuffner, J., Hodgins, J., & Kanade, T. (2005, April). Footstep planning for the honda asimo humanoid. In *Proceedings of the 2005 IEEE international conference on robotics and automation* (pp. 629-634). IEEE. 10.1109/ROBOT.2005.1570188

Chestnutt, J., Michel, P., Kuffner, J., & Kanade, T. 2007, October. Locomotion among dynamic obstacles for the honda asimo. In *2007 IEEE/RSJ International Conference on Intelligent Robots and Systems* (pp. 2572-2573). IEEE. 10.1109/IROS.2007.4399431

Deits, R., & Tedrake, R. (2014, November). Footstep planning on uneven terrain with mixed-integer convex optimization. In *2014 IEEE-RAS International Conference on Humanoid Robots* (pp. 279-286). IEEE. 10.1109/HUMANOIDS.2014.7041373

Deits, R., & Tedrake, R. (2015). Computing large convex regions of obstacle-free space through semidefinite programming. In *Algorithmic foundations of robotics XI* (pp. 109–124). Cham: Springer. doi:10.1007/978-3-319-16595-0_7

Deits, R., & Tedrake, R. (2015, May). Efficient mixed-integer planning for UAVs in cluttered environments. In *2015 IEEE international conference on robotics and automation (ICRA)* (pp. 42-49). IEEE.

Feng, S., Xinjilefu, X., Atkeson, C. G., & Kim, J. (2015, November). Optimization based controller design and implementation for the atlas robot in the darpa robotics challenge finals. In *2015 IEEE-RAS 15th International Conference on Humanoid Robots (Humanoids)* (pp. 1028-1035). IEEE. 10.1109/HUMANOIDS.2015.7363480

Focchi, M., Orsolino, R., Camurri, M., Barasuol, V., Mastalli, C., Caldwell, D. G., & Semini, C. (2018). *Heuristic Planning for Rough Terrain Locomotion in Presence of External Disturbances and Variable Perception Quality*. arXiv preprint arXiv:1805.10238

Gupta, A., Bhargava, P., Agrawal, S., Deshmukh, A., & Kadam, B. (2018, April). Comparative Study of Different Approaches to Inverse Kinematics. In *International Conference on Advances in Computing and Data Sciences* (pp. 556-563). Springer. 10.1007/978-981-13-1813-9_55

Hanger, M., Johansen, T. A., Mykland, G. K., & Skullestad, A. (2011, December). Dynamic model predictive control allocation using CVXGEN. In *Control and Automation (ICCA), 2011 9th IEEE International Conference on* (pp. 417-422). IEEE. 10.1109/ICCA.2011.6137940

Hirukawa, H., Hattori, S., Harada, K., Kajita, S., Kaneko, K., Kanehiro, F., . . . Morisawa, M. (2006, May). A universal stability criterion of the foot contact of legged robots-adios zmp. In *Robotics and Automation, 2006. ICRA 2006. Proceedings 2006 IEEE International Conference on* (pp. 1976-1983). IEEE. 10.1109/ROBOT.2006.1641995

Hirukawa, H., Hattori, S., Kajita, S., Harada, K., Kaneko, K., Kanehiro, F., . . . Nakaoka, S. 2007, April. A pattern generator of humanoid robots walking on a rough terrain. In *Robotics and Automation, 2007 IEEE International Conference on* (pp. 2181-2187). IEEE. 10.1109/ROBOT.2007.363644

Hutter, M., Gehring, C., Jud, D., Lauber, A., Bellicoso, C. D., Tsounis, V., ... Diethelm, R. (2016, October). Anymal-a highly mobile and dynamic quadrupedal robot. In *2016 IEEE/RSJ International Conference on Intelligent Robots and Systems (IROS)* (pp. 38-44). IEEE. 10.1109/IROS.2016.7758092

Hutter, M., Gehring, C., Lauber, A., Gunther, F., Bellicoso, C. D., Tsounis, V., ... Kolvenbach, H. (2017). ANYmal-toward legged robots for harsh environments. *Advanced Robotics*, *31*(17), 918–931. doi:10.1080/01691864.2017.1378591

Jatsun, S., Savin, S., & Yatsun, A. (2017, September). Footstep planner algorithm for a lower limb exoskeleton climbing stairs. In *International Conference on Interactive Collaborative Robotics* (pp. 75-82). Springer. 10.1007/978-3-319-66471-2_9

Jatsun, S., Savin, S., & Yatsun, A. (2018, September). Harmonic Function-Based ZMP Trajectory Generation for Nonlinear Motion of Walking Robots. In *2018 International Russian Automation Conference (RusAutoCon)* (pp. 1-6). IEEE. 10.1109/RUSAUTOCON.2018.8501657

Johnson, M., Shrewsbury, B., Bertrand, S., Calvert, D., Wu, T., Duran, D., ... Smith, J. (2017). Team IHMC's Lessons Learned from the DARPA Robotics Challenge: Finding Data in the Rubble. *Journal of Field Robotics*, *34*(2), 241–261. doi:10.1002/rob.21674

Johnson, M., Shrewsbury, B., Bertrand, S., Wu, T., Duran, D., Floyd, M., ... Carff, J. (2015). Team IHMC's lessons learned from the DARPA robotics challenge trials. *Journal of Field Robotics*, *32*(2), 192–208. doi:10.1002/rob.21571

Kajita, S., Kanehiro, F., Kaneko, K., Fujiwara, K., Harada, K., Yokoi, K., & Hirukawa, H. (2003, September). Biped walking pattern generation by using preview control of zero-moment point. In ICRA (Vol. 3, pp. 1620-1626). doi:10.1109/ROBOT.2003.1241826

Kanoulas, D., Stumpf, A., Raghavan, V. S., Zhou, C., Toumpa, A., Von Stryk, O., ... Tsagarakis, N. G. (2018, May). Footstep Planning in Rough Terrain for Bipedal Robots Using Curved Contact Patches. In *2018 IEEE International Conference on Robotics and Automation (ICRA)* (pp. 1-9). IEEE. 10.1109/ICRA.2018.8460561

Karumanchi, S., Edelberg, K., Baldwin, I., Nash, J., Reid, J., Bergh, C., ... Newill-Smith, D. (2017). Team RoboSimian: Semi-autonomous mobile manipulation at the 2015 DARPA robotics challenge finals. *Journal of Field Robotics*, *34*(2), 305–332. doi:10.1002/rob.21676

Kuindersma, S., Deits, R., Fallon, M., Valenzuela, A., Dai, H., Permenter, F., ... Tedrake, R. (2016). Optimization-based locomotion planning, estimation, and control design for the atlas humanoid robot. *Autonomous Robots*, *40*(3), 429–455. doi:10.100710514-015-9479-3

Kuindersma, S., Permenter, F., & Tedrake, R. (2014, May). An efficiently solvable quadratic program for stabilizing dynamic locomotion. In *Robotics and Automation (ICRA), 2014 IEEE International Conference on* (pp. 2589-2594). IEEE. 10.1109/ICRA.2014.6907230

LaValle, S. M. (1998). *Rapidly-exploring random trees: A new tool for path planning*. Academic Press.

LaValle, S. M., & Kuffner Jr, J. J. (2000). *Rapidly-exploring random trees: Progress and prospects*. Academic Press.

Long, X. (2017). *Optimization-based Whole-body Motion Planning for Humanoid Robots* (Doctoral dissertation). Northeastern University.

Mason, S., Rotella, N., Schaal, S., & Righetti, L. (2016, November). Balacing and walking using full dynamics LQR control with contact constraints. In *2016 IEEE-RAS 16th International Conference on Humanoid Robots (Humanoids)* (pp. 63-68). IEEE. 10.1109/HUMANOIDS.2016.7803255

Mastalli, C., Havoutis, I., Focchi, M., Caldwell, D., & Semini, C. (2018). *Motion planning for quadrupedal locomotion: coupled planning, terrain mapping and whole-body control*. Academic Press.

Mattingley, J., & Boyd, S. (2012). CVXGEN: A code generator for embedded convex optimization. *Optimization and Engineering*, *13*(1), 1–27. doi:10.100711081-011-9176-9

McGeer, T. (1990). Passive bipedal running. *Proceedings of the Royal Society of London. Series B, Biological Sciences*, *240*(1297), 107–134. doi:10.1098/rspb.1990.0030 PMID:1972987

McGeer, T. (1990). Passive dynamic walking. *The International Journal of Robotics Research*, *9*(2), 62–82. doi:10.1177/027836499000900206

Meredith, M., & Maddock, S. (2004). *Real-time inverse kinematics: The return of the Jacobian*. Technical Report No. CS-04-06, Department of Computer Science, University of Sheffield.

Panovko, G. Y., Savin, S. I., Yatsun, S. F., & Yatsun, A. S. (2016). Simulation of exoskeleton sit-to-stand movement. *Journal of Machinery Manufacture and Reliability*, *45*(3), 206–210. doi:10.3103/S1052618816030110

Playter, R., Buehler, M., & Raibert, M. (2006, May). BigDog. In *Unmanned Systems Technology VIII* (Vol. 6230, p. 623020). International Society for Optics and Photonics. doi:10.1117/12.684087

Pratt, G. A. (2000). Legged robots at MIT: What's new since Raibert? *IEEE Robotics & Automation Magazine*, *7*(3), 15–19. doi:10.1109/100.876907

Radford, N. A., Strawser, P., Hambuchen, K., Mehling, J. S., Verdeyen, W. K., Donnan, A. S., ... Berka, R. (2015). Valkyrie: Nasa's first bipedal humanoid robot. *Journal of Field Robotics*, *32*(3), 397–419. doi:10.1002/rob.21560

Raibert, M., Blankespoor, K., Nelson, G., & Playter, R. (2008). Bigdog, the rough-terrain quadruped robot. *IFAC Proceedings Volumes, 41*(2), 10822-10825. 10.3182/20080706-5-KR-1001.01833

Sakagami, Y., Watanabe, R., Aoyama, C., Matsunaga, S., Higaki, N., & Fujimura, K. (2002). The intelligent ASIMO: System overview and integration. In *IEEE/RSJ international conference on intelligent robots and systems* (Vol. 3, pp. 2478–2483). IEEE. doi:10.1109/IRDS.2002.1041641

Savin, S. (2017, June). An algorithm for generating convex obstacle-free regions based on stereographic projection. In *2017 International Siberian Conference on Control and Communications (SIBCON)* (pp. 1-6). IEEE. 10.1109/SIBCON.2017.7998590

Savin, S., & Ivakhnenko, A. (2019). Enhanced Footsteps Generation Method for Walking Robots Based on Convolutional Neural Networks. In Handbook of Research on Deep Learning Innovations and Trends (pp. 16-39). IGI Global. doi:10.4018/978-1-5225-7862-8.ch002

Savin, S., Jatsun, S., & Vorochaeva, L. (2017, November). Modification of Constrained LQR for Control of Walking in-pipe Robots. In *2017 Dynamics of Systems, Mechanisms and Machines (Dynamics)* (pp. 1-6). IEEE.

Savin, S., Khusainov, R., & Klimchik, A. (2019) Admissible region ZMP trajectory generation for bipedal robots walking over uneven terrain. *Zavalishin's Readings.* (forthcoming)

Savin, S., Vorochaev, A., & Vorochaeva, L. (2018). Inverse Kinematics for a Walking in-Pipe Robot Based on Linearization of Small Rotations. The Eurasia Proceedings of Science, Technology. *Engineering & Mathematics, 4*, 50–55.

Savin, S., & Vorochaeva, L. (2017, May). Nested quadratic programming-based controller for pipeline robots. In *2017 International Conference on Industrial Engineering, Applications and Manufacturing (ICIEAM)* (pp. 1-6). IEEE. 10.1109/ICIEAM.2017.8076142

Savin, S., Yatsun, A., & Loktionova, O. (2019) Footstep Planning for Bipedal Robots and Lower Limb Exoskeletons Moving Through Narrow Doors. *Zavalishin's Readings.* (forthcoming)

Short, A., & Bandyopadhyay, T. (2018). Legged motion planning in complex three-dimensional environments. *IEEE Robotics and Automation Letters, 3*(1), 29–36. doi:10.1109/LRA.2017.2728200

Thompson, C. M., & Raibert, M. H. (1990). Passive dynamic running. In *Experimental Robotics I* (pp. 74–83). Berlin: Springer. doi:10.1007/BFb0042513

Tsagarakis, N. G., Caldwell, D. G., Negrello, F., Choi, W., Baccelliere, L., Loc, V. G., ... Natale, L. (2017). Walk-man: A high-performance humanoid platform for realistic environments. *Journal of Field Robotics, 34*(7), 1225–1259. doi:10.1002/rob.21702

Vukobratović, M., & Borovac, B. (2004). Zero-moment point—thirty five years of its life. *International Journal of Humanoid Robotics, 1*(1), 157-173.

Xia, Z., Xiong, J., & Chen, K. (2011). Global navigation for humanoid robots using sampling-based footstep planners. *IEEE/ASME Transactions on Mechatronics, 16*(4), 716–723. doi:10.1109/TMECH.2010.2051679

Yi, S. J., McGill, S. G., Vadakedathu, L., He, Q., Ha, I., Han, J., ... Yim, M. (2015). Team thor's entry in the darpa robotics challenge trials 2013. *Journal of Field Robotics*, *32*(3), 315–335. doi:10.1002/rob.21555

Zheng, Y., Lin, M. C., Manocha, D., Adiwahono, A. H., & Chew, C. M. (2010, October). A walking pattern generator for biped robots on uneven terrains. In *2010 IEEE/RSJ International Conference on Intelligent Robots and Systems* (pp. 4483-4488). IEEE. 10.1109/IROS.2010.5653079

ADDITIONAL READING

Horn, J. C., Mohammadi, A., Hamed, K. A., & Gregg, R. D. (2019). Hybrid Zero Dynamics of Bipedal Robots Under Nonholonomic Virtual Constraints. *IEEE Control Systems Letters*, *3*(2), 386–391. doi:10.1109/LCSYS.2018.2888571

Kanoulas, D., Stumpf, A., Raghavan, V. S., Zhou, C., Toumpa, A., Von Stryk, O., ... Tsagarakis, N. G. (2018, May). Footstep Planning in Rough Terrain for Bipedal Robots Using Curved Contact Patches. In *2018 IEEE International Conference on Robotics and Automation (ICRA)* (pp. 1-9). IEEE. 10.1109/ICRA.2018.8460561

Katić, D., & Vukobratović, M. (2003). Survey of intelligent control techniques for humanoid robots. *Journal of Intelligent & Robotic Systems*, *37*(2), 117–141. doi:10.1023/A:1024172417914

Knight, R., Nehmzow, U., & Sq, C. C. (2002). Walking robots-a survey and a research proposal. Univ. Essex, Essex, UK, Technical Report CSM-375.

Mombaur, K., & Berns, K. (Eds.). (2013). *Modeling, simulation and optimization of bipedal walking*. Springer. doi:10.1007/978-3-642-36368-9

Nelson, G., Saunders, A., & Playter, R. (2019). The PETMAN and Atlas Robots at Boston Dynamics. Humanoid Robotics: A Reference, 169-186.

Rai, A., Antonova, R., Meier, F., & Atkeson, C. G. (2019). Using Simulation to Improve Sample-Efficiency of Bayesian Optimization for Bipedal Robots. *Journal of Machine Learning Research*, *20*(49), 1–24.

Siciliano, B., & Khatib, O. (Eds.). (2016). *Springer handbook of robotics*. Springer. doi:10.1007/978-3-319-32552-1

Wu, Q., Liu, C., Zhang, J., & Chen, Q. (2009). Survey of locomotion control of legged robots inspired by biological concept. *Science in China Series F: Information Sciences*, *52*(10), 1715–1729.

KEY TERMS AND DEFINITIONS

Anthropomorphic Robot: A robot with a shape and morphology generally similar to that of a human body.

Bipedal Robot: A robot with two legs.

Quadratic Program: A convex optimization problem with a quadratic cost function, linear equality constrains and linear constraints inequality constraints. Is usually solved with numeric algorithms.

Robot Feet: Elements of the robot's structure designed to remain in periodic contact with the supporting surface during walking, in a similar manner as do feet of humans.

Vertical Stability: A potential ability of a walking robot to remain from making unplanned contacts with the supporting surface or uncontrollably losing contacts with the supporting surface (falling).

Zero-Moment Point: A point on the supporting surface relative to which the reaction forces' torques are balanced. Used to find stable trajectories for the center of mass of the robot.

Chapter 8
The Problem of Using Landmarks for the Navigation of Mobile Autonomous Robots on Unknown Terrain

Oleksandr Vasilievich Poliarus
Kharkiv National Automobile and Highway University, Ukraine

Evhen Oleksandrovych Poliakov
Kharkiv National Automobile and Highway University, Ukraine

ABSTRACT

Navigation of mobile autonomous robots on unknown terrain in the absence of GPS is extremely difficult. The general aim of the chapter is to analyze the possibilities of reliable detecting landmarks and determining their coordinates for navigation purposes. It is shown that the method of solving such a problem is the complex use of the meters operating on different physical principles. The main attention is paid to the radar method of measuring the angular coordinates of the landmarks by an antenna with a small size. For a radar with a wide antenna pattern, the possibility of angular resolving of two or more closely spaced landmarks is estimated. The most reliable method providing angular resolution is the creation of a synthesized aperture of antennae in the process of linear movement of a robot. The possibilities of such antennae are analyzed, considering random phase distortions and errors.

DOI: 10.4018/978-1-5225-9924-1.ch008

INTRODUCTION

Navigation of mobile autonomous robots (MAR) in the unknown terrain is most easily provided using GPS. However, in some cases, such navigation is significantly complicated by the lack of satellites visibility. In the close vicinity of the buildings, hills, mountains, etc. GPS accuracy drops down due to unavailability of useful signal. In such situations, the landmarks for MAR navigation may be single concentrated objects that occur on the ground. As a rule, the non-contact remote methods, which are realized using of radar, ultrasound, infrared and laser systems, as well as systems of technical vision, are used for the landmarks detection by mobile autonomous robots. Each system has limitations due to various factors, such as weather conditions, daily changes, landmark type, presence of noise, etc. Thus, the most rational approach to MAR navigation is the complex use (fusion) of systems built on various physical principles or systems that differ significantly in their characteristics. An alternative approach is to use systems that have proven well in other branches. One such system is presented in the chapter of the book. After detecting a single landmark, its coordinates are determined. If, within the width of the antenna pattern, there are two landmarks that are not resolved by the range, the interference field of the reflected waves has a non-spherical fluctuating phase front. The errors of measuring the azimuth of these landmarks can increase significantly. Consequently, there is a problem of angular resolution of landmarks. The particular purpose of the chapter is to estimate the possibilities of angular resolution of landmarks and determining their coordinates for robotic navigation tasks and to suggest ways of increasing this resolution by radar methods using frequency range VHF or UHF.

BACKGROUND

Autonomous robot navigation is important due to the increasing interest in self-driving vehicles. The chapter deals with outdoor environment. The robot doesn't know its coordinates and path-planning step is not performed over previously known map. Many robots collide with unpredicted obstacles on their way if they do not have appropriate detection systems (Ferreira et al., 2008). The collision-free optimal path of a robot is selected using agent-based architecture in a virtual reality environment (Popirlan and Dupac, 2009). In (Borenstein et al., 1997) seven approaches for positioning systems are analyzed: odometry, inertial navigation, magnetic compasses, active beacons, GPS, landmarks navigation and model matching. Landmarks navigation demands a starting point. If such point isn't known, the robot must scan the environment for searching a landmark. The navigation task can be defined as the combination of localization, path planning and vehicle control (Dhanasingaraja et al., 2014). A robot localization is the robot's function to estimate its location, first of all, relative location (relative to landmarks, features of environments and so on). A robot must be able to achieve the aim without direct external control information. The localization of a robot is accomplished using beacons and computational triangulation (Melo et al., 2013). For robot localization a single-webcam distance measurement technique is proposed in (Li et al., 2014) and there is other technique based on system of optical sensors (Lee and Song, 2004). In (Howard et al., 2003) relative localization method for mobile robot team doesn't require GPS and landmarks. It uses measurements data from the nearby robot positions. In (Sergiyenko et al., 2016) the idea of data transferring in the group of robots during their movement on the terrain with a high density of obstacles is proposed. The authors used the behavioristic models and fuzzy logic for the choice of appropriate data for transferring models. Authors of (Ort et al., 2018) have presented a method for autonomous robot movement without detailed prior maps using a LIDAR-based trajectory algorithm. The method is effective in rural environments. The authors of (Jin et al., 2006) proposed a localization of mobile robots using the images of distributed networked devices which create an intelligent space and the authors of (Mester, Aleksandar Rodić, 2010) presented sensor-based intelligent mobile robot navigation using Sun SPOT technology in unknown environments. Laser scanning for autonomous robot navigation is described in (Andersen et al., 2006). The methods of rational use of different approaches to navigating robots are important. The odometric information from wheels encoders of a robot is combined with GPS data in Kalman filter (Vaz et al., n. d.). In (Surrécio et al., 2005) measuring information from the odometers is fused with data from magnetic markers. Kalman filtering used stabilized stereo camera for distance measurement to obstacles and removing noise from images (Budiharto et al., 2011). In (Lukin et al., 2013) a combination of 2D aperture synthesis and wideband noise radar with high resolution is analyzed and tomographic imaging experiments carried out in S-band.

Consequently, autonomous navigation of robots uses different physical principles, approaches and technical systems. Now the practice of navigation is beginning to be implemented methods that have long been used in other industries, for example, in aviation. One of these methods is the radar with a synthesized aperture antenna. The basic principles of synthesized aperture radar (SAR) are presented in (Chan and Koo, 2008), (Jean and Rouse, 1983). In (Corany et al., 2018) the authors presented synthesized phased array imaging of a known and unknown areas using narrowband radio frequency signals. Such antenna array produces an image of uncertain terrain. In (Ali et al., 2011) synthesized an aperture imaging radar produces a hologram using a small omnidirectional antenna mounted on a rotating robot's platform. In (Kidera, 2010) a synthesized aperture is formed for ultra-wide band pulse radar which expands its capabilities in conditions of poor visibility for optical waves.

1 Problems of Landmarks Detection and Estimation of Their Coordinates

The method of determining the position of a landmark using a radar, which may have small sizes and therefore can easily be installed on most MARs, is important in practical applications. The radar transmitter emits electromagnetic waves (EMW) in the space near the surface of the earth. These waves are reflected from local objects, as well as other objects in the surrounding area. Some objects, such as metal pillars, can be used as benchmarks for a robot. If the width of the radar antenna pattern (AP) is small, then determining the angular coordinates of the landmark relative to the robot does not cause any difficulties. These coordinates are stored in the computer memory and then used for autonomous navigation of the robot. However, measuring the angular coordinates by the antenna with a narrow AP does not guarantee reliable detection of landmarks, since most landmarks are immovable. Hence, the radar receiver does not differ in Doppler frequency most of the echo signals from landmarks and the signals reflected from the environment. In (Poliarus et al., 2018) a system for detecting jumps of echo signals amplitude from the landmarks is proposed and an optimal scheme of their detection is given. For jumps of signals amplitude exceeding half of that amplitude, the probability of detecting jumps is 0.9 and more, but there is a large dependence on the background signals that are reflected from the forest, urban structures, and so on. Unfortunately, the use of energy characteristics of signals for their detection does not always ensure the high reliability of this procedure.

Consequently, increasing the size of the antenna and reducing the wavelength ensure creation of a narrow AP. The size of the antenna is limited by the dimensions and other characteristics of the robot itself. Reducing the wavelength is definitely feasible, but there is a problem of increasing the intensity of the echo signals from small-scale irregularities of the surface, vegetation, and so on. At high wavelengths, due to features of terrain, the mirror reflection of EMW prevails. Electromagnetic waves are reflected in the direction from the radar, which often reduces the level of unintentional noise and facilitates the detection of landmarks. In addition, on such EMW there is a greater probability of resonance scattering of waves on the landmarks.

The electromagnetic wave reflected from a single landmark has a phase front, which is almost spherical, and in place of the robot position, it is close to the plane one. During a robot movement, the characteristics of the phase front vary slightly. Since the antenna of the radar determines the direction of the landmark in the normal to the phase front, the angular position of the landmark is stable for the radar. The robot measures the position of the landmark in the azimuthal plane. The presence of direct and reflected waves from the earth's surface has a little effect on the accuracy of the azimuth measurement. Another situation arises if there are two landmarks within the range of the unresolved volume of space (Figure 1). The radar antenna now receives two spherical waves, which create a phase front that differs from the spherical one. The antenna determines the direction of the EMW arrival, which may not coincide with the direction to the landmark (or the center between the two landmarks). Moreover, a random change of the phases of the reflected signals during the robot's motion leads to phase front fluctuation and to errors in the measurement of the azimuth of the landmark. In the general case, a system of two landmarks, which are not distinguished by angular coordinates, cannot be considered as a stable system for use in navigation. Consequently, the possibilities of landmarks resolution using other characteristics must be estimated. The known mathematical relations for the signals reflected from landmarks 1 and 2 (Figure 1) are given below.

The first landmark creates signal at the point O (Figure 1)

$$S_1(t) = A_1 \cos(\omega t - \varphi_1),$$ (1)

Figure 1. Scheme of location of two landmarks that do not resolve in angular coordinates

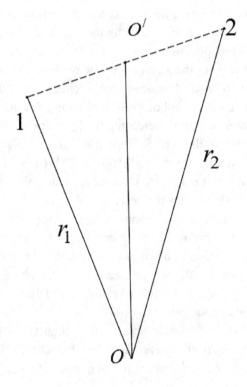

where A_1 - the amplitude of the echo signal depending on the directional properties of the radar transmitting antenna, the transmitter power, radar cross section (RCS) of the landmark, etc., $\omega=2\pi f$- angular frequency of the signal, t - time, φ_1 - the phase of the reflected wave, which depends on the distance r_1 between the landmark and the radar antenna, the initial phase δ_1 of the echo signal and its frequency f, moreover

$$\varphi_1 = \frac{2\pi f}{c} r_1 + \delta_1 ,$$ where c - the speed of light. Similarly for the second landmark

$$S_2(t) = A_2 \cos(\omega t - \varphi_2) . \tag{2}$$

At the point of reception O, an interference field of two echo-signals from the landmarks is created, the total amplitude of which is

$$A = \sqrt{A_1^2 + A_2^2 + 2A_1 A_2 \cos(\varphi_1 - \varphi_2)} , \tag{3}$$

and the phase of EMW at the input of the radar receiver

$$\varphi = arctg \frac{A_1 \sin\varphi_1 + A_2 \sin\varphi_2}{A_1 \cos\varphi_1 + A_2 \cos\varphi_2} . \tag{4}$$

The simple formulas (3) and (4) are useful for estimating the sensitivity of total amplitudes and phases of signals to the change of partial parameters, for example, frequency. The examples of the results of mathematical modeling of the dependence of total amplitudes in relative units and phases from frequency for the following data: $c = 3 \cdot 10^8 \frac{m}{s}$, r_1=123m, r_1=128m, δ_1=2,154, δ_2=1.154, A_1=2, A_2=1 are presented at Figures 2…7.

An example of dependencies of the total amplitudes and phases of the signal at the receiver input from the frequency, i. e. $A(f)$ and $\varphi(f)$ for given data are shown in Figure 2.

At equal amplitudes of echo signals at some frequencies there is a deep fading (Figure 3).

The frequency dependence of the total phase of the signal at the receiver input is shown in the figure 4.

The sensitivity of the total amplitude of the signal to the frequency change with unequal and equal amplitudes of the echo signals from the landmarks is shown in the figures 5 and 6 accordingly. This sensitivity is expressed in derivatives in frequency from the total amplitudes of echo signals from landmarks.

Figure 2. Frequency dependence of the total amplitude with unequal amplitudes of the echo signals from the landmarks

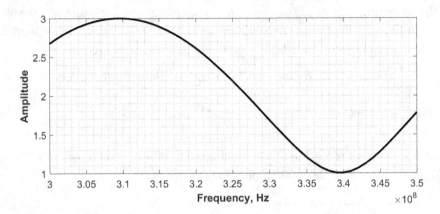

Figure 3. Frequency dependence of the total amplitude with equal amplitudes of the echo signals from the landmarks

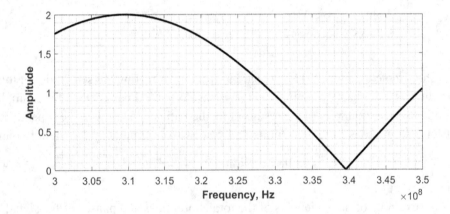

Similar dependence for the derivative of the total phase at unequal amplitudes of the echo signals from the landmarks is shown in the figure 7.

The figures are made for modeling conditions, in which the radar cross section of landmarks did not depend on the frequency. The reflexivity of landmarks of artificial origin and, in some cases, of natural landmarks, is frequency dependent. Only RCS of ball are not frequency dependent. The RCS of a circular cylindrical metal pillar with radius r and length L is determined by the known formula (Rajyalakshmi and Raju, 2011)

Figure 4. Frequency dependence of the total phase of the echo signals from the landmarks

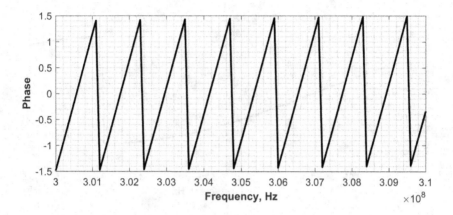

Figure 5. Dependence of the derivative in frequency of the total amplitude of the signal on frequency at unequal amplitudes of the echo signals from the landmarks

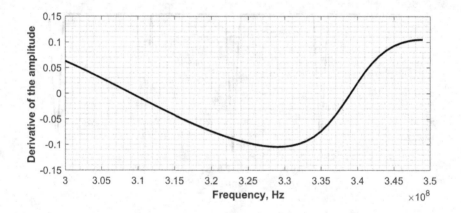

$$\sigma_{max} = \frac{2\pi f \cdot rL^2}{c},\tag{5}$$

For r=0.2m, L=4m in the frequency range 300 ... 350 MHz σ_{max} as the energy characteristic varies by about 15%. Similarly, the amplitude of the echo signal is changed. As it follows from the figures 2, ..., 7, due to the change of the carrier frequency of a signal the total amplitude can vary by several times. Consequently, the sensitivity to frequency variation is a sign that there are two or more landmarks in an unresolved volume of space, but in the latter case (many landmarks), the sensitivity

Figure 6. Dependence of the derivative in frequency of the total amplitude of the signal on frequency at equal amplitudes of the echo signals from the landmarks

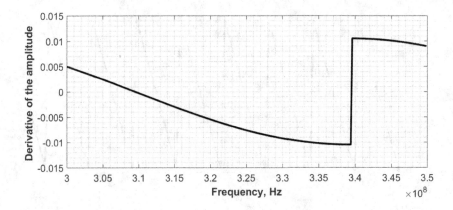

Figure 7. The dependence of the derivative of the total phase in frequency from the signal frequency at unequal amplitudes of the echo signals from the landmarks

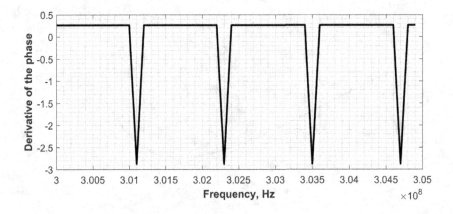

to frequency change is much lower. In the presence of one landmark in the angular range of the AP, the frequency sensitivity of an amplitude and phase is absent at all.

The largest amplitude of the EMW scattering by the landmarks is achieved in the resonant wavelength ranges, when the wavelength is commensurate with one of the dimensions of a landmark. This means that for detection of landmarks, it is necessary to use EMW with a relatively long length, for example, a wavelength of a meter range (VHF).

Thus, the use a few transmitters or transmitter with rebuilding frequency allows to form a dependence of the number of landmarks in an unresolved volume of space from frequency. Unfortunately, the sensitivity of the radar cross section of landmarks to the change of frequency in some cases reduces the reliability of such a dependence. Hence there is a need to develop a different approach to resolving landmarks.

It should also be noted that the binding of a robot to a landmark and the determination of its coordinates is impossible only with the use of one landmark. The navigation practice requires two or more landmarks that are spaced apart at an angle that is substantially larger than the AP width. As shown in (Poliarus et al., 2018), the spatial errors of the determination of the robot coordinates depend essentially on the distance ρ_i between the robot and the landmark and decrease significantly at small distances. For example, at a distance $\rho_i \approx 300$m the coordinates of the landmark are determined with errors that do not exceed 10 m, and then the measurement error of the azimuth is less than one degree that can satisfy the requirements of near navigation. A higher accuracy is obtained when the robot is near a landmark.

2 Increasing the Angular Resolution of Landmarks by Antennas With Synthesized Aperture

Due to the small size of the antenna installed on a mobile autonomous robot, the antenna beam width at half power $2\theta_{0.5P}$ in some plane (horizontal, vertical, or both) is large, and, as result, this leads to the radar resolution loss at angular coordinates. For example, with the maximum antenna size L, the linear resolution is

$$\Delta l \approx 2\theta_{0.5P}R_0 \approx \frac{\lambda}{L}R_0, \qquad (6)$$

where λ is the wavelength, R_0 - the minimum distance between the landmark and the robot antenna.

If, for example, $L=0.1$m, $\lambda=1$m, then $\Delta l \approx 0.1R_0$, that is, at a distance of only 100 m, it is impossible to resolve landmarks with angular coordinates, when the distance between these landmarks is 10 m or less. It does not make sense to increase the antenna size from a technical point of view, and reducing the wavelength is inappropriate because of the decrease in the amplitude of the reflected waves towards the radar and increasing external noise.

Consequently, the only approach of increasing the angular resolution of a robot antenna is the transition to a synthesized aperture of antenna. In conventional antenna array, the reception of reflected EMW occurs simultaneously by all the receiving elements of the array. The signals from the outputs of these elements are added in phase, creating a large amplitude of the total signal in the direction of maximum antenna pattern.

Figure 8. The geometry of the problem: the top view of the robot trajectory relative to a landmark

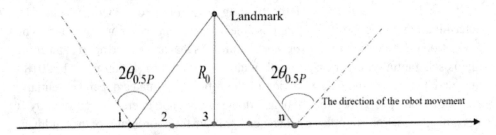

The synthesized antenna aperture is created with a uniform movement of the robot along the path in the detection region of the landmark (Figure 8).

During the robot movement the radar irradiates the surrounding area, and only when it is at point 1 (Figure 8), the EMW is reflected from the landmark and the signal enters the antenna of the radar receiver. Similarly, from the points 2, 3, .., the pulses are also emitted and reflected from the landmark. The EMW returns to the receiving element when the robot has practically not moved from the place. This is due to a small robot speed relative to the speed of the EMW and a relatively small range to the landmark. Consequently, now it is possible to assume that at points 1, 2, ..., there are emitters of a fictitious antenna array, in which the information is not processed in parallel (signals from the outputs of the emitters are in phase), but sequentially. However, if the signals from the outputs of receiving elements are added, but it is practically impossible to get the main lobe of the AP, because these signals are spaced apart in time. For phase-to-phase signals, their phases should be the same or different, for example, no more than $\pi/2$. This requirement can be met by two methods. In the first method, it is necessary to impose the requirements for the range from the receiving elements to the landmark. The EMW phase during its propagation from the radar to the landmark and in the reverse direction should not differ by more than $\pi/4$ that corresponds to the length of the path $\lambda/8$. If the distance between the points 1 and n (for Figure 8 $n=5$) is the length of the antenna with synthesized aperture L_{sa}, then from Figure 8 we have:

$$\left(\frac{L_{sa}}{2}\right)^2 = \left(R_0 + \frac{\lambda}{8}\right)^2 - R_0^2, \tag{7}$$

where $R_0 + \dfrac{\lambda}{8}$ - the length of the hypotenuse of a rectangular triangle (Figure 8), and the distance between the points 1 and 3 in Figure 8 is $L_{sa}/2$.

From the relation (7) we have

$$L_{sa} = \sqrt{R_0 \lambda}, \tag{8}$$

that is, for distance, for example, R_0=100m and λ=1m we get L_{sa}=10m.

The linear resolution of the radar using formula (6)

$$\Delta l \approx \frac{\lambda}{2L_{sa}} R_0 = \frac{1}{2}\sqrt{R_0 \lambda}. \tag{9}$$

In the above example (R_0=100m and λ=1m) the linear resolution $\Delta l \approx$5m.

So, the using of an unfocused antenna with a synthesized aperture allows to increase the linear resolving ability approximately in two times. However, the real value Δl will actually exceed the given one due to the instability of the frequency of the transmitter signals, non-linearity and unevenness of the robot movement due to the features of the terrain in which it moves, changing the conditions of reflection of the EMW from the landmarks at different angles, random errors of the reflected signals phases measurement during their processing and other factors. Such additional phase distortions $\Delta \varphi_e$ of the field are equivalent to changing the propagation distance Δr_e of the EMW from the receiving element of the antenna array to the landmark

$$\Delta r_e = \frac{\lambda}{2\pi} \Delta \varphi_e. \tag{10}$$

Addition this distance to the hypotenuse of the triangle (Figure 8) means that due the phase measurement error, the receiving element 1 in Figure 8 measures the range to the landmark with an error Δr_e that is greater than zero. On the contrary, we subtract Δr_e from the length R_0 of the triangle leg, further worsening the measurement situation, in which the length of the triangle leg is measured with an error Δr_e that is negative. Then equation (7) will have the form

$$\left(\frac{L_{sa}^*}{2}\right)^2 = \left(R_0 + \frac{\lambda}{8} + \Delta r_e\right)^2 - (R_0 - \Delta r_e)^2, \tag{11}$$

where L_{sa}^* - the new value of the synthesized aperture antenna length, which appeared due to errors in the measurement of the phase. Since $\lambda \ll R_0$, we obtain from equation (11)

$$L_{sa}^* \approx \sqrt{R_0 \lambda} \sqrt{1 + \frac{16 \Delta r_e}{\lambda}} \approx L_{sa} \sqrt{1 + \frac{16 \Delta r_e}{\lambda}}. \tag{12}$$

From formula (12) it follows that in the presence of phase measurement errors, the length of the antenna with the synthesized aperture increases $(L_{sa}^* > L_{sa})$, and this should not be, since with the increase of the aperture, some emitters will not receive reflected signals. Hence, the requirements for measurement errors of the phase $\frac{16 \Delta r_e}{\lambda} \ll 1$ or after simple transformations $\Delta \varphi \ll \frac{\pi}{8}$, are difficult to achieve in practice.

Consequently, due to the inequality of distances from the elements of the antenna with the synthesized aperture (SA) to the landmark along this array, a quadratic phase distribution of the field is formed, to which distortions of the phases described above are added. If all phase distortion is eliminated, then the focused SA antenna will have a length

$$L_{saf} = 2\theta_{0.5 P_b} R_0 = \frac{\lambda}{d} R_0, \tag{13}$$

where $2\theta_{0.5 P_b}$ - the width of the AP for radar, which is on MAR board, d - the maximum size of this antenna.

Linear resolution of the focused antenna with SA

$$\Delta l \approx 2\theta_{0.5 P_{saf}} R_0 = \frac{d}{2}. \tag{14}$$

In the presence of phase distortions of signals that cannot be compensated during signal processing, the formula (14) is not applicable to estimate the resolution. The calculating the AP with SA, considering phase distortions, and then already determining the width of this AP $2\theta_{0.5 P_{saf}}^*$ must be carried out. Then the linear resolution of the antenna with SA is determined as

$$\Delta l^* = 2\theta_{0.5 P_{saf}}^* R_0. \tag{15}$$

In case of large phase distortions, it may appear that $\Delta l^* \gg \Delta l$. Consequently, the accuracy of determining phase distortions of signals for further their compensation plays an important role in the creation of antennas with SA that can distinguish even near-placed landmarks.

The geometric length of antenna with synthesized aperture without considering phase distortions is

$$L_s = 2R_0 tg\left(\frac{1}{2}2\theta_{0.5b}\right),$$
(16)

where in brackets it is indicated half of the width of the AP on the MAR $\phi = \frac{1}{2}2\theta_{0.5b}$. This length differs from the similar length given in the formula (8), since phase distortions of the echo signals from the landmark, which are due to different range from the radiators to the landmark and phase measurement errors, are not yet considered here. The number of radiators m, which ensure the unity of the main lobe of AP for antenna with synthesized aperture (APSA), should be determined considering that the distance between them should not exceed half the wavelength, that is

$$2R_0 tg\phi = m\frac{\lambda}{2}.$$
(17)

Hence, the number of radiators of SA antenna is determined

$$m = \frac{4R_0 \cdot tg\phi}{\lambda}.$$
(18)

In the one-half part of the antenna with SA a quadratic phase distribution is created (in suggestion of excluding different distance between the emitters and the landmark)

$$\varphi_i = \frac{2\pi}{\lambda}R_i \approx \frac{2\pi}{\lambda}R_0\left(1 + \left(\frac{\lambda \cdot \left(i - \frac{m}{2}\right)}{2R_0}\right)^2\right),$$
(19)

where $i = 0 \ldots \frac{m}{2}$. Similar phase distribution is created in the other half of the antenna.

$$\varphi_i = \frac{2\pi}{\lambda} R_i \approx \frac{2\pi}{\lambda} R_0 \left(1 + \left(\frac{\lambda \cdot \left(\frac{m}{2} - i\right)}{2R_0}\right)^2\right), \tag{20}$$

where $i = \frac{m}{2} \dots m$. First, we are interested in the phase difference between the signals of emitters, which determines the shape of the AP. At the edge of the antenna with SA, the phase difference between the last and the central emitters will be

$$\Delta\varphi_m = \frac{2\pi}{\lambda} R_0 \left(\frac{\lambda \cdot m}{4R_0}\right)^2. \tag{21}$$

Let the distance from the robot to the landmark $R_0 = 100$m, and the width of the AP $2\theta_{0.5b} = 30°$. Then $L_s = 55.6$m, and the number of conditional radiators is $m = 107$. Without considering the quadratic phase distribution, that is, with a uniform phase and amplitude distribution, the width of the APSA is $2\theta_{05P}^\circ \approx 51\frac{\lambda}{L_s} = 0.92°$ at the wavelength $\lambda = 1$m.

We calculate the APSA, if the amplitude distribution is uniform and normalized to unity, and the phase distribution is determined by the formula

$$\varphi(z) = \frac{\pi}{\lambda} R_0 \left(\frac{z}{2R_0}\right)^2, \tag{22}$$

where discrete coordinates with numbers i are replaced by continuous coordinates z, in the antenna center (origin of coordinates) $z = 0$, and at the edges of a linear antenna $z = \pm\frac{m\lambda}{4}$. Such a transformation of a discrete antenna array to a continuous linear antenna can be made without significant change in the antenna, since the distance between the radiators of the antenna array does not exceed half the length waves. Then the antenna pattern of a linear continuous antenna is

$$F(\theta) = \int_{-\frac{m\lambda}{4}}^{\frac{m\lambda}{4}} e^{j\left[\frac{2\pi}{\lambda}\frac{R_0}{2}\left(\frac{z}{2R_0}\right)^2 + \frac{2\pi}{\lambda}z\sin\theta\right]} \, dz. \tag{23}$$

300

Figure 9. The pattern of an antenna with synthesized aperture: solid line - in absence of phase distortions; dotted line – in their presence

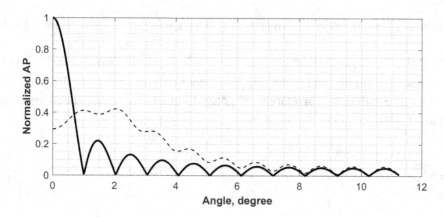

For distances R_0=100m and at wavelengths λ=1m, we obtain two APSA (Figure 9): the solid line indicates a normalized AP for antenna with SA in the absence of phase distortions due to the difference in distance from the radiators to the landmark, and the dotted line indicates the AP of this antenna, taking into consideration these distortions.

From the Figure 9 it follows that in the absence of compensation of phase distortions, the APSA significantly changes, which is displayed in the collapse of the main lobe and the loss of the necessary resolving ability of the robot's radar. These phase distortions can be compensated if the distance between the emitter and the landmark is known. It is more difficult to do this when the signal phase, measured by each emitter, is random. Random factors may cause instabilities of the radar transmitter frequency, the influence of atmospheric factors, the dependence of the reflection coefficient of the EMW on the landmark, the impact of noise, etc. If these factors do not significantly affect the phase distortion of signals, then the main factor will be the error of measuring the phase of the signals, which form the random phase distribution in the antenna with SA. The main parameters characterizing this phase distribution are the variance of phase fluctuations and their correlation coefficient.

The APSA by power with random phase distribution and length L_s is determined by the formula (Shifrin, 1971)

$$\overline{F^2(\psi)} = \frac{L_s^2}{4} e^{-\alpha} \int_{-1}^{1} \int_{-1}^{1} e^{\alpha r} e^{j\psi(x-x_1)} dx_1 dx, \tag{24}$$

where $\psi = \dfrac{\pi L_s}{\lambda} \sin\theta$ - a generalized angle, and $\theta = \arcsin\left(\dfrac{\psi\lambda}{\pi L_s}\right)$. Relative

coordinates x, x_1 are determined by the formulas: $x = \dfrac{2z}{L_s}$, $x_1 = \dfrac{2z_1}{L_s}$, where z, z_1 - the

coordinates of the point along the antenna with SA. The parameter $\alpha = \sigma_\varphi^2$, that is, the variance of phase fluctuations, is considered constant along the length of the antenna. The radius of correlation of phase fluctuations is calculated from the formula (Shifrin, 1971)

$$r = e^{-\frac{(x-x_1)^2}{c^2}}, \tag{25}$$

where $c = \dfrac{2\rho}{L_s}$ - the relative radius of correlation, and ρ - the radius of correlation

of phase errors, which characterizes the distance along the antenna, within which the phase errors can be considered interrelated (correlated).

Assume that the conditions for the reflection of the EMW from the landmark are the same for all angular directions from each emitter to this landmark. Then the coefficient of correlation of phase fluctuations can be considered large, for example,

$\rho = \dfrac{L_s}{2}$. Let's estimate the influence of the phase variance on the antenna angle and

linear resolution using formula (24).

The Figure 10 shows non-normalized APSA by power in the absence of phase fluctuations of the EMW (solid line) and, if they exist (dotted line, phase-fluctuation variance $\sigma_\varphi^2 = 2$). As can be seen, the maximum power value in the main direction decreased by 2.4 times, which leads to a decrease in the signal-to-noise ratio at the input of the radar receiver of the MAR. The side lobes also disappeared, and a large side background was created. The level of this background has increased by about 10 dB (Figure 11).

The disappearance of the main lobe of the APSA means the loss of the antenna resolution by angular coordinates. Consequently, it is necessary to limit the level of phase fluctuations in the antenna array radiators.

If the conditions for the EMW reflection from the landmark differ significantly for the different angles to the emitter, the correlation radius of the phase fluctuations of the EMW decreases. The Figures 12, 13 are represented for the radius of the

correlation of the EMW phase fluctuations $\rho = \dfrac{L_s}{5}$. They are like the graphs depicted

in the figures 10, 11.

Figure 10. Unnormalized pattern of antenna with synthesized aperture at a large radius of phase correlations: solid line - absence of phase fluctuations; dotted line - variance of phase fluctuations is 2

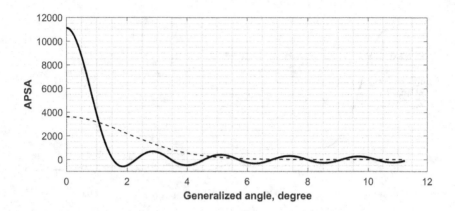

Figure 11. The normalized pattern of antenna for the antenna with the synthesized aperture for a large radius of correlation of phase fluctuations: solid line - the absence of phase fluctuations; dotted line - variance of phase fluctuations is 2

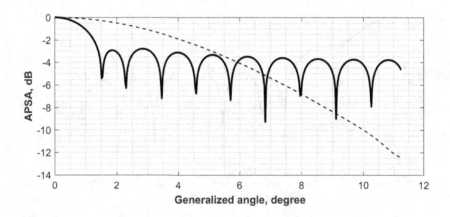

The reduction of the radius of correlation of phase fluctuations by 2.5 times more distorted the APSA. The function of angular resolution of landmarks is absent. The radiation power in the direction of the main lobe of APSA decreases by 5.1 times. However, even with a small radius of correlation of the phase fluctuations of EMW, an angular resolution of the landmark is possible, but this occurs at small values of the variance of phase fluctuations of the EMW. Consequently, for the angular resolution of the landmark, it is necessary that the scattering of the EMW in the

Figure 12. Unnormalized pattern of antenna with synthesized aperture at low radius of phase correlations: solid line - absence of phase fluctuations; dotted line - variance of phase fluctuations is 2

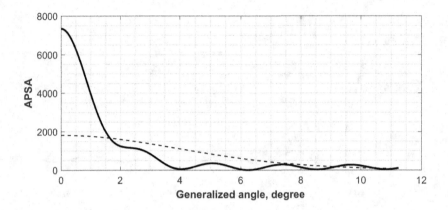

Figure 13. Normalized pattern of antenna direction with synthesized aperture for a small radius of correlation of phase fluctuations: solid line - absence of phase fluctuations; dotted line - variance of phase fluctuations is 2

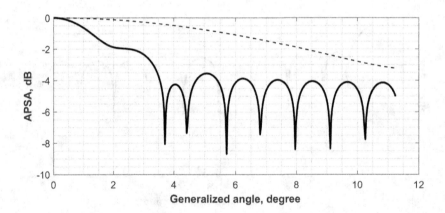

directions on the emitter of the antenna with SA is identical. For landmarks in the form of, for example, columns, this requirement is fulfilled.

Analysis of the resolution of the antenna with SA at large correlation radii of phase fluctuations of EMW shows that under these conditions, the main factor affecting the AP antenna is the variance of phase fluctuations. In a variety of mobile robot navigation conditions, this variance may vary. Let's choose a single factor that affects the variance level in any navigation environment. This factor is the measurement

Figure 14. Dependence of the APSA width on the variance of phase fluctuations at high their correlation within the antenna length

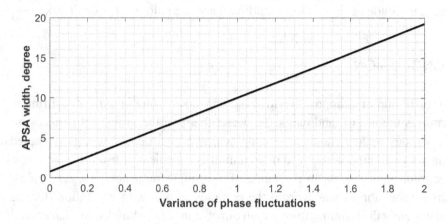

error of the antenna phase and the radar receiver. The magnitude of these errors is precisely determined by the level of variance of phase fluctuations. Let's determine the allowable level of variance, in which the width of the APSA will increase no more than μ times. To do this, we construct a graph of the dependence of the antenna width in degrees on the variance of phase fluctuations, using the formula (24). This graph is represented in Figure 14.

The linear dependence in Figure 14 was obtained by processing the results of calculating the width of the AP by the method of least squares. As it follows from the figure 14, the increase of phase fluctuations in 2 times leads to the expansion of the AP width approximately by twice, that is, the decrease of the angular resolution at the same number of times. This is an approximate estimate, since in case of a large variance of phase fluctuations, the AP in general "collapses" and robot's radar does not have the need angular resolution ability if to speak correctly. By setting the permissible level of the AP width, it is easy to obtain an acceptable level of variance of phase fluctuations. All results are obtained for the width of the board antenna of the robot equal to 30°. For different widths, they differ, which should be considered during process of designing an antenna with SA.

FUTURE RESEARCH DIRECTIONS

All existing approaches to navigating mobile autonomous robots have their advantages and disadvantages that do not always overlap. This is the basis for the simultaneous use of different approaches at the same time, which is relevant in the future with the growth of robot speeds. It is also possible to synthesize several measuring channels of the robot considering the influence of the internal and external factors. For example, such channels may include channels for detecting and measuring the jump amplitude of signals from landmark and the color of this landmark.

CONCLUSION

One of the important approaches to navigating mobile autonomous robots in unknown terrain is the use of landmarks. To detect them on the ground, it is necessary to use various remote methods, among which a significant place is occupied by radars. Depending on the type of terrain, it is expedient to use different ranges of electromagnetic waves. In the range of very high frequencies (meter range), the mirror reflection of waves from the complex terrain with vegetation dominates, and wavelength is commensurate with dimensions of landmarks. In such conditions resonance scattering of waves on the landmark is possible, which increases the amplitude of the echo signals. The width of the small antenna pattern in the range of very high frequencies is large, which reduces the accuracy of the measurement of the landmark azimuth and eliminates the possibility of resolving two near-placed landmarks. The use of a synthesized aperture antenna, created during the robot's movement, solves these problems. The simulation results indicate the possibility of implementation such synthesized apertures into practice in areas of the straightforward movement of mobile robots.

This research received no specific grant from any funding agency in the public, commercial, or not-for-profit sectors.

REFERENCES

Ali, F., Urban, A., & Vossiek, M. A. (2011). Short Range Synthetic Aperture Imaging Radar with Rotating Antenna. *INTL Journal of Electronics and Telecommunications*, *57*(1), 97–102. doi:10.2478/v10177-011-0014-y

Andersen, J. C., Blas, M. R., Ravn, O., Andersen, N. A., & Blanke, M. (2006). Traversable terrain classification for outdoor autonomous robots using single 2D laser scans. *Integrated Computer-Aided Engineering*, *13*(3), 222–232. doi:10.3233/ICA-2006-13303

Borenstein, J., Everett, H. R., Feng, L., & Wehe, D. (1997). Mobile Robot Positioning: Sensors and Techniques. *Journal of Robotic Systems*, *14*(4), 231–249. doi:10.1002/(SICI)1097-4563(199704)14:4<231::AID-ROB2>3.0.CO;2-R

Budiharto, W., Santoso, A., Purwanto, D., & Jazidie, A. (2011). A Navigation System for Service Robot Using Stereo Vision and Kalman Filtering. *Proceedings of 11th International Conference on Control, Automation and Systems*, 1771-1776.

Chan, Y. K., & Koo, V. C. (2008). An introduction to synthetic aperture radar (SAR). *Progress in Electromagnetics Research*, *2*, 27–60. doi:10.2528/PIERB07110101

Corany, B., Karanam, C. R., & Mostofi, Y. (2018). Adaptive Near-Field Imaging with Robotic Arrays. *IEEE 10ᵗʰ Sensor Array and Multichannel Signal Processing Workshop (SAM)*, 1-5.

Dhanasingaraja R., Kalaimagal S. & Muralidharan G. (2014). Autonomous Vehicle Navigation and Mapping System. *International Journal of Innovative Research in Science, Engineering and Technology*, *3*(3), 1347-1350.

Ferreira,, A., Pereira,, F. G., Vassalo,, R. F., & Teodiano,, F., Bastos, & Sarcinelli, F. M. (2008). An approach to avoid obstacles in mobile robot navigation: The tangential escape. *Revista Controle & Automação*, *19*(4), 395–405.

Howard, A., Matarić, M. J., & Sukhatme, G. S. (2003). Cooperative relative localization for mobile robot teams: an egocentric approach. *Proceedings of the Naval Research Laboratory Workshop on Multirobot Systems*, 1-12.

Jean, B. R., & Rouse, J. W. (1983). A Multiple Beam Synthetic Aperture Radar Design Concept for Geoscience Applications. *IEEE Transactions on Geoscience and Remote Sensing*, *GE-21*(2), 201–207. doi:10.1109/TGRS.1983.350489

Jin, T.-S., Lee, J.-M., & Hashimoto, H. (2006). Position Control of Mobile Robot for Human-Following in Intelligent Space with Distributed Sensors. *International Journal of Control, Automation, and Systems*, *4*(2), 204–216.

Kidera, S. (2010). Shadow region imaging algorithm using array antenna based on aperture synthesis of multiple scattered waves for UWB radars. *Proceedings of International Geoscience and Remote Sensing Symposium*, 1-4. 10.1109/IGARSS.2010.5651782

Lee, S., & Song, J.-B. (2004). Mobile Robot Localization Using Optical Flow Sensors. *International Journal of Control, Automation, and Systems*, *2*(4), 485–493.

Li, I.-H., Chen, M.-C., Wang, W.-Y., Su, S.-F., & Lai, T.-W. (2014). Mobile Robot Self-Localization System Using Single Webcam Distance Measurement Technology in Indoor Environments. *Sensors (Basel)*, *14*(2), 2089–2109. doi:10.3390140202089 PMID:24473282

Lukin, K. A., & .... (2013). Tomographic imaging using noise radar and 2D aperture synthesis. *Applied Radio Electronics*, *12*(1), 152–156.

Melo, L. F., Rosário, J. R., & Silveira, A. F. (2013). Mobile Robot Indoor Autonomous Navigation with Position Estimation Using RF Signal Triangulation. *Positioning*, *4*(01), 20–35. doi:10.4236/pos.2013.41004

Mester, G., & Rodić, A. (2010). Sensor-Based Intelligent Mobile Robot Navigation in Unknown Environments. *International Journal of Electrical and Computer Engineering Systems*, *1*(2), 1–8.

Ort, T., Paull, L., & Rus, D. (2018). *Autonomous Vehicle Navigation in Rural Environments without Detailed Prior Maps*. MIT CSAIL. doi:10.1109/ICRA.2018.8460519

Poliarus, O., Poliakov, Y., & Lindner, L. (2018). Determination of landmarks by mobile robot's vision system based on detecting abrupt changes of echo signals parameters. *The 44th Annual Conference of the IEEE Industrial Electronics Society*, 3165-3170.

Popirlan, C., & Dupac, M. (2009). An Optimal Path Algorithm for Autonomous Searching Robots. *Annals of University of Craiova. Math. Comp. Sci. Ser.*, *36*(1), 37–48.

Rajyalakshmi, P., & Raju, G. S. N. (2011). Characteristics of Radar Cross Section with Different Objects. *International Journal of Electronics and Communication Engineering*, *4*(2), 205-216.

Sergiyenko, O. Y. (2016). Data transferring model determination in robotic group. *Robotics and Autonomous Systems*, 1–10.

Shifrin, J. S. (1971). Statistical antenna theory. Golem Press.

Surrécio, A., Nunes, U., & Araújo, R. (2005). *Using Kalman Filters and Augmented System Models for Mobile Robot Navigation*. Dubrovnic, Croatia: IEEE ISIE.

Vaz, D., Serralherio, A., & Gerald, J. (n. d.). *Navigation System for a Mobile Robot Using Kalman Filters*. Academic Press.

Chapter 9
Image Compression Technique Based on Some Principal Components Periodicity

Wilmar Hernandez
iD https://orcid.org/0000-0003-4643-8377
Universidad de Las Americas, Ecuador

Alfredo Mendez
Universidad Politecnica de Madrid, Spain

ABSTRACT

In this chapter, the almost periodicity of the first principal components is used to carry out the reconstruction of images. First, the principal component analysis technique is applied to an image. Then, the periodicity of its principal components is analyzed. Next, this periodicity is used to build periodic vectors of the same length of the original principal components, and finally, these periodic vectors are used to reconstruct the original image. The proposed method was compared against the JPEG (Joint Photographic Experts Group) compression technique. The mean square error and peak signal-to-noise ratio were used to perform the above-mentioned comparison. The experimental results showed that the proposed method performed better than JPEG, when the original image was reconstructed using the principal components modified by periodicity.

DOI: 10.4018/978-1-5225-9924-1.ch009

INTRODUCTION

In order to represent information, different data sets are available. Therefore, sometimes the information through redundant data sets is represented. Image compression methods are procedures that are used to reduce redundant data and represent these images profitably (Gonzalez et al. 2008).

This type of methods is having a great growth and a great reception for several years, and the perspectives of the development of digital communications indicate that image compression is a field of study with a great projection. As an example, some data extracted from (Gonzalez et al. 2008, p. 525) are shown below:

"Think about the volume of data that is necessary to store a two-hours long standard-definition television movie, by using arrays of 720×480×24 bit pixel arrays. Due to the fact that the digital movie is a sequence of video frames, in which a frame is a photo in full color and 30 frames per second are used, 2.24×10^{11} bytes are required to store the digital movie, which approximately is equal to 224 GB (gigabytes) of data. Therefore, it is necessary to use 27 dual-layer DVDs of 8.5 GB to store the movie. In order to have a two-hour movie on a single DVD, the user must compress each frame by a factor of 26.3, on average" (Gonzalez et al. 2008, p. 525).

Image compression methods are also used in the following cases: digital cameras, high-definition television, medical images that are reconstructed in order to help doctors and researchers interpret diagnosis of rare illnesses and improve it, surveillance systems based on video processing techniques, traffic control in highways and inside cities, and interpretation of satellite images, among other applications.

Due to the large number of people who use communication devices today, large volumes of images are shared over the internet. To store these images, which often have high resolution, large amounts of bits are needed. In addition, the transmission of said images is done through networks that have limited bandwidth. What has been said above, entails excessive bandwidth consumption and justifies the fact that it is necessary to have procedures to compress images using few bits. Therefore, it is necessary to have procedures for compressing and transmitting images through the network in a fast and efficient way.

The image compression process is based on reducing the amount of data that is necessary to be able to represent it. These compression processes eliminate data that does not provide relevant information about the content of the image and cause losses in terms of their visual quality. However, this loss due to image degradation is not significant compared to the decrease in the size of the file that contains the image, while this is tolerable.

When there was no internet, people worked with high resolution and carrying out the compression of the files was only a requirement to take into account to carry out the transfer of data from one site to another. At that time, the storage media did not have great capacity. With the widespread use of the Internet, in 1986 a committee of experts, called the Joint Photographic Experts Group, set to work to create a standard procedure for compression and coding of images, and the JPEG format emerged in 1992 (JPEG, 2019).

The JPEG format is based on the DCT (discrete cosine transform) (CCIT, 1992) and is a format, in general, with losses, which means that each and every pixel that forms the bitmap is not saved. When the compressed image is reopened, the deleted pixels are plotted based on their resemblance to the surrounding pixels. This procedure supports different levels of compression; that is to say, it carries a very high quality, if little is compressed, and instead the quality decreases if it is highly compressed.

Due to the fact that compression always implies the loss of information, if an image that is stored with the JPEG format is opened and then saved again with this format, after having performed this operation several times it will be observed that said image will be degraded (Lifewire, 2019).

In (Hernandez et al., 2018), a compression technique based on the almost periodicity of the first principal components is presented. Also, in (Hernandez et al., 2019) this technique is explained in detail. However, a comparison of this technique with the JPEG format has not yet been made. Therefore, it is important to have scientific literature that demonstrates how this new compression technique, introduced by the authors, has benefits that under certain criteria is better than JPEG.

In order to avoid redundancies in the explanation of the above-mentioned technique and to try not to repeat words already said in (Hernandez et al., 2018) and (Hernandez et al., 2019), this chapter is focused on using the image compression technique based on taking advantage of the almost periodicity of the first principal components - to carry out the compression of a particular image - and to perform the comparison of said compression with that performed by using the JPEG technique.

The chapter is structured as follows: A summary of the JPEG compression method is given in Section II. The methodology used to perform image compression by using the almost periodicity of the first principal components is explained in Section III. Comparisons between both compression methods for different cases are performed in Sections IV and V. And, finally, the conclusion is given in Section VI.

JPEG Compression

In the late 80s of the twentieth century, an expert committee was conceived to create an image compression and coding standard, Joint Photographic Experts Group, which resulted in the acronym JPEG, its website is http://www.jpeg.org.

The procedure that follows is flexible enough so that most images can be treated. In this chapter, although the procedure is suitable for color images, only grayscale images will be analyzed where each pixel, representing the brightness, is written with 8 bits of information (that is, 0 to 255 different values) and, in accordance with (Martin-Marcos, 1999; Khedr et al., 2016; Raid et al., 2014), is structured as follows:

1. First, the image is subdivided into blocks of 8×8 non-overlapping pixels.
2. Because the DCT concentrates the energy in a few coefficients, then 128 is subtracted so that many values close to zero come out and thus the values are in the range [-128,127].
3. The DCT is applied to each block obtaining blocks of coefficients of 8×8.
4. These coefficients are quantified, that is, many similar coefficients are assigned the same value. This is done with a quantization matrix, Q, which is added to the header of the file so that the image can be decompressed later.
5. Then, these values are encoded, with different methods, to encode very frequent values with few bits.
6. A file is constructed where, in the headers, the quantization matrix and the coding table are included.
7. With the transmitted headers, the file is decoded obtaining quantified indexes.
8. The discrete inverse cosine transform (IDCT) is applied to the indexes and 128 is added to obtain the reconstructed image.

In this process, the discrete cosine transform is a transformation that has the inverse version that returns the original data and the coding, which is used to represent some data in one way or another. Compression with losses is reflected when using quantization matrices; that is, when unifying coefficients with similar values and when performing the inverse process, which consists in assigning a coefficient to the index considered.

To analyze the results, three families of quantization matrices have been used in this chapter:

-   JPEG standard quantization matrix (Khedr et al., 2016):

$$Q_{50} = \begin{bmatrix} 16 & 11 & 10 & 16 & 24 & 40 & 51 & 61 \\ 12 & 12 & 14 & 19 & 26 & 58 & 60 & 55 \\ 14 & 13 & 16 & 24 & 40 & 57 & 69 & 56 \\ 14 & 17 & 22 & 29 & 51 & 87 & 80 & 62 \\ 18 & 22 & 37 & 56 & 68 & 109 & 103 & 77 \\ 24 & 35 & 55 & 64 & 81 & 104 & 113 & 92 \\ 49 & 64 & 78 & 87 & 103 & 121 & 120 & 101 \\ 72 & 92 & 95 & 98 & 112 & 100 & 103 & 99 \end{bmatrix} \tag{1}$$

For $0 < n < 100$,

$$Q_n = \begin{cases} \left(\dfrac{100-n}{50}\right)Q_{50}, & n > 50 \\[4mm] \left(\dfrac{50}{n}\right)Q_{50}, & n < 50 \end{cases} \tag{2}$$

- Parametric quantization matrices (Kornblum, 2008) defined from Q_n:

$$S(n) = \begin{cases} \left(\dfrac{5000}{n}\right), & n < 50 \\[4mm] 200 - 2n, & n > 50 \end{cases} \tag{3}$$

$$T_S[i,j] = \left[\dfrac{S \cdot Q_{50} + 50}{100}\right] \tag{4}$$

where $n \in [1,100]$ is the quality factor, S is the scale factor, and $[\bullet]$ is the floor function.

- In addition, the following family has been included (Salomon et al., 2010):

$$R_S[i,j] = 1 + (i+j) \cdot (100 - S) \tag{5}$$

where S is a quality parameter.

In the above-mentioned three families, increasing the quality parameter increases the image quality. In addition, as the peak signal-to-noise ratio (PSNR) increases, the mean square error (MSE) and compression rate decrease. The PSNR, the MSE and the compression rate at 64 equally spaced points of the interval [1,100] for the Barbara image are shown below. The Barbara image is shown in Figure 1, the PSNR of the JPEG compression applied to the Barbara image is shown in Figure 2, the MSE of the JPEG compression applied to the Barbara image is shown in Figure 3, and the compression rate of the JPEG compression applied to the Barbara image is shown in Figure 4.

As it can be seen, the graphs corresponding to the first two families related to the quantization matrices produce very similar results. In fact, the curves that appear in the graphs overlap from a low value. In addition, the results for these families

Figure 1. Gray image considered: Barbara

Figure 2. PSNR of the JPEG compression: Barbara

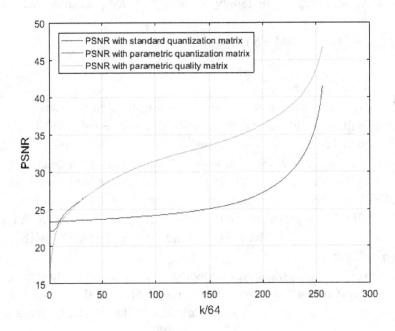

Figure 3. MSE of the JPEG compression: Barbara

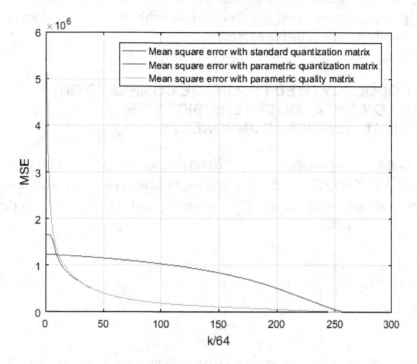

Figure 4. Compression rate of the JPEG compression: Barbara

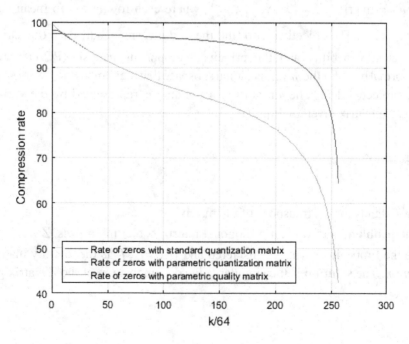

of matrices are much better, because the PSNR grows rapidly, the MSE decreases sharply, and the compression rate decreases smoothly, to produce a steep decline as the quality parameter increases greatly.

METHODOLOGY USED FOR IMAGE COMPRESSION BASED ON THE ALMOST PERIODICITY OF THE FIRST PRINCIPAL COMPONENTS

In accordance with (Hernandez et al., 2018; Hernandez et al., 2019), the gray scale version of size 512×512 (i.e., $2^9 \times 2^9$) of an image will be considered. Next, the image is partitioned into blocks of size $2^h \times 2^h = k \times k$, A_{ij}, and $2^{9-h} \cdot 2^{9-h} = l^2$ blocks are obtained (See Eq. (6)).

$$\Theta = \begin{bmatrix} A_{1,1} & \cdots & A_{1,l} \\ \vdots & \ddots & \vdots \\ A_{l,1} & \cdots & A_{l,l} \end{bmatrix} \tag{6}$$

Now, each matrix $A_{i,j}$ is stored in a data vector of dimension k^2, $a_{(i-1)l+j} = vec\left(A_{ij}\right)$, $i,j=1,\ldots,l$, which contains the elements of the matrix by rows. The data is arranged in an $l^2 \times k^2$ matrix, $X = \left(x_1; x_2; \ldots; x_{l^2}\right)$, where each row is a zero mean vector, $x_p = a_p - \overline{a_p}$. The objective is to find the m-dimensional subspace that captures the greatest variability and the principal component analysis (PCA) finds an orthonormal basis for the m-dimensional subspace that maximizes the sum of squares of the projected data. The variability of the data is represented by the variance/covariance matrix given by Eq. (7)

$$Z = \frac{1}{l^2 - 1} X^T X \tag{7}$$

where T stands for the transpose of the matrix.

The **problem** is to find an orthogonal matrix U that minimizes $\left\| Z - ZUU^T \right\|_F^2$, where the Frobenius norm is considered and $U^T U = I_m$, I_m is the identity matrix of size $m \times m$. The **solution** is the singular value decomposition of the Z matrix given by Eq. (7).

The k^2 pairs of eigenvalues and eigenvectors, (λ_i, e_i), are ordered according to the eigenvalues from highest to lowest. Then, the spectral decomposition of Z (Eq. (7)) is given by

$$Z = P \rangle P^T = \lambda_1 e_1 e_1^T + \ldots + \lambda_{k^2} e_{k^2} e_{k^2}^T \tag{8}$$

where Λ is a diagonal matrix that consists of the eigenvalues of Z, $\left(\lambda_1, \ldots, \lambda_{k^2}\right)$, and $P = \left[e_1, \ldots, e_{k^2} \right]$.

Therefore, the k^2 principal components given by Eq. (9)

$$y_j = e_j^T x_q = e_{1,j} x_1 + \ldots + e_{k^2,j} x_{k^2} \tag{9}$$

with $j=1,\ldots,k^2$ and x_q (for $q=1,\ldots,k^2$) the q-column of the above-defined X matrix, have been built and consequently the orthonormal basis $B' = \left\{ e_1, \ldots, e_{k^2} \right\} \subset \mathbb{R}^{k^2}$ is obtained.

Given a vector v with coordinates $\left(x_1, \ldots, x_{k^2} \right)$ with respect to the canonical base, and with coordinates $\left(y_1, \ldots, y_{k^2} \right)$ with respect to the base B', the relation between these coordinates is given by Eq. (10)

$$\left(x_1, \ldots, x_{k^2} \right)^T = P \left(y_1, \ldots, y_{k^2} \right)^T \tag{10}$$

In addition, due to the fact that P (Eq. (10)) is orthogonal, Eq. (10) can also be given by Eq.(11)

$$\left(y_1, \ldots, y_{k^2} \right) = \left(x_1, \ldots, x_{k^2} \right) P \tag{11}$$

As a result, the coordinates of the k^2 vectors that form the observation matrix have $y=xP$ as coordinates. If the B' vectors are kept, the data matrix can be reconstructed, because $y=xP$ implies that $x=yP^{-1}=yP^T$.

In order to compress the image, it has to be considered that the first vectors of the base B' are going to be used, each of them with k^2 components, but that only m, $m<k^2$, vectors are going to be used. Then, the T_m matrix of size $k^2 \times k^2$ given by Eq. (12) is defined.

$$T_m = \begin{bmatrix} I_m & 0 \\ 0 & 0 \end{bmatrix} \tag{12}$$

where I_m is the identity matrix. Therefore, the dimension of $y_m = yT_m = xPT_m = xU$ is $l^2 \times k^2$, and its last (k^2-m) columns are all equal to zero. So, the dimension $l^2 \times k^2$ has been reduced to $l^2 \times m$, and the other part has been filled with zeros.

The compressed image is obtained by Eq. (13)

$$b_p = a_p PT_m P^T + \overline{a_p} \tag{13}$$

with $p=1,\dots,l^2$, and the reconstruction is performed by reordering each of the vectors in the $k \times k$ matrix $B_{ij} = matrix\left(b_{(i-1)l+j} \right)$.

Finally, the compressed image matrix given by Eq. (14) is built. The dimension of Θ_c is 512×512, and it consists of $k \times k$ dimensional blocks, each of them being a B_{ij} matrix.

$$\Theta_c = \begin{bmatrix} B_{1,1} & \cdots & B_{1,l} \\ \vdots & \ddots & \vdots \\ B_{l,1} & \cdots & B_{l,l} \end{bmatrix} \tag{14}$$

At this point, it is important to highlight the dependency of a pixel in with respect to its neighbors. It can be observed that when there are $k \times k$-dimensional submatrices, stored by rows in vectors, the value of one pixel of one submatrix is quite similar to the value of another pixel that is separated of this by k elements of the vector that has been built. In short, there is a periodic behavior in that the first k pixels are adjacent to the next k pixels, and so on, up to k times. Consequently, most of the vectors are close to be periodic of period k. Since the first principal components (PC) collect a large part of the characteristics of the vectors, it is plausible that they also reflect the periodicity of the vectors.

Therefore, if the first PC are replaced with their periodic version, the original image can be reconstructed. In order to do this, the first PC are replaced with periodic series with tendency (Box et al., 2016). First, the trend is eliminated using low-degree polynomials. Then, these k^2-dimensional vectors are replaced with others whose components are periodic. Next, the trend is added to the periodic vector. Finally, it has already been replaced the first PC with their periodic versions.

Comparison Between Compression by Principal Components Modified by Periodicity and JPEG Compression: 8×8 Case

This section is aimed at performing a comparative analysis between the two compression methods that have been analyzed. That is, the results obtained by compressing the image using the JPEG method and the results obtained by using periodic functions constructed from the first PC.

To perform the comparisons, three methods will be used. Specifically, the MSE and PSNR will be used to measure image quality in terms of the differences between the modified and the original image. Then, the third method will carry out the comparison in terms of the number of zeros that contain the modified image.

Assuming that the pixels of the original image are $\{x_n, n=1,\ldots,N\}$, where N the number of rows multiplied by the number of columns of the image, the pixels of the modified image are $\{y_n, n=1,\ldots,N\}$. Then, the MSE is given by Eq. (15)

$$MSE = \frac{1}{N}\sum_{i=1}^{N}\left(x_n - y_n\right)^2 \tag{15}$$

In order to make the subjective assessment of the intensity value independent, a measure is established based on comparing the error with the maximum possible value. This value is the PSNR and is given by (16)

$$PSNR = 10log_{10}\left(\frac{\left(2^8-1\right)^2}{MSE}\right) = 10log_{10}\left(\frac{255^2}{MSE}\right) = 10log_{10}\left(\frac{255^2 \cdot N}{\sum_{i=1}^{N}\left(x_n - y_n\right)^2}\right) \tag{16}$$

To measure the compression rate in images modified with JPEG, the zero rate after dequantization has been considered. On the other hand, to measure the compression rate in images modified by periodized PC, with the first PC modified by periodicity, the percentage of zeros and PC that are not used has been considered.

The following figures show the PSNR of the decomposition applying PCA, the reconstructions using PCA but modifying from the first to the sixth main component and keeping the others intact, the values using the JPEG method with the type-II discrete cosine transform, and the families of quantization matrices given by Eqs. (2), (4) and (5). Figure 5 shows the PSNR, Figure 6 shows the MSE, and the compression rate is shown in Figure 7.

Figure 5. PSNR of the compressions: Barbara

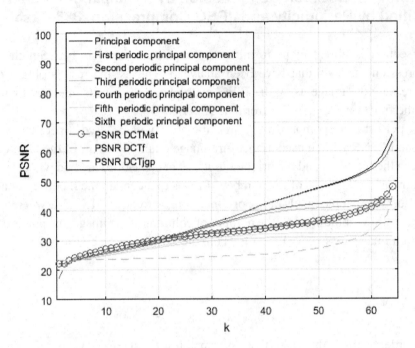

Figure 6. MSE of the compressions: Barbara

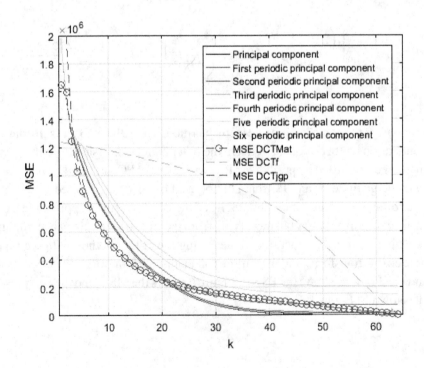

Figure 7. Compression rate of the compressions: Barbara

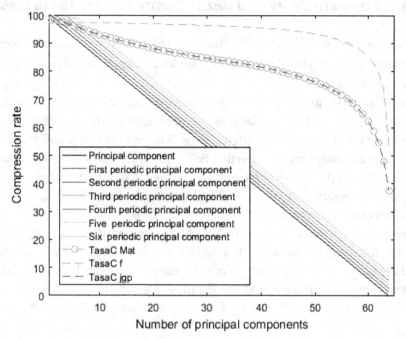

From Figure 5, it can be seen that the PSNR for JPEG compressions using the matrices given by Eqs. (2) and (4), and when the first four PC are replaced by their periodic versions, are analogous until more than 20 PC are used. After this, the behavior is better in reconstructions using four PC modified by periodicity and keeping the other intact.

Something similar happens with the MSE (see Figure 6). In this case, there is good behavior using JPEG with the families of matrices given by Eqs. (2) and (4), and when the first three PC are replaced by their periodic version and the remaining ones are kept intact, up to more than 20. From that point, compressions that are made by using PCA with the first three PC replaced by their periodic versions behave better.

However, from Figure 7 it can be seen that when more than 2 PC modified by periodicity are used, the zero rate behavior is worse for PCA reconstructions with the first PC replaced by periodic functions, which is linear, than for those reconstructions performed by JPEG.

Comparison Between Compression by Principal Components Modified by Periodicity and JPEG Compression: 16×16 Case

If the images are of size 512×512 and blocks of size 8×8 are made, then 4096 vectors are obtained in $\mathbb{R}^{64}$. Therefore, if working with blocks of size 16×16, then 1024 vectors are obtained in $\mathbb{R}^{256}$.

The following figures show the PSNR of the reconstructions using PCA, modifying from the first to the sixth PC and keeping the others intact. In addition, the values are shown using the JPEG method with the Type-II discrete cosine transform, and the families of quantization matrices given by Eqs. (2) and (4) considering reconstructions with 16 vectors of the complete base of 256, where the first PC have been modified by periodicity. Figure 8 shows the PSNR, Figure 9 shows the MSR and Figure 10 shows the compression rate of the compressions, respectively, using 16 PC.

The following figures show the PSNR of the decomposition applying PCA, the reconstructions using PCA but modifying from the first to the sixth PC and keeping the others intact. In addition, the values are shown using the JPEG method with the Type-II cosine transform, and the families of quantization matrices given by Eqs. (2), (4), and (5) are also shown. It is made with 16 x 16 blocks. The PSNR is

Figure 8. PSNR of the compressions using 16 PC: Barbara

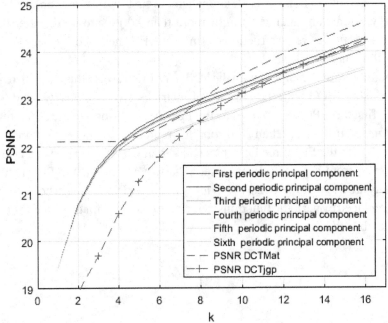

Figure 9. MSE of the compressions using 16 PC: Barbara

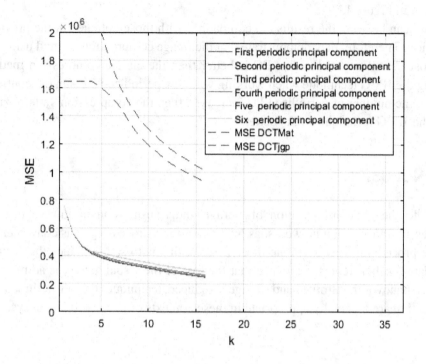

Figure 10. Compression rate of the compressions using 16 PC: Barbara

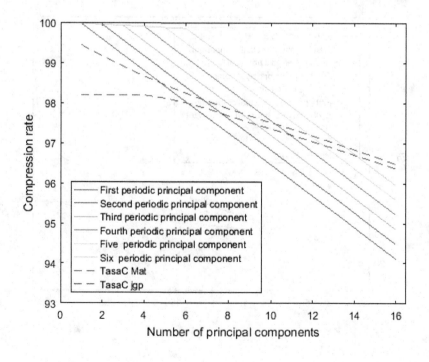

shown in Figure 11, the MSE is shown in Figure 12, and the compression rate is shown in Figure 13.

As can be seen, the results are compatible with those obtained in the previous section. Therefore, they demonstrate that the image compression method using the almost periodicity of the first PC is better than the classic compression method using the JPEG technique when working with few periodized PC. On the contrary, when the number of periodized PC increases, then the compression system based on the JPEG technique works better.

CONCLUSION

Due to the great proliferation of storage and transmission of images through different transmission networks, as well as for their processing, it is convenient to have procedures to concentrate those images, that is to say to reduce the number of data for their representation so that the loss of the visual quality is admissible. When bandwidth is limited and storage is reduced, techniques that allow dimension reduction are especially appropriate in these contexts of integrated vision systems.

Figure 11. PSNR of the compressions: Barbara

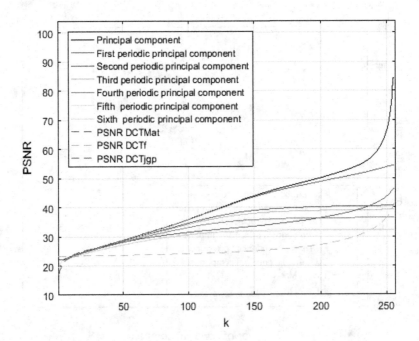

Figure 12. MSE of the compressions: Barbara

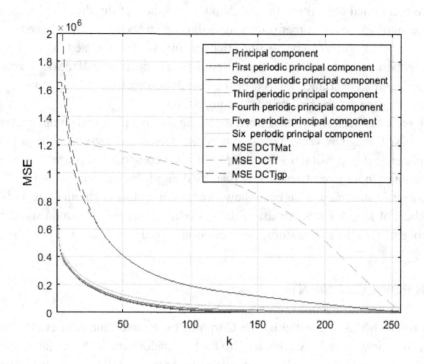

Figure 13. Compression rate of the compressions: Barbara

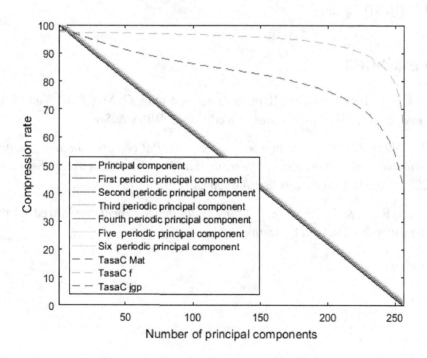

The analysis of principal components has been a technique widely used for the above-mentioned objectives. In this chapter, a method of dividing the image into squares and to decompose them into principal components has been explained. Given a pixel of an image, it can be assumed that this pixel is quite related to the pixels of its neighborhood, which implies that the ellipsoid concentration of the observations, defined by the covariance matrix, or the correlation matrix, is lengthened by the directions given by the first principal components.

Here, it has been shown that when replacing the first principal components of a compressed image by vectors consisting of the periodized version of the principal components, it is possible to carry out the image compression process of images under study in a more efficient way than by doing it by using the classical JPEG technique. In summary, it has been shown that when vectors of length k^2 are replaced by others of length k that consist of the periodized version of them, a significant improvement in the image storage process is achieved.

ACKNOWLEDGMENT

This research was supported by the Corporacion Ecuatoriana para el Desarrollo de la Investigacion y la Academia (CEDIA), Ecuador [under the research project CEPRA XII-2018-13)]; the Universidad de Las Americas (UDLA), Ecuador [under the research Project ERa.ERI.WHP.18.01], and the Universidad Politecnica de Madrid (UPM), Spain.

REFERENCES

Box, G. E. P., Jenkins, G. M., Reinsel, G. C., & Ljung, G. M. (2016). *Time Series Analysis: Forecasting and Control* (5th ed.). John Wiley & Sons.

CCIT. (1992). *T.81 - information technology – digital compression and coding of continuous – tone still images – requirements and guidelines*. Retrieved September 22, 2019, from https://www.w3.org/Graphics/JPEG/itu-t81.pdf

Gonzalez, R. C., & Woods, R. E. (2008). *Digital Image Processing* (3rd ed.). Upper Saddle River, NJ: Pearson Education.

Hernandez, W., & Mendez, A. (2018). Application of Principal Components to Image Compression. In T. Göksel (Ed.), *Statistics - Growing Data Sets and Growing Demand for Statistics* (pp. 107-136). IntechOpen. Retrieved September 22, 2019, from https://www.intechopen.com/books/statistics-growing-data-sets-and-growing-demand-for-statistics/application-of-principal-component-analysis-to-image-compression

Hernandez, W., Mendez, A., Quezada-Sarmiento, P. A., Jumbo-Flores, L. A., Mercorelli, P., Tyrsa, V., . . . Cevallos Cevallos, W. B. (2019, October). *Image compression based on periodic principal components.* Paper presented at the 45th Annual Conference of the IEEE Industrial Electronics Society, Lisbon, Portugal.

JPEG. (2019). *Overview of JPEG.* Retrieved September 22, 2019, from https://jpeg.org/jpeg/index.html

Khedr, W. M., & Abdelrazek, M. (2016). Image Compression Using DCT Upon Various Quantization. *International Journal of Computers and Applications, 137*(1), 11–13. doi:10.5120/ijca2016908648

Kornblum, J. D. (2008). Using JPEG Quantization Tables to Identify Imagery Processed by Software. *Digital Investigation, 5,* S21-S25.

Lifewire. (2019). *JPEG File Format Myths and Facts.* Retrieved September 24, 2019, from https://www.lifewire.com/jpeg-myths-and-facts-1701548

Martin-Marcos, A. (1999). *Compresión de Imágenes: Norma JPEG.* Madrid: Editorial Ciencia 3 S.L.

Raid, A. M., Khedr, W. M., El-dosuky, M. A., & Ahmed, W. (2014). Jpeg Image Compression Image Compression Using Discrete-Cosine Transform – A Survey. *International Journal of Computer Science & Engineering Survey, 5*(2), 39–47. doi:10.5121/ijcses.2014.5204

Salomon, D., & Motta, G. (2010). *Handbook of Data Compression* (5th ed.). London: Springer-Verlag. doi:10.1007/978-1-84882-903-9

Chapter 10
Obstacle Classification Based on Laser Scanner for Intelligent Vehicle Systems

Danilo Caceres Hernandez
Universidad Tecnológica de Panamá, Panama

Laksono Kurnianggoro
University of Ulsan, South Korea

Alexander Filonenko
ABBYY, Russia

Kang-Hyun Jo
University of Ulsan, South Korea

ABSTRACT

In the field of advanced driver-assistance and autonomous vehicle systems, understanding the surrounding vehicles plays a vital role to ensure a robust and safe navigation. To solve detection and classification problem, an obstacle classification strategy based on laser sensor is presented. Objects are classified according the geometry, distance range, reflectance, and disorder of each of the detected object. In order to define the best number of features that allows the algorithm to classify these objects, a feature analysis is performed. To do this, the set of features were divided into four groups based on the characteristic, distance, reflectance, and the entropy of the object. Finally, the classification task is performed using the support vector machines (SVM) and adaptive boosting (AdaBoost) algorithms. The evaluation indicates that the method proposes a feasible solution for intelligent vehicle applications, achieving a detection rate of 87.96% at 48.32 ms for the SVM and 98.19% at 79.18ms for the AdaBoost.

DOI: 10.4018/978-1-5225-9924-1.ch010

INTRODUCTION

A combination of environmental issues and demand for safety factors influence the automotive industry. The automotive industry has currently shown robust growth which has been reflected in the increase of access to data and information resulting in development of sensor solutions. Advance driver assistance systems (ADAS) and autonomous ground vehicle navigation (AGVN) are still facing important challenges in the field of robotics and automation. Essentially due to the uncertain nature of the environments, moving obstacles, and sensor fusion accuracy. In that sense, the systems must be able to recognize features in which the method should be able to detected lanes, objects on the road, e.g vehicle, pedestrian or animal (Zhang, F. et al.2016; Ibarra-Arenado et al. 2017; Wang, T et al. 2003). Zhang, F et al. proposed the use of LiDAR to develop a vehicle detection method. The authors presented a probability hypothesis density filter which is a multiple-target filter and for the case of hypothesis verification the authors used SVM. To evaluate their proposed idea the authors used the KITTI dataset (Geiger, A et al. 2013). Although the result shows a good performance, there are still issues with processing time due to the 3D Velodyne point clouds. Authors in (Ibarra-Arenado et al. 2017) proposed a vision-based vehicle detection method, which is comprised of two steps: hypotheses generation (HG) and hypotheses verification (HV). The authors are focuses on the HG strategy. In HG vehicles are localized by using a shadow-based vehicle detection method, distinguishing the difference on intensity between the shadow under the vehicle and road surface. As a results this method overcome the problems given by lateral shadows, asphalt stains and traffic markings on the road. However, the method still encounters problems of false positives rate mostly due to the ego-vehicle trajectory. Wang, T. et al. (2003) presented a radar-vision fusion based strategy. The idea focuses on the radar target detection of candidate objects. Once the candidate is detected, the coordinate information is used to define the searching are within the image where the object is located. The idea is to design a robust object detection; however, in order to be used in open road scenarios the ongoing idea need to be improved. Current research efforts in the field of obstacle detection and recognition strive to understand and improve the problem of features extraction stage by leaping between different sensors and strategies. Regarding the above paragraph, the most utilized sensors for achieving ADAS and AGVN using feature-based strategy (color, edges, shape, geometric) are commonly cameras and laser.

Therefore, the use of laser sensor is increasing in the automotive industry due to high accuracy and precision of measurement, and the cost reduction. Laser-based vehicles application is becoming relevant in the automotive industry, examples include detection (Premebida, C. et al. 2007; Hsu, C.W. et al. 2012; Liu, Z. et al. 2013), and classification (Zewei, X. et al. 2015; Lehtomäki, M. et al. 2015; Wang,

H., et al. 2014). Premebida, C. et al. propose a laser-vision based application for pedestrian and vehicle detection. In case of the laser, the authors use geometrical information; a total of five features were used in total to perform the classification task using Gaussian mixture model classifier, while in the vision task the AdaBoost classifier was proposed. The vehicle classification rate was 84%, while the pedestrian classification was 82.9% at approximately 13 frames per second, using AdaBoost processing performance as a reference since the authors do not provide information related to the processing time. Hsu, C.W. et al. 2012 proposed a car-following system based on laser. The laser is used to detect the vehicles; the main feature is the vehicle width (in a range of [1.45 -1.97 m]) using random sample consensus algorithm, while the camera is used for verification. The proposed car-following idea provides good results. Although the algorithm was verified with a speed range of 15 to 35 km/h, the authors do not provide detailed information about the feature detection as well as processing time. Lehtomäki, M. et al. 2015, and Wang, H., et al. 2014 use laser scanning point cloud data to deal with the task of classification. Authors in Zewei, X. et al. 2007 use geometrical properties such as the top-length ratio, top-height ratio, front-rear ratio, and variance of the top height. To classify the vehicle, authors use the genetic algorithm-back-propagation as the classifier. The point distribution to classify objects such as trees, lamp recognition was over 90% for the group of vehicles (saloon, passenger, truck). The authors do not provide information about the processing time. Lehtomäki, M. et al. 2015 propose an object classification and recognition using a mobile laser scanning point posts, and traffic clouds. The authors propose to use three sets of features, the local descriptor histograms, spin images, and general shape and signs, cars, pedestrians, and hoardings. The best performance was given by the task of vehicle recognition, which shows results of 96.9%. The processing time of the proposed method was in a range of 19 to 24 min per frame. The features that are typically used while developing laser-based applications are based on geometrical properties and statistical analysis of the distance (Arras, K.O.). The width, height, centroid, and shape estimation (square, circle and polygon approximation) are the examples of the first group (Nashashibi, F. et al 2008; Mertz, C. et at. 2013). The mean, variance, standard deviation, kurtosis, and skewness of the distance of each beams related to each object are examples of the second group (Azim, A. et al. 2014; Kim, B. et al. 2015). In that sense and looking forward improvements of the classification taking into account the detection rate as well as the processing time as a key in real-time intelligent vehicles applications, a group of additional features is considered, the reflectance and the spatial disorder respectively. The reflectance is related to the angle incidence, surface and material. The spatial disorder feature is considered since objects on the road can be described in terms of geometrical shapes while vegetation (bushes, grasses) is not consistent with the geometrical shapes assumption.

As part of the ongoing research in (Hernández, D.C. et al. 2016) to increase the robustness of scene understanding in dynamic outdoor environments a laser-based obstacle detection is proposed. The purpose of this paper is three-fold:

1. To introduce an object classification method based on laser sensor for intelligent vehicle systems in applications such as collision avoidance, navigation, and localization.
2. To propose a real-time object classification strategy combining a set of features, which uses traditional features of objects; geometrical and statistical features, and includes a new set of features based on the reflectance value, and the disorder within the object dataset.
3. To propose a low cost solution for vehicle intelligent systems. Although the proposed classification method shows a good performance, there will be future improvements. These improvements will be made by developing a laser-camera hybrid method.

SYSTEM OVERVIEW

The laser was mounted on the roof at a height of 1.7 meters at the geometric center line of the vehicle, tilted down at 7.5° and 2.3 meters of the front most part of the vehicle. The maximum range of the LMS was 18 meters with an aperture angle of 270° and angle resolution of 0.50°. The distance between the point P(x,y,z) located in the road surface and sensor is about 13.1 meters with respect to the laser. The distance with respect to the ground is approximately 13 meters, as shown in Figure 1. The vehicle speed was acquired by a National Instruments PXI Express module.

Figure 1. Vehicle platform and laser scanning model
Note: *the image is not scaled*

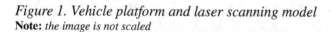
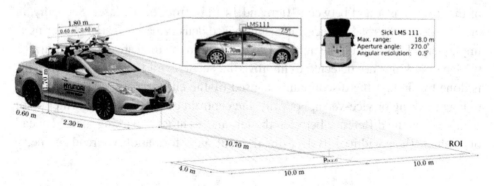

The laser sensor (SICK LMS 111) was mounted on the vehicle roof at the geometric center line of the vehicle, tilted down 7.5o. The image shows schematic top view of laser scanning, and the rectangle in black shows the region of interest, that consists of the space of 10 meters to the left and right side from the point P(x, y, z) and length of approximately 13 meters

METHODOLOGY

Figure 2 shows the flow diagram of the proposed approach in (Hernández, D.C. et al. 2016; Hernández, D.C. et al. 2015). The research proposed a method focus on the exploitation of lane marking surface given the laser point data set. For given data of the scanned road surface, it had been noticed that there was a significant difference in distance between road and marking surface, explained as a result of the retro-reflective properties of the marker materials. Once the lanes have been disclosed, the vehicle angle heading is computed by considering that the vehicle is moving in circular motion toward the center of the circle as; known as Instantaneous Center of Rotation. To determine the number of lane markings presented in the lane surface, the lane marking recognition method uses the discontinuous behavioral responses of the laser beam in presence of reflective surfaces. To do this end, the peak values, the lane road as well as the lane marking where able to be detected as a cluster due to the discontinuity formed by the difference between the road and lane marking surface. The laser scanning data collected are affected by the presence of paint marking material. To identify the crosswalk by finding its representative pattern, a square wave template matching was used. For detecting the objects within the previous describe lane region, the aim of this step is to extract the set of point which have not been considered during the lane cluster formation step.

The algorithm is briefly described as follows: firstly, for a laser line-scanner input range data, the set of points within the region of interest (ROI) are extracted. The ROI is defined as $20x4$ m^2, the square box is located 10 meters to the left and right of the laser sensor and between 10 m and 14 m in front of the laser. Secondly, in order to detect the road surface, an unsupervised density-based spatial clustering is implemented. As a result the number of lanes within the road are detected. Thirdly, the lane markings are detected to improve the accuracy of the previous step. This is done by finding the discontinuity caused by the change between surface, road to lane marking or vice-versa. Fourthly, the centroid of the lane is the center point computed as the difference between the left and right lane marking located ahead of a vehicle (lane width). Fifthly, using the centroid information, the road geometry

Figure 2. Main flowchart. lane surface, heading angle detection and object detection based on laser scanner

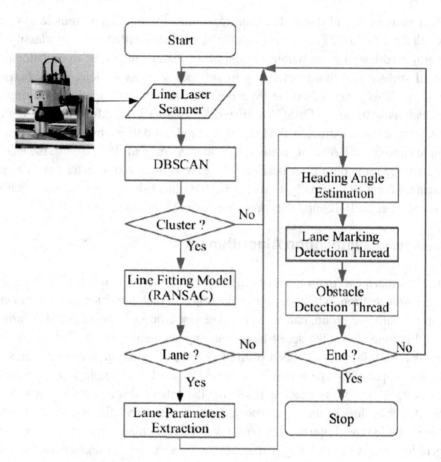

parameters are extracted; the deflection angle, the chord length, and the vehicle heading angle are computed. Once the lane region was define, the algorithm is able to sense the presence of an obstacle within the ROI.

Hence, the collision risk region is defined by taking into consideration both the vehicle speed and the vehicle stopping distance. Then, objects within the collision risk region are detected. At this step, the algorithm is able to define the lane region located ahead of the vehicle, the collision risk region, and objects within the region.

OBSTACLE CLASSIFICATION

In this section, the obstacle detection algorithm for intelligent vehicle systems through the use of a 2D laser sensor is described. The algorithm works by classifying the projected laser beams within objects in two main categories: those belonging to road surface and those belonging to surfaces such as vehicle, grass, barrier, poles, etc. This process is done by using the density based spatial clustering of applications with noise, DBSCAN (Ester, M. et al. 1996). After the lane region cluster candidate is extracted, the lane fitting model is defined by using the random sample consensus, RANSAC method (Fischler, M.A. et al. 1981). Once the lane is detected, the remaining laser beam information is analyzed to extract the objects presented along ROI. Finally, to recognize the object the support vector machine (Boser, B.E et al. 1992) and the adaptive boosting method are used.

Obstacle Classification Algorithm

In this approach two main types of objects were proposed to be classified: vehicle and grass/vegetation, while the rest of the objects; such a barrier, pole, or human, are mixed into one group, named others. The last group will be analyzed in further study. Figure 3 shows the set of laser scanning segments samples for the objects to be detected. It can be noticed from the Figures that the gathered segments do not show a specific shape between the same classes. For example, in Figure 3(a) a sedan vehicle type was extracted, (b) shows the extracted segments for the vehicle located at both sides of the vehicle centerline, and (c) shows the extracted point for a large vehicle type. Figures 2 (d)-(f) show a set of segments extracted from grass/ vegetation areas. Consequently, in this approach these groups of segments are labeled as a non-car and non-grass/vegetation objects.

Features Description

Over the last years the use of laser range finder has been successfully increased in the field of intelligent systems, for example in areas such as agriculture, transport, mining, mapping, architecture. Researchers in the field of agricultural science are using new techniques and technologies (Sugiura, R. et al. 2003; Ehlert, D. et al. 2009). Ehlert, D. et al. presented a vehicle based laser application to measure crop parameters for agricultural production. To do that, researchers focused on distance and reflection for measuring crop plants, as following: the range depending on medium under static conditions, the distribution of the light intensity, the properties for multiple reflection levels inside of the beam, and the properties for variable velocities of target medium and measuring distances. In the field of intelligent transport system several methods have been developed which allows autonomous vehicle to self navigate (Liu, Z et al. 2003; Li, Q. et al. 2014). For example, Li, Q. et al. proposed a sensor-fusion strategy to detect the drivable region for structured and non structured road. In this paper, authors uses the elevation as the angle bisector, the X,Y,Z variance

of a set of points as a feature to detect the curb points. In case of human detection, Budzan, S. et al. proposed a laser-vision/infrared fusion system. Authors take into account the number of points, different shapes of the objects, and some noise based on distance-based method using the Euclidean distance between points. In case of detection using 3D laser researchers applied voxel-based modeling and/or clustering strategies to solve the detection problem (Oh, S.I. et al. 2017; Yu, Y. et al. 2015). Yu, Y. et al. 2015 proposed a method that generated a perpendicular profile along the trajectory. Then the profile is divided to extract a set of points in which the slope and elevation between two consecutive points are the most important feature to detect the curbs region. Then, the set of points belong the surface are converting into a geo-referenced raster image. To solve the problem of the car and grass/vegetation detection a feature vector ($F = f_1, f_2,..., f_d$, where f_d is d-dimensional features) based on shape, physical property statistical measures, and the disorder of the object is proposed. The number of laser points, combined with object size (width and length) as well as object distance, are expected to distinguish car and vegetation since most of car sizes are constrained according to their design while vegetation are randomly shaped. Furthermore, some geometrical properties also useful to detect a man-made object such as car since the laser points lies on that object will be detected as line instead of randomly scattered points. Thus, mean of angle constructed from 3 points are utilized.

Figure 3. Object-segments based laser scanning located near to the laser scanner. (a) Small vehicles, (b) Smalls vehicles to both left and right side of the vehicle. (c) Large vehicles. (d)-(f) Different grass/vegetation-segments based laser scanning.

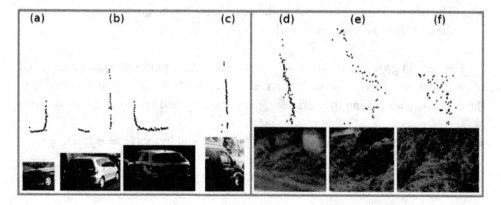

The first set of features is related to the characteristic of the objects (O) to be detected. This group consider the geometric and dimensions of each detected object (height, width, length, and shape). The first four features are related to the dimensions, the next six features are focused on the square shape, while the last two on the circle shape.

- The first feature is the total amount of laser beams within the detected object ($f_1 = n$).
- The second and the third features are the height ($f_2 = h_i$) and width ($f_3 = w_i$) of the object. This is done by assuming that the object can be described using a square shape. The farthest points along the x and y axes are used as the corners of a virtual red-dashed rectangle shown in Fig. 3(b).
- The fourth feature is related to the Euclidean distance ($f_4 = d_{j(max)}$) between the centroid of the virtual bonding box ($c_{j(x,y)}$) and the laser beam (p_j) located farthest from c_j, see Fig. 3(b).
- The fifth feature is the standard deviation of the distance between the virtual centroid c_j and the set of points within the object p_j ($f_5 = \sigma_d$).
- The sixth feature is the deviation from the mean of the laser beams ($f_6 = \sigma_m$).
- The seventh and eighth features are related to the boundary. The first one is the length of the boundary of the object ($f_7 = b$), while the second one is the standard deviation ($f_8 = \sigma_b$).
- The ninth feature is the mean of the angle formed between three consecutive points ($f_9 = \phi$).
- The tenth feature is related to the shape analysis to distinguish rectangular objects. This feature is the sum of the distance differences ($f_{10} = d_{jbe}$) between the laser beams (p_j) and the projection of the point p_j into the virtual bonding box. Figure 4(c) shows the features diagram for a single point p_j in yellow.
- The eleventh and twelfth features are related to the shape analysis to distinguish circular objects. The best fit model is obtained using least square circle ($f_{11} = r_j$).
- The twelve feature is related to the error between the circle fitting model and the laser beams position ($f_{11} = d_{jre}$).

The second group of features is the statistics related to the raw laser range data of each detected object. This set of features are related to the distribution of the data, the dependencies among the variables, the symmetry and the presence of outliers

- The thirteenth and fourteenth are the variance along x ($f_{13} = \sigma_x^2$) and y-axes ($f_{14} = \sigma_y^2$) respectively.

Figure 4. Laser based object feature extraction. The first row shows the set of extracted features for a vehicle class, while the second row shows the set of vegetation features. (a) Detected object. (b) and (c) shows the set of extracted features using a square shape. Image (d) show the feature set using a circle shape.

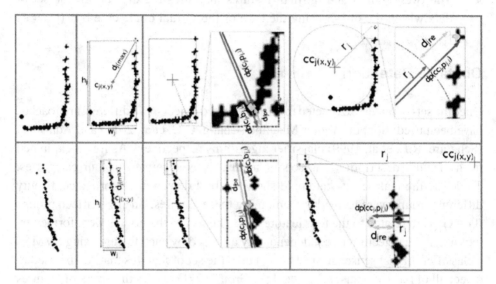

- To measure the input data linear dependence, the fifteenth feature is the covariance of the raw laser range data ($f_{15} = \text{cov}(x,y)$).
- To measure the asymmetry, the sixteenth and seventeenth features are the skewness of the input data along x ($f_{16} = s_x$) and y-axes ($f_{17} = s_y$) respectively.
- To measure the tailed of the input data, the eighteenth and nineteenth features are the kurtosis along x ($f_{18} = k_x$) and y-axes ($f_{19} = k_y$) respectively.

The third set of feature is related to the raw laser reflectance data. These features belonging to the material type and its surface properties.

- Features from twentieth to twenty-fourth are the mean, variance, standard deviation, skew, and kurtosis of the reflectance respectively ($f_{20} = \mu_r$, $f_{21} = \sigma_r^2$, $f_{22} = \sigma_r$, $f_{23} = s_r$, $f_{24} = k_r$).

The fourth set of features is related to the disorder of the laser raw data. These features related to the entropy of the raw data of each detected object.

- The twenty fifth and twenty-sixth features measure the disorder of the set of data along the x ($f_{25} = e_x$) and y ($f_{26} = e_y$) axes.

- The twenty-seventh and twenty-eighth features measure the disorder of the set of data within the object using the case of the virtual bounding box ($f_{27} = e_b$, $f_{28} = e_{sb}$).
- The twenty-ninth and thirtieth features measures the disorder of the set of data within the object using the case of the virtual circle case box ($f_{29} = e_r$, $f_{30} = e_{sr}$).

Obstacle Detection

Once the sets of features extracted from the detected objects, two different approaches have been used: Support Vector Machine (Chang, C.C. et al. 2011) and AdaBoost (Schapire, R.E. et al. 1999) classifier algorithms respectively. As it was explained in the feature description, the sets of features was separated in four categories. To define the feature set that could show the best detection performance, twenty different groups were formed by combining these features. The first fifteen groups (G_1 - G_{15}) are related to the four feature sets. The rest of the groups were formed by looking at the detection rate independently assessed by each feature using SVM as a classifier. In that sense, group 16 (G_{16}) uses the set of features that can be used to detect all of these objects; f_{16}, f_{18}, and f_{23}. Group 17 (G_{17}) uses the group of features for which the total detection average (vehicle, vegetation, others) is greater than or equal to 85% for at least two features; f_1, f_{10}, f_{15}, f_{21}, f_{27}, and f_{28}. Group 18 (G_{18}) is formed by using the features which can independently detect objects with accuracy over 85% detection; f_1, f_3, f_7, f_9, f_{17}, f_{19}, and f_{21}, f_{30}. Group 19 (G_{19}) is formed in a similar manner, with the difference that those features which solely detect a single object with 100% accuracy are not considered; f_1, f_3, f_5, f_7, f_9, f_{10}, f_{12}, f_{16}, f_{21}, f_{23}, and f_{25}, f_{30}. Finally, group 20 (G_{20}) uses the 50% of the each feature group which detected obstacles; f_3, f_5, f_7, f_9, f_{10}, f_{14}, f_{16}, f_{22}, f_{23}, and f_{25}, f_{27}.

Table 1 shows the feature vector extraction results for the two main objects: vehicle and grass. The first column shows the object within the sensor field of view, the camera sensor and laser scanner in Figure 5. For the object laser data set the second column shows the histogram along of the x-axis, while the third column shows the histogram for the y-axis. Figure 5(a) shows the vector for the case of vehicle within the field of view of both sensors (laser and camera). The vehicle is located ahead and on the left side of the platform vehicle. Figure 5(b) shows the vector that belongs to a vehicle located to the left lateral side of the vehicle. In this case, the vehicle is out of the field of view of the camera. The difference in shape of the detected object can be noticed, which is related to the position of the object with respect to the laser scanner. Figure 5(c) shows the vector for the case of grass located to the left side of the vehicle. The histogram shown in Figure 5 helps readers to visually distinguish between vehicles and vegetation objects. In this study, these

Figure 5. Obstacles detected by a laser sensor

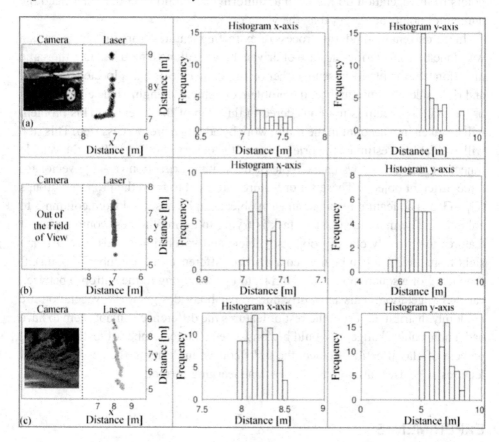

differences are used to classify objects by their similarities and difference by using machine learning. It should be noticed that all those differences are affected by the 2D object coordinates with respect to the sensor reference system, for example, the difference in range, frequency, and distribution.

In Figure 5, the first column show the object laser-based detection results, the second column shows the histogram of the raw data along the x-axis coordinate while the third column shows the histogram along the y-axis. The first row shows the obstacle detection results for a vehicle case, half observed by the camera sensor. The second row shows the vehicle detection result, the vehicle was located outside of the field of view of the camera. The third row shows the vegetation detection results

In the case of a driver traveling in the left lane, it can be observed for the height that a high group of point to the total set of the laser point are split within a short interval, 1 cm for the case of the points along the x-axis. In the case of the width the set of point lies along the y-axis while, more than 1.2 m. In contrast, the set of points from vegetation do not keep to uniform distribution, on the other hand the number of points are larger than the vehicles.

In the current method, the process of extracting features from the raw laser data was done by considering the case of drivers moving along a road heading the same direction that the platform setup either on the same lane or lanes located to the left and right side. In consequence, the problem of vehicle changing lane either drivers or the platform setup is not considered in this step of the research. This problem will be taken into account in the future work by adding a pre-process step. This step will evaluate and estimate the orientation of the observed objects along the vehicle trajectory. Table 1 shows an example of the feature extraction $(f_1 - f_{30})$ vector for three different objects. These set of features are used to form the proposed groups $(G_1 - G_{20})$. The feature response in each object can be observed from column 3 to column 5. The main idea of this task is trying to define the best combination of features to correctly classify objects, vehicle and vegetation in this case. At first sight from table 1 it can be noticed a marked difference in f_1 (number of point), f_8 (boundary of a square box), and from f_{25} to f_{30} (belonging to the entropy) between car and vegetation. This is mostly due to car body design while vegetation are randomly distributed. From the rest of features the difference can not seem to have had a noticeable change, it should be considered a large number of features for the process of classification. Hence, the tasks consist in defining the best number of features to be used, as well as, the machine learning method.

EXPERIMENTS

To validate and verify the feasibility of the proposed obstacle detection based on laser, the Kitti raw dataset from the Karlsruhe Institute of Technology was used (Geiger, A. et al. 2013) (http://www.cvlibs.net/datasets/kitti/raw_data.php). The program was built in C++ and compiled in Ubuntu16.04 using an Intel Core i7-6700HQ CPU, 2.60 GHz with 8GB of RAM. A total of 2,494 objects within the region reached by the laser beams were manually labeled. To do this test, the beams within a range of 3° with respect to the horizontal projection where extracted from the Velodyne points. For a total of 2,494 objects manually labeled, 876 belong to vehicles (35.12%), 1124 belong to vegetation (45.07%), while belong to other types of objects (19.81%). For both classifiers the trained data use a total of 40% of the total objects within the class (vehicles, vegetation, others), while the rest of the objects were used for testing. As previously mentioned in Section 3, the algorithm starts extracting from the laser dataset those points which are located within the region of interest. Then, by using both the DBSCAN and RANSAC methods, clusters belongs

Table 1. Objects features extraction results for Figure 5

Type	[a]Feature	Fig. 5(a)	Fig. 5(b)	Fig. 5(c)
f_1	n	41.00	45.00	94.00
f_2	w	0.54	0.10	0.58
f_3	h	1.95	2.01	3.96
f_4	d_{jmax}	1.01	1.00	2.00
f_5	σ_d	0.27	0.29	0.62
f_6	σ_m	17.85	21.48	14.89
f_7	b	12.26	12.92	57.04
f_8	σ_b	0.36	0.26	0.53
f_9	φ	21.20	49.47	11.80
f_{10}	d_{jbe}	21.98	2.98	382.50
f_{11}	r	1.15	2.35	1.20
f_{12}	d_{jre}	3.54	1.98	43.60
f_{13}	σ_x^2	0.02	0.00	0.02
f_{14}	σ_y^2	0.29	0.27	1.14
f_{15}	cov(x,y)	-0.04	0.01	0.14
f_{16}	s_x	1.33	0.11	-0.13
f_{17}	s_y	1.32	-0.04	0.37
f_{18}	k_x	0.47	-0.54	-1.03
f_{19}	k_y	0.94	-1.06	-0.72
f_{20}	μ_r	7.56	6.57	6.60
f_{21}	σ_r^2	0.29	0.27	1.13
f_{22}	σ_r	0.54	0.52	1.07
f_{23}	s_r	1.32	-0.04	0.37
f_{24}	k_r	0.94	-1.06	-0.72
f_{25}	e_x	253.40	268.80	705.80
f_{26}	e_y	272.50	242.30	511.50
f_{27}	e_b	4.38	3.33	-259.30
f_{28}	e_{sb}	-29.50	-1.41	-987.90
f_{29}	e_r	3.37	2.49	12.26
f_{30}	e_{sr}	-1.94	-0.59	-71.48

a Note: The set of features are defined in 4.1.1.

to the road surface are detected. Those detected cluster that are not belonging to the road surface class are used to classify objects as vehicle, vegetation and others.

From the cluster which define the lane ahead of the vehicle the road parameters geometry information is computed. The proposed method was tested using a real road traffic scene on a bright sunny day with the presence of strong shadows, see the bottom image in Figure 6 and 7. Figure 6 shows the result of the proposed method; a total of five objects were detected within the region of interest in the laser line scan, the first three belong to vehicles, the fourth belongs to a fence and the last one belongs to vegetation. Four objects were detected within the vehicle class while one within the vegetation class. The first three are correctly classify as vehicles, the fourth one is not correctly classified as a fence, the last one is correctly classified within the vegetation class. On the other hand, Figure 7 shows two objects classified into the same class. The first object was correctly classified as a vehicle, while the algorithm categorized incorrectly the second object as a vehicle been a grass and dry soil texture.

According to the above mentioned paragraph, to evaluate the performance of the proposed method in detecting and classifying multiple objects, the classification result of the proposed algorithm was compared against the objects ground truth data (manually labeled). The evaluation describes the feasibility of the presented classification method to classify correctly objects from traffic scenes using laser range sensor. The traffic scene data has the following characteristics: traffic jam scene, bright sunny day and presence of strong shadows. To evaluate the obstacle classification results, sensitivity (St), specificity (Sp), accuracy (A), the Matthews correlation coefficient (MCC), false positive per second (FPs), and negative per second (FNs) rate metrics were used, see from Equation 1 to Equation 6. The sensitivity measures the correct detection of the objects candidate as a vehicle, vegetation and other types, while the specificity measures the correct rejection of the false candidate as vehicles, vegetation or others objects. The accuracy describes how well the presented method correctly classifies the objects, while the MCC measures the quality of binary classification. FPs and FNs indicate the amount of false positives or negatives that occur within the frames. They are computed as follows:

$$St = \frac{TP}{TP + FN} \tag{1}$$

$$Sp = \frac{TN}{TN + FP} \tag{2}$$

$$A = \frac{TP + TN}{TP + TN + FP + FN} \qquad (3)$$

$$MCC = \frac{(TP \times TN) - (FP \times FN)}{\sqrt{(TP + FP)(TP + FN)(TP + FP)(TP + FN)}} \qquad (4)$$

$$FPs = \frac{FP \times PT}{TF} \qquad (5)$$

Figure 6. Obstacle Detectionand Classification

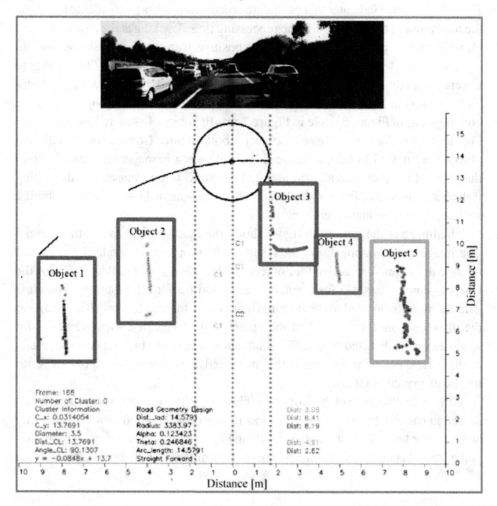

$$FNs = \frac{FN \times PT}{TF} \qquad (6)$$

The image at the top of the figure 6 shows the road scene ahead of the vehicle. The image a the bottom shows the result of the proposed method. The proposed method detect and categorize five objects within the region of interest, C1 for the case of vehicle class, C2 for vegetation class and C3 for other objects. The class data information located to the bottom right part of the image; class (C1 - C3) and the distance from the centroid of the object to the trajectory, uses indexed color specifies by the detected object. The square box in blue and green show the results of the correct classification, while the square red box magenta show the result for the incorrectly classified where TP (true positive) indicates that candidate objects are correctly detected, TN (true negative) indicates that candidates are correctly rejected, FN (false negative) indicates that candidate objects were incorrectly rejected, and FP (false positive) indicates that candidate objects were incorrectly identified, TF is the total amount of frames, PT is the processing time. Looking at the case of vehicle classification, Figure 6 and 7 show the true positive, true negative, false positive and false negative classification results. For example, as a true positive (TP) in Figure 6 were correctly classified the following objects 1, 2 and 3 and object 2 in Figure 7 (square box in blue). For the case of true negative (TN) was the object 5 (square box in green) in Figure 6 while in Figure 7 was the object 1 (square box in brown). For the case of false positive (FP), object 4 both Figures (square box in red), and object 3 in Figure 7 as false negative (FN) (square box in magenta). Table 2 shows the result of the proposed algorithm tested on the Kitti raw dataset. For the testing task a total amount of frames were used with an amount of 1494 objects combining vehicle, vegetation and others types.

The image at the top of the figure shows the road scene ahead of the vehicle. The image a the bottom shows the result of the proposed method. The proposed method detect and categorize four objects within the region of interest, C1 for the case of vehicle class, C2 for vegetation class and C3 for other objects. The class data information located to the bottom right part of the image; class (C1 - C3) and the distance from the centroid of the object to the trajectory, uses indexed color specifies by the detected object. The square box in blue and brown show the results of the correct classification, while the square red and magenta box show the result for the incorrectly classification

Although objects can be correctly classified, there were some incorrectly detection cases. Object 4 in Figure 6 was incorrectly classified as a vehicle being an barrier or fence. Likewise, object 4280 in Figure 7 being a region with moist soil and grass. This error occurs mostly due to the similarities between the features set

Figure 7. Obstacle detection and classification

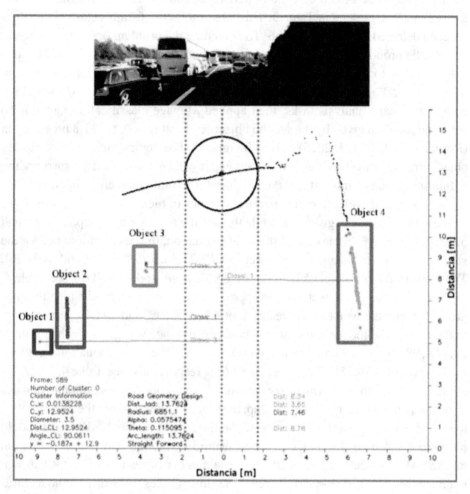

Table 2. Object Detection Rate for Different Types of Objects

[b]Types	TP[%]	TN[%]	FP[%]	FN[%]	St[%]	Sp[%]	A[%]	MCC[%]	FPs[μs]	FNs[μs]
S_c	83.40	85.00	14.90	16.50	83.40	85.00	84.30	0.68	7.10	7.10
S_v	84.70	86.90	13.20	15.20	84.70	86.70	86.10	0.71	8.60	3.90
S_o	86.30	87.20	12.70	13.60	86.30	87.20	87.00	0.73	8.80	3.00
A_c	99.50	99.60	0.39	0.43	99.50	99.60	99.50	0.99	0.10	0.10
A_v	95.60	99.40	0.86	4.30	95.60	99.10	98.10	0.94	0.60	1.40
A_o	88.88	98.91	1.08	11.10	88.80	98.90	96.40	0.88	0.80	2.40

b Note: S stands for SVM, A stands for AdaBoost, index c stands for car, index v stands for vegetation, index o stands for others, TP stands for true positive, FP stands for false positive, FN stands for false negative, TN stands for true negative, St stands for sensitivity, Sp stands for specificity, A stands for accuracy, FPs stand for false positive per second, and FNs stand for false negative per second.

used, previously described in section 4.1.1. These similarities between the fence and the vehicles can be seen in Table 3. This table shows the features extraction values for each detected object in Figure 6. To overcome the problem presented in Figure 6 it will be proposed an hybrid laser-vision based classification method. The idea behind will be focused on transforming the 3D coordinate object information (from laser) to the 2D image coordinate plane, it will be done to reduce the time searching within the image analysis tasks, then applied a image classification method. To solve the problem given by the second object (soil and grass), the idea presented in (Hernández, D.C. et al. 2015) will be integrated. The algorithm is able to classify objects belong to the road surface. Additional it will be considered the surrounding within the region of interest and the height information while creating cluster.

For the global evaluation, the best performance of the SVM classifier was given by the set of features in group 20, while the worst was given by group 4, 87.9% and 59.1% respectively. In the case of the local evaluation, the best performance for the vehicle, vegetation and other classes were 89.7% (G20), 95.2% (G16), and 94.9 (G6). The worst evaluations were 57.5% (G16), 85.9% (G8), and 8.1% (G1), see Table 4. In the case of the AdaBoost, the best performance was given by the feature set in group 1, while the worst was given by group 16, 98.19% and 79.04% respectively. In the case of the local evaluation, the best performance was given by the group 1, 99.77%, 98.23%, and 96.56% respectively). Similarly, the worst evaluation was given by the group 16, 76.25%, 78.21%, and 82.65% respectively, see Table 4.

Table 5 shows the performance of the proposed method. The first column evaluates the time spent on feature extraction step, the second column shows the time spent on the SVM while the third column shows the time spent on the AdaBoost. It should be noticed the time difference between feature extraction and the classification task (SVM and AdaBoost respectively). The number of features (f1− f30) chosen to create the groups (G1−G20) increase the time it takes to extract all the features values. Although the processing time shows a fast response, it can be observed that the processing time increases as the amount of features increases and vice-versa within a range of 5.37 µs to 75.51 µs. With respect to the classification task, the SVM shows a better performance than the AdaBoost classifier for the best detection results; however, AdaBoost shows a better a performance in detection.

Table 6 shows the performance of the proposed method per frame. To do this, the frame with the maximum (8) and minimum (1) number of detected objects within the field of view of the laser beams were used. It should be noted that Table 6 was created using the best performance of both classifiers, group 20 and group 1 in Table 5. In Table 6, to is the computing time per object shown in Table 5, tf is the computed time using the maximum and minimum numbers of objects detected in a laser line scan, ta is the computing time for all the process. As it can be noticed that to and tf are related to features (f1− f30), while ta is the computing time required by

Table 3. Objects features extraction results for figure 6

Type	Feature	Object 1	Object 2	Object 3	Object 4	Object 5
		Vehicle	Vehicle	Vehicle	Fence	Vegetation
f_1	n	62.00	26.00	58.00	26.00	87.00
f_2	w	1.21	0.19	1.62	0.15	1.10
f_3	h	3.11	3.25	1.97	1.28	4.01
f_4	d_{jmax}	1.67	1.63	1.26	0.64	2.08
f_5	σ_d	0.37	0.56	0.08	0.19	0.65
f_6	σ_m	1.68	4.62	12.78	13.88	62.71
f_7	b	24.77	20.64	6.23	3.33	67.52
f_8	σ_b	0.56	1.09	0.24	0.12	0.88
f_9	φ	30.31	15.39	1.36	7.84	8.44
f_{10}	d_{jbe}	82.96	7.34	105.91	20.54	343.07
f_{11}	r	0.89	8.22	1.07	2.10	1.36
f_{12}	d_{jre}	17.88	0.61	3.40	0.49	39.39
f_{13}	σ_x^2	0.02	0.00	0.25	0.00	0.06
f_{14}	σ_y^2	0.58	0.87	0.22	0.14	1.39
f_{15}	cov(x,y)	0.02	0.00	0.11	0.02	0.18
f_{16}	s_x	6.55	0.67	-0.17	-0.32	-0.74
f_{17}	s_y	0.39	-0.53	2.70	0.22	0.28
f_{18}	k_x	45.80	-0.34	-1.28	-1.66	0.31
f_{19}	k_y	-0.54	-0.55	6.12	-1.20	-1.03
f_{20}	μ_r	6.20	8.53	9.95	8.91	6.72
f_{21}	σ_r^2	0.58	0.87	0.22	0.14	1.39
f_{22}	σ_r	0.76	0.94	0.47	0.38	1.18
f_{23}	s_r	0.39	-0.53	2.70	0.22	0.28
f_{24}	k_r	-0.54	-0.55	6.12	-1.20	-1.03
f_{25}	e_x	439.06	60.99	60.65	88.47	646.28
f_{26}	e_y	305.89	206.98	576.65	220.26	488.19
f_{27}	e_b	-12.22	3.57	-30.28	0.42	-234.30
f_{28}	e_{sb}	-159.19	-6.35	-214.45	-26.95	-869.82
f_{29}	e_r	8.04	0.88	3.76	0.79	10.60
f_{30}	e_{sr}	-22.38	0.13	-1.81	0.15	-62.84

^{c} Note: Feature set are defined in 4.1.1.

Table 4. Object classification

Support Vector Machine		AdaBoost							
[d]G_i	F[#]	C[%]	V[%]	[O%]	T[%]	C[%]	V[%]	[O%]	T[%]
G_1	30	84.36	86.03	87.03	86.08	99.77	98.23	96.56	98.19
G_2	12	81.50	87.61	68.62	79.25	98.39	96.95	94.08	96.47
G_3	7	88.01	89.20	60.73	79.32	95.88	95.18	89.32	93.46
G_4	5	81.62	87.46	8.09	59.06	92.42	92.51	87.70	90.87
G_5	6	66.67	87.14	83.80	79.21	93.55	95.25	91.61	98.19
G_6	19	72.94	88.25	94.23	85.38	98.68	97.86	94.56	97.09
G_7	17	76.59	88.25	82.79	82.87	99.08	97.86	95.99	97.46
G_8	18	83.10	85.87	89.06	86.02	98.84	97.30	95.40	97.06
G_9	12	87.10	89.04	69.23	86.02	97.10	96.67	91.61	95.13
G_{10}	13	82.65	86.50	80.16	83.10	99.06	96.14	94.28	96.49
G_{11}	11	67.69	86.50	84.21	79.47	99.08	96.75	95.04	96.96
G_{12}	24	86.41	87.77	85.62	86.61	99.09	97.91	96.18	97.73
G_{13}	25	84.70	86.35	91.90	87.65	99.53	98.20	95.04	97.59
G_{14}	23	81.05	86.50	90.08	85.88	99.08	97.62	96.09	97.60
G_{15}	18	79.90	88.41	83.60	83.97	98.62	97.92	95.23	97.26
G_{16}	3	57.53	95.24	95.71	82.83	76.25	78.22	82.65	79.04
G_{17}	6	71.92	88.25	82.39	80.85	97.00	94.02	89.23	93.42
G_{18}	26	84.24	86.51	93.32	88.02	98.85	98.21	95.32	97.46
G_{19}	21	84.58	86.98	94.53	88.70	99.32	98.20	95.13	97.55
G_{20}	14	89.73	86.50	87.65	87.96	98.61	97.89	94.38	96.95

[d] Note: Gi stands for Group, F stands for the number of selected features, C stands for car, V stands for vegetation, O stands for others, T stands for total respectively. From G1 to G20 are the groups formed using the proposed feature sets described in 4.1.1. Numerical results are the percentage of correct detection of the objects.

the algorithm shows in Figure 2 to complete the loop. In the first column, row one and three belong to the maximum and minimum number of detected objects within the dataset. The second row is estimated using Figure 6 as an example.

CONCLUSION

In this approach an obstacle type detection method based on laser sensor has been presented, to detect obstacles such as vehicles, grass and others (barrier, pole, fence or human) respectively. To solve this problem a set of features divided in four

Table 5. Classification performance

ᵉGᵢ	F[#]	Time[µs]	Support Vector Machine		AdaBoost	
			Time[µs]	Detection[%]	Time[%]	Detection[%]
G_1	30	75.51	0.59	86.08	3.67	98.1
G_2	12	50.7	0.67	79.25	3.82	96.47
G_3	7	17.65	0.66	79.32	3.25	93.43
G_4	5	5.37	0.65	59.06	3.01	90.88
G_5	6	6.07	0.63	79.21	3.3	93.47
G_6	19	65.48	0.68	85.38	3.57	97.07
G_7	17	59.28	0.64	82.87	3.47	97.46
G_8	18	53.61	0.61	86.02	3.43	97.06
G_9	12	27.27	0.68	81.79	3.57	95.13
G_{10}	13	23.98	0.69	83.1	3.58	96.46
G_{11}	11	16.13	0.76	79.47	3.66	96.96
G_{12}	24	70.9	0.61	86.61	3.53	97.73
G_{13}	25	72.15	0.66	87.65	3.55	97.59
G_{14}	23	71.41	0.67	85.88	3.79	97.6
G_{15}	18	32.99	0.66	83.88	4	97.25
G_{16}	3	7.73	0.68	81.83	2.63	79.04
G_{17}	6	27.74	0.65	80.85	2.93	93.42
G_{18}	26	74.12	0.65	88.02	4.06	97.46
G_{19}	21	62.09	0.74	88.7	3.67	97.55
G_{20}	14	47.69	0.63	87.96	3.54	96.96

ᵉ Note: Gi stands for Group, F stands for the number of selected features. From G1 to G20 are the groups formed using the proposed feature sets described in 4.1.1. Time is the processing time in µs for the feature extraction, and the machine learning task (SVM and AdaBoost).

Table 6. Computing time performance

ᶠNO	Supporting Vector Machine (G_{20})			AdaBoost (G_{20})		
	t_o[µs]	t_f[µs]	t_a[ms]	t_o[µs]	t_f[µs]	t_a[ms]
8	48.32	0.386	27.56	79.18	0.634	28.44
5	48.32	0.241	27.36	79.18	0.396	27.52
1	48.32	0.048	27.22	79.18	0.079	27.25

ᶠ Note: NO stands for number of detected object within the laser data set. t_o is the computing time per object, t_f is the computed time using the maximum and minimum numbers of objects detected in a laser line scan, t_a is the computing time for all the process related to the number of detected object.

groups was proposed: object characteristic, statistical analysis of range distance and reflectance, and entropy. To verify the reliability of the proposed idea, two different classifiers were used. The features were combined between them in order to find the best set of relevant features. Although the response of both classifiers showed a good performance (over 85%), it has been demonstrated that the use of AdaBoost provides better results for the task of object classification. As a result for a total of 28 features, the SVM classifier provides the best response-performance with a set of 14 features (87.96% at 48.32 μs), while with the AdaBoost the best result was given by the full set of features (98.19% at 79.18 μs). However, it should be noted that the result of the AdaBoost classifier using the group 20 shows a better response than the SVM, with a detection rate of 96.96% at 51.23 μs. Although the presented method shows good performance there is still the potential to reduce conflicts between objects classes by taken into consideration the results of the research results shown in Cáceres Hernández, D. et al. 2016 which detect the road geometric model using a vision based real-time lane understanding application and the hybrid lane detection results using camera and laser sensors.

ACKNOWLEDGMENT

The authors would like to thank the National Bureau of Science, Technology and Innovation of Panamá (SENACYT) and the Sistema Nacional de Investigación of Panamá, SNI 168-2017. They are also grateful for the support received from the Universidad Tecnológica de Panamá.

REFERENCES

Arras, K. O., Mozos, O. M., & Burgard, W. (2007). Using Boosted Features for the Detection of People in 2D Range Data. *IEEE International Conference on Robotics and Automation*, 3402–3407. 10.1109/ROBOT.2007.363998

Azim, A., & Aycard, O. (2014). Layer-based supervised classification of moving objects in outdoor dynamic environment using 3D laser scanner. *IEEE Intelligent Vehicles Symposium Proceedings*, 1408–1414. 10.1109/IVS.2014.6856558

Boser, B. E., Guyon, I. M., & Vapnik, V. N. (1992). A Training Algorithm for Optimal Margin Classifiers. In *Annual Workshop on Computational Learning Theory*. ACM. 10.1145/130385.130401

Budzan, S., Wyzgolik, R., & Ilewicz, W. (2018). Improved Human Detection with a Fusion of Laser Scanner and Vision/Infrared Information for Mobile Applications. *Applied Sciences*, *8*(10), 1–27. doi:10.3390/app8101967

Cáceres Hernández, D., Kurnianggoro, L., Filonenko, A., & Jo, K. H. (2016). Real-Time Lane Region Detection Using a Combination of Geometrical and Image Features. *Sensors (Basel)*, *16*(11), 1–19. doi:10.339016111935 PMID:27869657

Chang, C.C., & Lin, C.J. (2011). LIBSVM: A Library for Support Vector Machines. *ACM Trans. Intell. Syst. Technol.*, *2*(27), 1–27.

Ehlert, D., Adamek, R., & Horn, H. J. (2009). Vehicle Based Laser Range Finding in Crops. *Sensors (Basel)*, *9*(5), 3679–3694. doi:10.339090503679 PMID:22412333

Ester, M., Kriegel, H. P., Sander, J., & Xu, X. (1996). A Density-based Algorithm for Discovering Clusters a Density-based Algorithm for Discovering Clusters in Large Spatial Databases with Noise. International Conference on Knowledge Discovery and Data Mining. AAAI Press. *KDD: Proceedings / International Conference on Knowledge Discovery & Data Mining. International Conference on Knowledge Discovery & Data Mining*, *96*, 226–231.

Fischler, M. A., & Bolles, R. C. (1981). Random Sample Consensus: A Paradigm for Model Fitting with Applications to Image Analysis and Automated Cartography. *Communications of the ACM*, *24*(6), 381–395. doi:10.1145/358669.358692

Geiger, A., Lenz, P., Stiller, C., & Urtasun, R. (2013). Vision meets Robotics: The KITTI Dataset. *The International Journal of Robotics Research*, *32*(11), 1231–1237. doi:10.1177/0278364913491297

Hernández, D. C., Filonenko, A., Hariyono, J., Shahbaz, A., & Jo, K. H. (2016). Laser based collision warning system for high con>ℓict vehicle-pedestrian zones. *IEEE International Symposium on Industrial Electronics*, 935–939. 10.1109/ISIE.2016.7745016

Hernández, D. C., Filonenko, A., Seo, D., & Jo, K. (2015). Laser scanner based heading angle and distance estimation. *IEEE International Conference on Industrial Technology*, 1718–1722. 10.1109/ICIT.2015.7125345

Hsu, C. W., Hsu, T. H., & Chang, K. J. (2012). Implementation of car-following system using LiDAR detection. *International Conference on ITS Telecommunications*, 165–169.

Ibarra-Arenado, M., Tjahjadi, T., Pérez-Oria, J., Robla-Gómez, S., & Jiménez-Avello, A. (2017). Shadow-Based Vehicle Detection in Urban Traffic. *Sensors*, *17*(5), 1-19.

Kim, B., Choi, B., Park, S., Kim, H., & Kim, E. (2016). Pedestrian/Vehicle Detection Using a 2.5-D Multi-Layer Laser Scanner. *IEEE Sensors Journal, 16*(2), 400–408. doi:10.1109/JSEN.2015.2480742

Lehtomäki, M., Jaakkola, A., Hyyppä, J., Lampinen, J., Kaartinen, H., Kukko, A., ... Hyyppä, H. (2015). Object Classification and Recognition From Mobile Laser Scanning Point Clouds in a Road Environment. *IEEE Transactions on Geoscience and Remote Sensing, 54*(2), 1226–1239. doi:10.1109/TGRS.2015.2476502

Li, Q., Chen, L., Li, M., Shaw, S., & Nüchter, A. (2014). A Sensor-Fusion Drivable-Region and Lane-Detection System for Autonomous Vehicle Navigation in Challenging Road Scenarios. *IEEE Transactions on Vehicular Technology, 63*(2), 540–555. doi:10.1109/TVT.2013.2281199

Liu, Z., Liu, D., & Chen, T. (2013). Vehicle detection and tracking with 2D laser range finders. *International Congress on Image and Signal Processing, 2*, 1006–1013. 10.1109/CISP.2013.6745203

Liu, Z., Wang, J., & Liu, D. (2003). A New Curb Detection Method for Unmanned Ground Vehicles Using 2D Sequential Laser Data. *Sensors (Basel), 13*(1), 1102–1120. doi:10.3390130101102 PMID:23325170

Mertz, C., Navarro-Serment, L. E., MacLachlan, R., Rybski, P., Steinfeld, A., Suppé, A., ... Gowdy, J. (2013). Moving Object Detection with Laser Scanners. *Journal of Field Robotics, 30*(1), 17–43. doi:10.1002/rob.21430

Nashashibi, F., & Bargeton, A. (2008). Laser-based vehicles tracking and classification using occlusion reasoning and confidence estimation. *IEEE Intelligent Vehicles Symposium*, 847–852. 10.1109/IVS.2008.4621244

Oh, S.I., & Kang, H.B (2017). Object Detection and Classification by Decision-Level Fusion for Intelligent Vehicle Systems. *Sensors, 17*(1), 1-21.

Premebida, C., Monteiro, G., Nunes, U., & Peixoto, P. (2007). A Lidar and Vision-based Approach for Pedestrian and Vehicle Detection and Tracking. *IEEE Intelligent Transportation Systems Conference*, 1044–1049. 10.1109/ITSC.2007.4357637

Schapire, R. E. (1999). A Brief Introduction to Boosting. In *International Joint Conference on Artificial Intelligence*. Morgan Kaufmann Publishers Inc.

Sugiura, R., Fukagawa, T., Noguchi, N., Ishii, K., Shibata, Y., & Toriyama, K. (2003). Field information system using an agricultural helicopter towards precision farming. *Proceedings IEEE/ASME International Conference on Advanced Intelligent Mechatronics, 2*, 1073–1078. 10.1109/AIM.2003.1225491

Wang, H., Wang, C., Luo, H., Li, P., Cheng, M., Wen, C., & Li, J. (2014). Object Detection in Terrestrial Laser Scanning Point Clouds Based on Hough Forest. IEEE Geoscience and Remote Sensing Letters, 11, 1807–1811.

Wang, T., Zheng, N., Xin, J., & Ma, Z. (2011). Integrating Millimeter Wave Radar with a Monocular Vision Sensor for On-Road Obstacle Detection Applications. *Sensors (Basel), 11*(9), 8992–9008. doi:10.3390110908992 PMID:22164117

Yu, Y., Li, J., Guan, H., Jia, F., & Wang, C. (2015). Learning Hierarchical Features for Automated Extraction of Road Markings From 3-D Mobile LiDAR Point Clouds (2015). *IEEE Journal of Selected Topics in Applied Earth Observations and Remote Sensing, 8*(2), 709–726. doi:10.1109/JSTARS.2014.2347276

Zewei, X., Jie, W., & Xianqiao, C. (2015). Vehicle recognition and classification method based on laser scanning point cloud data. *International Conference on Transportation Information and Safety*, 44–49. 10.1109/ICTIS.2015.7232078

Zhang, F., & Knoll, A. (2016). Vehicle Detection Based on Probability Hypothesis Density Filter. *Sensors, 16*(4), 1-13.

Compilation of References

Ackerman, E. (2012). Boston dynamics sand flea robot demonstrates astonishing jumping skills. *IEEE Spectrum Robotics Blog, 2*(1).

Ahmad, N., Ghazilla, R. A. R., Khairi, N. M., & Kasi, V. (2013). Reviews on various inertial measurement unit (IMU) sensor applications. *International Journal of Signal Processing Systems, 1*(2), 256–262. doi:10.12720/ijsps.1.2.256-262

Akkar & Najim (2016). Kinematics Analysis and Modeling of 6 Degree of Freedom Robotic Arm from DFROBOT on Labview. *Research Journal of Applied Sciences, Engineering and Technology, 13*(7), 569–575. doi:10.19026/rjaset.13.3016

Alexander, C., Erenskjold Moeslund, J., Klith Bøcher, P., Arge, L., & Svenning, J.-C. (2013, July). Airborne laser scanner (LiDAR) proxies for understory light conditions. *Remote Sensing of Environment, 134*, 152–161. doi:10.1016/j.rse.2013.02.028

Ali, F., Urban, A., & Vossiek, M. A. (2011). Short Range Synthetic Aperture Imaging Radar with Rotating Antenna. *INTL Journal of Electronics and Telecommunications, 57*(1), 97–102. doi:10.2478/v10177-011-0014-y

American, S., America, N., & American, S. (2017). *Robot Systems.* Nature America, Inc. Retrieved from http://www.jstor.org/stable/24950284

Andersen, J. C., Blas, M. R., Ravn, O., Andersen, N. A., & Blanke, M. (2006). Traversable terrain classification for outdoor autonomous robots using single 2D laser scans. *Integrated Computer-Aided Engineering, 13*(3), 222–232. doi:10.3233/ICA-2006-13303

Antoshchenkov, R. V. (2017). *Dynamics and energy of motion of multi-element machine-tractor aggregates: a monograph.* Kharkiv: KhNTUA.

Apoorva, S., Prabhu, R. S., Shetty, S. B., & Souza, D. D. (2017). *Autonomous Garbage Collector Robot.* doi:10.5923/j.ijit.20170602.06

Arras, K. O., Mozos, O. M., & Burgard, W. (2007). Using Boosted Features for the Detection of People in 2D Range Data. *IEEE International Conference on Robotics and Automation,* 3402–3407. 10.1109/ROBOT.2007.363998

Artemov, N. P., Lebedev, A. T., Podrygalo, M. A., Polyansky, A. S., & Klets, D. M. (2012). The method of partial accelerations and its applications in the dynamics of mobile machines: a monograph. Kharkiv: Urban Press.

Artyomov, N. P., & Shuliak, M. L. (2015). Use of a filter to increase the accuracy of the study of the dynamics of mobile machines. *Scientific Bulletin of NUBIP of Ukraine, 226*, 290–295.

Atkeson, C. G., Babu, B. P. W., Banerjee, N., Berenson, D., Bove, C. P., Cui, X., . . . Gennert, M. (2015, November). No falls, no resets: Reliable humanoid behavior in the DARPA robotics challenge. In *2015 IEEE-RAS 15th International Conference on Humanoid Robots (Humanoids)* (pp. 623-630). IEEE.

Atkeson, C. G., Babu, B. P. W., Banerjee, N., Berenson, D., Bove, C. P., Cui, X., ... Gennert, M. (2016). What happened at the darpa robotics challenge, and why. *Journal of Field Robotics, 1*.

Atkinson, P. (2012). Feedback Control Theory for Engineers. Springer Science & Business Media. doi:10.1007/978-1-4684-7453-4

Avalos-Gonzalez, D., Hernandez-Balbuena, D., Tyrsa, V., Kartashov, V., Kolendovska, M., Sheiko, S., . . . Murrieta-Rico, F. N. (2018). Application of Fast Frequency Shift Measurement Method for INS in Navigation of Drones. In *IECON 2018-44th Annual Conference of the IEEE Industrial Electronics Society* (pp. 3159-3164). IEEE. 10.1109/IECON.2018.8591377

Azim, A., & Aycard, O. (2014). Layer-based supervised classification of moving objects in outdoor dynamic environment using 3D laser scanner. *IEEE Intelligent Vehicles Symposium Proceedings*, 1408–1414. 10.1109/IVS.2014.6856558

Basaca-Preciado, L., Sergiyenko, O., Rodriguez-Quinonez, J., & Rivas-Lopez, M. (2012). *Optoelectronic 3D Laser Scanning Technical Vision System based on Dynamic Triangulation. In Photonics Conference (IPC) in Burlingame California* (pp. 648–649). IEEE. Retrieved from http://ieeexplore.ieee.org/xpl/articleDetails.jsp?arnumber=6358788

Bass, M., DeCusatis, C., Enoch, J., Lakshminarayanan, V., Li, G., MacDonald, C., & Stryland, E. (2009). *Handbook of Optics* (D. M. Bass, Ed.; 3rd ed.; Vol. 1). McGraw-Hill Professional.

Borenstein, J., Everett, H. R., Feng, L., & Wehe, D. (1997). Mobile Robot Positioning: Sensors and Techniques. *Journal of Robotic Systems, 14*(4), 231–249. doi:10.1002/(SICI)1097-4563(199704)14:4<231::AID-ROB2>3.0.CO;2-R

Boser, B. E., Guyon, I. M., & Vapnik, V. N. (1992). A Training Algorithm for Optimal Margin Classifiers. In *Annual Workshop on Computational Learning Theory*. ACM. 10.1145/130385.130401

Box, G. E. P., Jenkins, G. M., Reinsel, G. C., & Ljung, G. M. (2016). *Time Series Analysis: Forecasting and Control* (5th ed.). John Wiley & Sons.

Boyd, S., & Vandenberghe, L. (2004). *Convex optimization*. Cambridge University Press. doi:10.1017/CBO9780511804441

Bruggemann, T. S., Greer, D. G., & Walker, R. A. (2011). GPS fault detection with IMU and aircraft dynamics. *IEEE Transactions on Aerospace and Electronic Systems*, *47*(1), 305–316. doi:10.1109/TAES.2011.5705677

Buccella, T. (1997). *Servo Control of a DC Brush Motor (AN532)*. Microchip.

Budiharto, W., Santoso, A., Purwanto, D., & Jazidie, A. (2011). A Navigation System for Service Robot Using Stereo Vision and Kalman Filtering. *Proceedings of 11th International Conference on Control, Automation and Systems*, 1771-1776.

Budzan, S., Wyzgolik, R., & Ilewicz, W. (2018). Improved Human Detection with a Fusion of Laser Scanner and Vision/Infrared Information for Mobile Applications. *Applied Sciences*, *8*(10), 1–27. doi:10.3390/app8101967

Buss, S. R. (2004). Introduction to inverse kinematics with jacobian transpose, pseudoinverse and damped least squares methods. *IEEE Journal of Robotics and Automation, 17*(1-19), 16.

Cáceres Hernández, D., Kurnianggoro, L., Filonenko, A., & Jo, K. H. (2016). Real-Time Lane Region Detection Using a Combination of Geometrical and Image Features. *Sensors (Basel)*, *16*(11), 1–19. doi:10.339016111935 PMID:27869657

Calinon, S., Guenter, F., & Billard, A. (2007). On learning, representing, and generalizing a task in a humanoid robot. *IEEE Transactions on Systems, Man, and Cybernetics. Part B, Cybernetics*, *37*(2), 286–298. doi:10.1109/TSMCB.2006.886952 PMID:17416157

CCIT. (1992). *T.81 - information technology – digital compression and coding of continuous – tone still images – requirements and guidelines*. Retrieved September 22, 2019, from https://www.w3.org/Graphics/JPEG/itu-t81.pdf

Chandrasekara, C., Rathnapriya, U., Handwriting, S., & Recognition, C. (2012). *Automatic Garbage Disposal System*. doi:10.13140/RG.2.2.12124.18566/1

Chang, C.C., & Lin, C.J. (2011). LIBSVM: A Library for Support Vector Machines. *ACM Trans. Intell. Syst. Technol., 2*(27), 1–27.

Chang. (2001). A new encycle algorithm for image cryptosystems. *Journal of Systems and Software, 58*, 83-91.

Chan, Y. K., & Koo, V. C. (2008). An introduction to synthetic aperture radar (SAR). *Progress in Electromagnetics Research*, *2*, 27–60. doi:10.2528/PIERB07110101

Chestnutt, J., Lau, M., Cheung, G., Kuffner, J., Hodgins, J., & Kanade, T. (2005, April). Footstep planning for the honda asimo humanoid. In *Proceedings of the 2005 IEEE international conference on robotics and automation* (pp. 629-634). IEEE. 10.1109/ROBOT.2005.1570188

Chestnutt, J., Michel, P., Kuffner, J., & Kanade, T. 2007, October. Locomotion among dynamic obstacles for the honda asimo. In *2007 IEEE/RSJ International Conference on Intelligent Robots and Systems* (pp. 2572-2573). IEEE. 10.1109/IROS.2007.4399431

Corany, B., Karanam, C. R., & Mostofi, Y. (2018). Adaptive Near-Field Imaging with Robotic Arrays. *IEEE 10ᵗʰ Sensor Array and Multichannel Signal Processing Workshop (SAM)*, 1-5.

Čupera, J., & Sedlak, P. (2011). The use of CAN-BUS messages of an agricultural tractor for monitoring its operation. *Research in Agricultural Engineering*, *57*(4), 117–127. doi:10.17221/20/2011-RAE

Dautenhahn, K. (2017). *Socially intelligent robots: dimensions of human-robot interaction.* doi:10.1098/rstb.2006.2004

Debain, C. A., Chateau, T., Berducat, M., Martinet, P., & Bonton, P. (2000). Guidance-assistance system for agricultural vehicles. *Computers and Electronics in Agriculture*, *25*(1-2), 29–51. doi:10.1016/S0168-1699(99)00054-X

Deergha Rao, K. (2011). A New and Secure Cryptosyce for Image Encryption and Decryption. *IETE Journal of research*, *57*(2), 165-171.

Deits, R., & Tedrake, R. (2015, May). Efficient mixed-integer planning for UAVs in cluttered environments. In *2015 IEEE international conference on robotics and automation (ICRA)* (pp. 42-49). IEEE.

Deits, R., & Tedrake, R. (2014, November). Footstep planning on uneven terrain with mixed-integer convex optimization. In *2014 IEEE-RAS International Conference on Humanoid Robots* (pp. 279-286). IEEE. 10.1109/HUMANOIDS.2014.7041373

Deits, R., & Tedrake, R. (2015). Computing large convex regions of obstacle-free space through semidefinite programming. In *Algorithmic foundations of robotics XI* (pp. 109–124). Cham: Springer. doi:10.1007/978-3-319-16595-0_7

Design, E. (2015). *Group 7's BOE-BOT Stair Climbing Robot.* Academic Press.

Dhanasingaraja R., Kalaimagal S. & Muralidharan G. (2014). Autonomous Vehicle Navigation and Mapping System. *International Journal of Innovative Research in Science, Engineering and Technology*, *3*(3), 1347-1350.

Dorobantu, R., & Zebhauser, B. (1999). *Field evaluation of a low-coststrapdown IMU by means GPS. In Ortung und Navigation* (pp. 51–65). Bonn: DGON.

Drenkow, G. (2006). LXI – A New Generation of Measuring Systems. *Electronics NTB - Scientific and Technical Journal. Control and Measurement*, *6*, 13–16.

Ehlert, D., Adamek, R., & Horn, H. J. (2009). Vehicle Based Laser Range Finding in Crops. *Sensors (Basel)*, *9*(5), 3679–3694. doi:10.339090503679 PMID:22412333

Eling, C., Klingbeil, L., & Kuhlmann, H. (2015). Real-time single-frequency GPS/MEMS-IMU attitude determination of lightweight UAVs. *Sensors (Basel)*, *15*(10), 26212–26235. doi:10.3390151026212 PMID:26501281

El-Sheimy, N., Nassar, S., & Noureldin, A. (2004). Wavelet de-noising for IMU alignment. *IEEE Aerospace and Electronic Systems Magazine, 19*(10), 32–39. doi:10.1109/MAES.2004.1365016

Endevco, M. (2019). *773 Triaxial Variable Capacitance Accelerometer.* March, 03, 2019, from MEGGITT Endevco Website: https://buy.endevco.com/773-accelerometer-1.html

Engineering, F. O. F. (2013). *Challenges and opportunities in solid waste management in Zimbabwe's urban councils.* Academic Press.

Eremenko, A. V., Maloletov, A. V., & Skakunov, V. N. (2010). Microprocessor control system of robotic manipulators. News of VolgGTU, 3, 88–94.

Ester, M., Kriegel, H. P., Sander, J., & Xu, X. (1996). A Density-based Algorithm for Discovering Clusters a Density-based Algorithm for Discovering Clusters in Large Spatial Databases with Noise. International Conference on Knowledge Discovery and Data Mining. AAAI Press. *KDD: Proceedings / International Conference on Knowledge Discovery & Data Mining. International Conference on Knowledge Discovery & Data Mining, 96,* 226–231.

Featherstone, R. (2014). *Rigid body dynamics algorithms.* Springer.

Feng, S., Xinjilefu, X., Atkeson, C. G., & Kim, J. (2015, November). Optimization based controller design and implementation for the atlas robot in the darpa robotics challenge finals. In *2015 IEEE-RAS 15th International Conference on Humanoid Robots (Humanoids)* (pp. 1028-1035). IEEE. 10.1109/HUMANOIDS.2015.7363480

Ferreira,, A., Pereira,, F. G., Vassalo,, R. F., & Teodiano,, F., Bastos, & Sarcinelli, F. M. (2008). An approach to avoid obstacles in mobile robot navigation: The tangential escape. *Revista Controle & Automação, 19*(4), 395–405.

Fischler, M. A., & Bolles, R. C. (1981). Random Sample Consensus: A Paradigm for Model Fitting with Applications to Image Analysis and Automated Cartography. *Communications of the ACM, 24*(6), 381–395. doi:10.1145/358669.358692

Flores-Fuentes, W., Rivas-Lopez, M., Sergiyenko, O., Rodriguez-Quinonez, J. C., Hernandez-Balbuena, D., & Rivera-Castillo, J. (2014, March 6). Energy Center Detection in Light Scanning Sensors for Structural Health Monitoring Accuracy Enhancement. *IEEE Sensors Journal, 17*(7), 2355–2361. doi:10.1109/JSEN.2014.2310224

Focchi, M., Orsolino, R., Camurri, M., Barasuol, V., Mastalli, C., Caldwell, D. G., & Semini, C. (2018). *Heuristic Planning for Rough Terrain Locomotion in Presence of External Disturbances and Variable Perception Quality.* arXiv preprint arXiv:1805.10238

Fu, G., Menciassi, A., & Dario, P. (2012, October). Development of a low-cost active 3D triangulation laser scanner for indoor navigation of miniature mobile robots. *Robotics and Autonomous Systems, 60*(10), 1317–1326. doi:10.1016/j.robot.2012.06.002

Gao, Z., Ge, M., Shen, W., Zhang, H., & Niu, X. (2017). Ionospheric and receiver DCB-constrained multi-GNSS single-frequency PPP integrated with MEMS inertial measurements. *Journal of Geodesy, 91*(11), 1351–1366. doi:10.100700190-017-1029-7

Gebre-Egziabher, D., Hayward, R. C., & Powell, J. D. (1998). A Low-Cost GPS/Inertial Attitude Heading Reference System (AHRS) for General Aviation Application. *Proc. of IEEE Position Location and Navigation Symp.*, 518–525. 10.1109/PLANS.1998.670207

Geiger, A., Lenz, P., Stiller, C., & Urtasun, R. (2013). Vision meets Robotics: The KITTI Dataset. *The International Journal of Robotics Research, 32*(11), 1231–1237. doi:10.1177/0278364913491297

Gilani, O., & Ben-Tzvi, P. (2011, January). Bioinspired Jumping Mobility Concepts for Rough Terrain Mobile Robots. In *ASME 2011 International Mechanical Engineering Congress and Exposition* (pp. 207-214). American Society of Mechanical Engineers. 10.1115/IMECE2011-64050

Gomez-Gil, J., Alonso-Garcia, S., Gómez-Gil, F. J., & Stombaugh, T. (2011). A Simple Method to Improve Autonomous GPS Positioning for Tractors. *Sensors (Basel), 11*(6), 5630–5644. doi:10.3390110605630 PMID:22163917

Gonzalez, R. C., & Woods, R. E. (2008). *Digital Image Processing* (3rd ed.). Upper Saddle River, NJ: Pearson Education.

Gupta, A., Bhargava, P., Agrawal, S., Deshmukh, A., & Kadam, B. (2018, April). Comparative Study of Different Approaches to Inverse Kinematics. In *International Conference on Advances in Computing and Data Sciences* (pp. 556-563). Springer. 10.1007/978-981-13-1813-9_55

Haddad, N. (2011, June). From ground surveying to 3D laser scanner: A review of techniques used for spatial documentation of historic sites. *Journal of King Saud University - Engineering and Science, 23*(2), 109–118. doi:10.1016/j.jksues.2011.03.001

Han. (2005). An Asymmetric Image Encryption Based on Matrix Transformation. *Ecti Transactions on Computer and Information Technology, 1*(2), 126-133.

Hanger, M., Johansen, T. A., Mykland, G. K., & Skullestad, A. (2011, December). Dynamic model predictive control allocation using CVXGEN. In *Control and Automation (ICCA), 2011 9th IEEE International Conference on* (pp. 417-422). IEEE. 10.1109/ICCA.2011.6137940

Hanshar, F. (2013). *Pick and place robot. Dynamic vehicle routing using Genetic Algorithms* (2nd ed.). McGraw Hill.

Hayawi, M. J. (2000). *The Closed Form Solution of the Inverse Kinematics of a 6-DOF Robot.* Academic Press.

Hernandez, W., & Mendez, A. (2018). Application of Principal Components to Image Compression. In T. Göksel (Ed.), *Statistics - Growing Data Sets and Growing Demand for Statistics* (pp. 107-136). IntechOpen. Retrieved September 22, 2019, from https://www.intechopen.com/books/statistics-growing-data-sets-and-growing-demand-for-statistics/application-of-principal-component-analysis-to-image-compression

Hernandez, W., Mendez, A., Quezada-Sarmiento, P. A., Jumbo-Flores, L. A., Mercorelli, P., Tyrsa, V., . . . Cevallos Cevallos, W. B. (2019, October). *Image compression based on periodic principal components*. Paper presented at the 45th Annual Conference of the IEEE Industrial Electronics Society, Lisbon, Portugal.

Hernandez-Balbuena, D., Sergiyenko, O., Rosas-Méndez, P. L., Tyrsa, V., & Rivas-Lopez, M. (2012). Fast method for frequency measurement by rational approximations with application in mechatronics. In *Modern Metrology Concerns*. IntechOpen. doi:10.5772/23225

Hernández, D. C., Filonenko, A., Hariyono, J., Shahbaz, A., & Jo, K. H. (2016). Laser based collision warning system for high con>ℓict vehicle-pedestrian zones. *IEEE International Symposium on Industrial Electronics*, 935–939. 10.1109/ISIE.2016.7745016

Hernández, D. C., Filonenko, A., Seo, D., & Jo, K. (2015). Laser scanner based heading angle and distance estimation. *IEEE International Conference on Industrial Technology*, 1718–1722. 10.1109/ICIT.2015.7125345

Hirukawa, H., Hattori, S., Harada, K., Kajita, S., Kaneko, K., Kanehiro, F., . . . Morisawa, M. (2006, May). A universal stability criterion of the foot contact of legged robots-adios zmp. In *Robotics and Automation, 2006. ICRA 2006. Proceedings 2006 IEEE International Conference on* (pp. 1976-1983). IEEE. 10.1109/ROBOT.2006.1641995

Hirukawa, H., Hattori, S., Kajita, S., Harada, K., Kaneko, K., Kanehiro, F., . . . Nakaoka, S. 2007, April. A pattern generator of humanoid robots walking on a rough terrain. In *Robotics and Automation, 2007 IEEE International Conference on* (pp. 2181-2187). IEEE. 10.1109/ROBOT.2007.363644

Honeywell. (2019). *HG4930 IMU*. March, 03, 2019, from Honeywell Website: https://aerospace.honeywell.com/en/products/navigation-and-sensors/hg4930

Howard, A., Matarić, M. J., & Sukhatme, G. S. (2003). Cooperative relative localization for mobile robot teams: an egocentric approach. *Proceedings of the Naval Research Laboratory Workshop on Multirobot Systems*, 1-12.

Hsu, C. W., Hsu, T. H., & Chang, K. J. (2012). Implementation of car-following system using LiDAR detection. *International Conference on ITS Telecommunications*, 165–169.

Huang, L., Yang, H., Gao, Y., Zhao, L., & Liang, J. (2013). Design and implementation of a micromechanical silicon resonant accelerometer. *Sensors (Basel)*, *13*(11), 15785–15804. doi:10.3390131115785 PMID:24256978

Hutter, M., Gehring, C., Jud, D., Lauber, A., Bellicoso, C. D., Tsounis, V., ... Diethelm, R. (2016, October). Anymal-a highly mobile and dynamic quadrupedal robot. In *2016 IEEE/RSJ International Conference on Intelligent Robots and Systems (IROS)* (pp. 38-44). IEEE. 10.1109/IROS.2016.7758092

Hutter, M., Gehring, C., Lauber, A., Gunther, F., Bellicoso, C. D., Tsounis, V., ... Kolvenbach, H. (2017). ANYmal-toward legged robots for harsh environments. *Advanced Robotics, 31*(17), 918–931. doi:10.1080/01691864.2017.1378591

I, A. P., Much, R., Faithful, M., You, T., & Palumbo, D. (2017). *International Association for the Fantastic in the Arts Alex Proyas's I, Robot : Much More Faithful to Aslmov Than You Think.* Academic Press.

Ibarra-Arenado, M., Tjahjadi, T., Pérez-Oria, J., Robla-Gómez, S., & Jiménez-Avello, A. (2017). Shadow-Based Vehicle Detection in Urban Traffic. *Sensors, 17*(5), 1-19.

Ihn, J. B., & Chang, F. K. (2008). Pitch-catch active sensing methods in structural health monitoring for aircraft structures. *Structural Health Monitoring, 7*(1), 5–19. doi:10.1177/1475921707081979

Jamelske, E. M. (2005). Assessing the support for the switch to automated collection of solid waste with single stream recycling in Madison. doi:10.1177/1087724X05283676

Jatsun, S. F., Volkova, L. Yu., & Vorochaev, A. V. (2013). Investigation of acceleration modes of the four-link jumping device. *Izvestia Volgograd State Technical University, 24*(127), 86–92.

Jatsun, S., Loktionova, O., Volkova, L., & Yatsun, A. (2014). Investigation into the influence of the foot attachment point in the body on the four-link robot jump characteristics. In *Advances on Theory and Practice of Robots and Manipulators* (pp. 159–166). Cham: Springer. doi:10.1007/978-3-319-07058-2_18

Jatsun, S., Savin, S., & Yatsun, A. (2017, September). Footstep planner algorithm for a lower limb exoskeleton climbing stairs. In *International Conference on Interactive Collaborative Robotics* (pp. 75-82). Springer. 10.1007/978-3-319-66471-2_9

Jatsun, S., Savin, S., & Yatsun, A. (2018, September). Harmonic Function-Based ZMP Trajectory Generation for Nonlinear Motion of Walking Robots. In *2018 International Russian Automation Conference (RusAutoCon)* (pp. 1-6). IEEE. 10.1109/RUSAUTOCON.2018.8501657

Jean, B. R., & Rouse, J. W. (1983). A Multiple Beam Synthetic Aperture Radar Design Concept for Geoscience Applications. *IEEE Transactions on Geoscience and Remote Sensing, GE-21*(2), 201–207. doi:10.1109/TGRS.1983.350489

Jin, T.-S., Lee, J.-M., & Hashimoto, H. (2006). Position Control of Mobile Robot for Human-Following in Intelligent Space with Distributed Sensors. *International Journal of Control, Automation, and Systems, 4*(2), 204–216.

Johansson, S. (2006). New frequency counting principle improves resolution. In *Proceedings of the 20th European Frequency and Time Forum* (pp. 139-146). IEEE.

Johnson, M., Shrewsbury, B., Bertrand, S., Calvert, D., Wu, T., Duran, D., ... Smith, J. (2017). Team IHMC's Lessons Learned from the DARPA Robotics Challenge: Finding Data in the Rubble. *Journal of Field Robotics, 34*(2), 241–261. doi:10.1002/rob.21674

Johnson, M., Shrewsbury, B., Bertrand, S., Wu, T., Duran, D., Floyd, M., ... Carff, J. (2015). Team IHMC's lessons learned from the DARPA robotics challenge trials. *Journal of Field Robotics, 32*(2), 192–208. doi:10.1002/rob.21571

JPEG. (2019). *Overview of JPEG*. Retrieved September 22, 2019, from https://jpeg.org/jpeg/index.html

Kadochnikov, G. N. (2006). *Test report No. 07-06-2006 (1200012). Information measuring system IP-256M.* FGU: Kuban MIS.

Kajita, S., Kanehiro, F., Kaneko, K., Fujiwara, K., Harada, K., Yokoi, K., & Hirukawa, H. (2003, September). Biped walking pattern generation by using preview control of zero-moment point. In ICRA (Vol. 3, pp. 1620-1626). doi:10.1109/ROBOT.2003.1241826

Kalisz, J. (2003). Review of methods for time interval measurements with picosecond resolution. *Metrologia, 41*(1), 17–32. doi:10.1088/0026-1394/41/1/004

Kanoulas, D., Stumpf, A., Raghavan, V. S., Zhou, C., Toumpa, A., Von Stryk, O., ... Tsagarakis, N. G. (2018, May). Footstep Planning in Rough Terrain for Bipedal Robots Using Curved Contact Patches. In *2018 IEEE International Conference on Robotics and Automation (ICRA)* (pp. 1-9). IEEE. 10.1109/ICRA.2018.8460561

Karumanchi, S., Edelberg, K., Baldwin, I., Nash, J., Reid, J., Bergh, C., ... Newill-Smith, D. (2017). Team RoboSimian: Semi-autonomous mobile manipulation at the 2015 DARPA robotics challenge finals. *Journal of Field Robotics, 34*(2), 305–332. doi:10.1002/rob.21676

Kawatsuma, S., Fukushima, M., & Okada, T. (2012). Emergency response by robots to Fukushima-Daiichi accident: Summary and lessons learned. *IndustrialRobot: AnInternational Journal, 39*(5), 428–435.

Kessler, S. S., Jugenheimer, K. A., Size, A. B., & Dunn, C. T. (2008). *U.S. Patent No. 7,469,595.* Washington, DC: U.S. Patent and Trademark Office.

Khedr, W. M., & Abdelrazek, M. (2016). Image Compression Using DCT Upon Various Quantization. *International Journal of Computers and Applications, 137*(1), 11–13. doi:10.5120/ijca2016908648

Khoroshko, V. O. (2003). *Methods and means of information protection: Teaching. Academic Press.*

Kidera, S. (2010). Shadow region imaging algorithm using array antenna based on aperture synthesis of multiple scattered waves for UWB radars. *Proceedings of International Geoscience and Remote Sensing Symposium*, 1-4. 10.1109/IGARSS.2010.5651782

Kim, B., Choi, B., Park, S., Kim, H., & Kim, E. (2016). Pedestrian/Vehicle Detection Using a 2.5-D Multi-Layer Laser Scanner. *IEEE Sensors Journal, 16*(2), 400–408. doi:10.1109/JSEN.2015.2480742

Kirianaki, N. V., Yurish, S. Y., & Shpak, N. O. (2001). Methods of dependent count for frequency measurements. *Measurement, 29*(1), 31–50. doi:10.1016/S0263-2241(00)00026-9

Klets, D. M. (2012). Improving the accuracy of the experimental evaluation of the performance properties of wheeled vehicles using Kalman filter. *Technological and technological aspects of development and development of new technical and technological technologies for the Ukrainian government: Zbirnik nauk.pr. DNU UkrNDIPVT im. L. Pogoriloy,* Doslidnitske. *UkrNDIPVT, 16*(30), 467–484.

Kodenko, M. N., & Lebedev, A. T. (1969). *Automation of tractor units.* Moskow: Mashinostroenie.

Korkishko, T. A. (2003). *Algorithms and Processors of Symmetric Block Encryption: Scientific Edition.* Baku: V.A. Melnik. - Lviv.

Kornblum, J. D. (2008). Using JPEG Quantization Tables to Identify Imagery Processed by Software. *Digital Investigation, 5,* S21-S25.

Kovac, M. (2010). *Bioinspired jumping locomotion for miniature robotics.* Academic Press.

Kovalchuk A. (2009). *Increasing the stability of the RSA system when encrypting images.* Academic Press.

Krasilenko, V. G. (2006). A noise-immune crptographis information protection method for facsimile information transmission and the realization algorithms. *Proc. SIEE, 6241,* 316-322.

Krasilenko, V. G. (2006). Development of the method of cryptographic protection of information text-graphic type. Science and educational process: a scientific and methodical collection of scientific and practical conference, 73-74.

Krasilenko, V. G. (2008). Simulation of the modified algorithm for creating 2-D keys in cryptographic applications. *Scientific-methodical collection of the scientific-practical conference "Science and educational process",* 107-109.

Krasilenko, V. G. (2010). Modeling of Matrix Affine Algorithms for the Encryption of Color Images. Computer technologies: science and education: abstracts of reports v VseUkr. sci. conf., 120-124.

Krasilenko, V. G. (2012). Modifications of the RSA system for creation of matrix models and algorithms for encryption and decryption of images on its basis. Systems of information processing, 8, 102-106.

Krasilenko, V. G. (2012). Simulation of Blind Electronic Digital Signatures of Matrix Type on Confidential Text-Graphic Documentation. *International Scientific-Methodical Conference,* 103-107.

Krasilenko, V. G. (2013). Matrix models of cryptographic transformations of images with matrix-bit-map decomposition and mixing and their modeling. Materials of 68 NTC "Modern Information Systems and Technologies. Informational security", 139-143.

Krasilenko, V. G. (2014). Cryptographic transformations of images based on matrix models of permutations with matrix-bit-map decomposition and their modeling. *Bulletin of Khmelnitsky National University. Technical sciences, 1*, 74-79.

Krasilenko, V. G. (2016). Cryptographic transformations (CTs) of color images based on matrix models with operations on modules. In *Modern methods, information and software management systems for organizational and technical complexes: a collection of abstracts of reports of the All-Ukrainian scientific and practical Internet conference.* Lutsk: RVB of Lutsk National Technical University.

Krasilenko, V. G. (2016). Modeling and research of cryptographic transformations of images based on their matrix-bit-map decomposition and matrix models of permutations with verification of integrity. In *Electronics and Information Technologies: a collection of scientific works.* Lviv: Lviv Ivan Franko National University. Retrieved from http://elit.lnu.edu.ua/pdf/6_12.pdf

Krasilenko, V. G. (2016). Modeling cryptographic transformations of color images with verification of the integrity of cryptograms based on matrix permutation models. *Materials of the scientific and practical Internet conference "Problems of modeling and development of information systems",* 128-136. Retrieved from http://ddpu.drohobych.net/wp-content/uploads/2016/04/material_konf.pdf37

Krasilenko, V. G. (2016). Simulation of cryptographic transformations of color images based on matrix models of permutations with spectral and bit-map decompositions. *Computer-integrated technologies: education, science, 23,* 31-36. Retrieved from http://ki.lutsk-ntu.com.ua/node/132/section/9

Krasilenko, V. G. (2017). Modeling of multi-stage and multi-protocol protocols for the harmonization of secret matrix keys. *Computer-integrated technologies: education, science, production: scientific journal, 26,* 111-120. Retrieved from http://ki.lutsk-ntu.com.ua/node/134/section/27

Krasilenko, V. G. (2017). Modeling Protocols for Matching a Secret Matrix Key for Cryptographic Transformations and Matrix-type Systems. *Systems of information processing, 3*(149), 151-157.

Krasilenko, V.G. (2004). Algorithms and architecture for high-precision matrix-matrix multipliers based on optical four-digit alternating arithmetic. *Measuring and computing engineering in technological processes, 1,* 13-26.

Krasilenko, V.G. (2009). Modeling of Matrix Cryptographic Protection Algorithms. *Bulletin of the National University of Lviv Polytechnic "Computer Systems and Networks", 658,* 59-63.

Krasilenko, V.G. (2011). Matrix Affine Ciphers for the Creation of Digital Blind Signatures for Text-Graphic Documents. *Systems of information processing, 7*(97), 60 - 63.

Krasilenko, V.G. (2012). Algorithms for the formation of two-dimensional keys for matrix algorithms of cryptographic transformations of images and their modeling. *Systems of information processing, 8,* 107-110.

Krasilenko, V.G. (2012). Matrix affine and permutation ciphers for encryption and decryption of images. *Systems of information processing, 3*(101), 53-62.

Krasilenko, V. G. (2013). *Matrix models of permutations with matrix-bit decomposition for cryptographic transformations of images and their modeling. In Science and educational process: a scientific and methodical collection of materials of the NPC of all the Universities "Ukraine"* (pp. 90–92). Vinnytsya: Vinnytsia Socio-Economic Institute of the University of Ukraine.

Kring, J., & Travis, J. (2006). LabVIEW for Everyone: Graphical Programming Made Easy and Fun (3rd ed.). Prentice Hall.

Kuindersma, S., Permenter, F., & Tedrake, R. (2014, May). An efficiently solvable quadratic program for stabilizing dynamic locomotion. In *Robotics and Automation (ICRA), 2014 IEEE International Conference on* (pp. 2589-2594). IEEE. 10.1109/ICRA.2014.6907230

Kuindersma, S., Deits, R., Fallon, M., Valenzuela, A., Dai, H., Permenter, F., ... Tedrake, R. (2016). Optimization-based locomotion planning, estimation, and control design for the atlas humanoid robot. *Autonomous Robots, 40*(3), 429–455. doi:10.100710514-015-9479-3

Kundur, P., Balu, N. J., & Lauby, M. G. (1994). *Power system stability and control* (Vol. 7). New York: McGraw-Hill.

Kuvachev, V. P., Ayubov, A. M., & Kotov, O. G. (2007). Improvement of the method of registration of vertical vibrations of mobile energy means. *Proceedings of Tavria DATE. Melitopol. TDAATA, 7*(1), 139–145.

Lapin, A. (2005). New generation of Texas Instruments products for a controlled electric drive. *Electronics: Science, Technology Business, 7*, 56–59.

LaValle, S. M. (1998). *Rapidly-exploring random trees: A new tool for path planning*. Academic Press.

LaValle, S. M., & Kuffner Jr, J. J. (2000). *Rapidly-exploring random trees: Progress and prospects*. Academic Press.

Lee, S., & Song, J.-B. (2004). Mobile Robot Localization Using Optical Flow Sensors. *International Journal of Control, Automation, and Systems, 2*(4), 485–493.

Lehtomäki, M., Jaakkola, A., Hyyppä, J., Lampinen, J., Kaartinen, H., Kukko, A., ... Hyyppä, H. (2015). Object Classification and Recognition From Mobile Laser Scanning Point Clouds in a Road Environment. *IEEE Transactions on Geoscience and Remote Sensing, 54*(2), 1226–1239. doi:10.1109/TGRS.2015.2476502

Lifewire. (2019). *JPEG File Format Myths and Facts*. Retrieved September 24, 2019, from https://www.lifewire.com/jpeg-myths-and-facts-1701548

Li, I.-H., Chen, M.-C., Wang, W.-Y., Su, S.-F., & Lai, T.-W. (2014). Mobile Robot Self-Localization System Using Single Webcam Distance Measurement Technology in Indoor Environments. *Sensors (Basel), 14*(2), 2089–2109. doi:10.3390140202089 PMID:24473282

Lindner, L. (2016). Laser Scanners. In O. Sergiyenko, J. Rodriguez-Quinonez, O. Sergiyenko, & J. Rodriguez-Quinonez (Eds.), Developing and Applying Optoelectronics in Machine Vision (p. 38). Hershey, PA: IGI Global. doi:10.4018/978-1-5225-0632-4.ch004

Lindner, L., Sergiyenko, O., Rivas-Lopez, M., Ivanov, M., Rodriguez-Quinonez, J., Hernandez-Balbuena, D., . . . Mercorelli, P. (2017). Machine vision system errors for unmanned aerial vehicle navigation. In *Industrial Electronics (ISIE), 2017 IEEE 26th International Symposium on* (pp. 1615-1620). Edinburgh, UK: IEEE. doi:10.1109/ISIE.2017.8001488

Lindner, L., Sergiyenko, O., Rivas-Lopez, M., Rodriguez-Quinonez, J., Hernandez-Balbuena, D., Flores-Fuentes, W., . . . Kartashov, V. (2016). Issues of exact laser ray positioning using DC motors for vision-based target detection. In *2016 IEEE 25th International Symposium on Industrial Electronics (ISIE)* (pp. 929-934). Santa Clara, CA: IEEE.

Lindner, L., Sergiyenko, O., Rodriguez-Quinonez, J., Rivas-Lopez, M., Hernandez-Balbuena, D., Flores-Fuentes, W., ... Tyrsa, V. (2016). Mobile robot vision system using continuous laser scanning for industrial application. *The Industrial Robot*, *43*(4), 360–369. doi:10.1108/IR-01-2016-0048

Li, Q., Chen, L., Li, M., Shaw, S., & Nüchter, A. (2014). A Sensor-Fusion Drivable-Region and Lane-Detection System for Autonomous Vehicle Navigation in Challenging Road Scenarios. *IEEE Transactions on Vehicular Technology*, *63*(2), 540–555. doi:10.1109/TVT.2013.2281199

Liqing, N., & Qingjiu, H. (2012). Inverse Kinematics for 6-DOF Manipulator by the Method of Sequential Retrieval. *Proceedings of the 1st International Conference on Mechanical Engineering and Material Science, 1*, 255–258. 10.2991/mems.2012.157

Li, T., Zhang, H., Gao, Z., Niu, X., & El-sheimy, N. (2019). Tight Fusion of a Monocular Camera, MEMS-IMU, and Single-Frequency Multi-GNSS RTK for Precise Navigation in GNSS-Challenged Environments. *Remote Sensing*, *11*(6), 610. doi:10.3390/rs11060610

Li, T., Zhang, H., Niu, X., & Gao, Z. (2017). Tightly-coupled integration of multi-GNSS single-frequency RTK and MEMS-IMU for enhanced positioning performance. *Sensors (Basel)*, *17*(11), 2462. doi:10.339017112462 PMID:29077070

Liu, Z., Liu, D., & Chen, T. (2013). Vehicle detection and tracking with 2D laser range finders. *International Congress on Image and Signal Processing*, *2*, 1006–1013. 10.1109/CISP.2013.6745203

Liu, Z., Wang, J., & Liu, D. (2003). A New Curb Detection Method for Unmanned Ground Vehicles Using 2D Sequential Laser Data. *Sensors (Basel)*, *13*(1), 1102–1120. doi:10.3390130101102 PMID:23325170

Long, X. (2017). *Optimization-based Whole-body Motion Planning for Humanoid Robots* (Doctoral dissertation). Northeastern University.

Loughery, J. (2018). The Hudson Review. *Inc*, *48*(2), 301–307.

Lukin, K. A., & ..... (2013). Tomographic imaging using noise radar and 2D aperture synthesis. *Applied Radio Electronics*, *12*(1), 152–156.

Lu, T.-C., & Chen, S.-L. (2016). Genetic algorithm-based S-curve acceleration and deceleration for five-axis machine tools. *International Journal of Advanced Manufacturing Technology*, *87*(1-4), 219–232. doi:10.100700170-016-8464-0

Machale, S. (2015). Smart garbage collection system in residential area. *Solid Waste Management and Monitoring*, *5*(11), 13–17.

Madgwick. (2010). *An effcient orientation filter for inertial and inertial/magnetic sensor arrays.* Academic Press.

Madisetti, V. (2009). *Digital signal processing fundamentals.* CRC Press. doi:10.1201/9781420046076

Maklouf, O., Ghila, A., Abdulla, A., & Yousef, A. (2013). Low Cost IMU\GPS Integration Using Kalman Filtering for Land Vehicle Navigation Application. *International Journal of Electrical, Computer, Energetic Electronic and Communication Engineering*, *7*(2), 184–190.

Martin-Marcos, A. (1999). *Compresión de Imágenes: Norma JPEG.* Madrid: Editorial Ciencia 3 S.L.

Más, F. R., Zhang, Q., & Hansen, A. C. (2011). *Mechatronics and Intelligent Systems for Off-road Vehicles.* London: Springer-Verlag London.

Mason, S., Rotella, N., Schaal, S., & Righetti, L. (2016, November). Balacing and walking using full dynamics LQR control with contact constraints. In *2016 IEEE-RAS 16th International Conference on Humanoid Robots (Humanoids)* (pp. 63-68). IEEE. 10.1109/HUMANOIDS.2016.7803255

Mastalli, C., Havoutis, I., Focchi, M., Caldwell, D., & Semini, C. (2018). *Motion planning for quadrupedal locomotion: coupled planning, terrain mapping and whole-body control.* Academic Press.

Mattingley, J., & Boyd, S. (2012). CVXGEN: A code generator for embedded convex optimization. *Optimization and Engineering*, *13*(1), 1–27. doi:10.100711081-011-9176-9

Maxon Academy. (n.d.). Retrieved November 09, 2016, from www.maxonmotor.com

McGeer, T. (1990). Passive bipedal running. *Proceedings of the Royal Society of London. Series B, Biological Sciences*, *240*(1297), 107–134. doi:10.1098/rspb.1990.0030 PMID:1972987

McGeer, T. (1990). Passive dynamic walking. *The International Journal of Robotics Research*, *9*(2), 62–82. doi:10.1177/027836499000900206

Melo, L. F., Rosário, J. R., & Silveira, A. F. (2013). Mobile Robot Indoor Autonomous Navigation with Position Estimation Using RF Signal Triangulation. *Positioning*, *4*(01), 20–35. doi:10.4236/pos.2013.41004

Meredith, M., & Maddock, S. (2004). *Real-time inverse kinematics: The return of the Jacobian.* Technical Report No. CS-04-06, Department of Computer Science, University of Sheffield.

Mertz, C., Navarro-Serment, L. E., MacLachlan, R., Rybski, P., Steinfeld, A., Suppé, A., ... Gowdy, J. (2013). Moving Object Detection with Laser Scanners. *Journal of Field Robotics, 30*(1), 17–43. doi:10.1002/rob.21430

Mester, G., & Rodić, A. (2010). Sensor-Based Intelligent Mobile Robot Navigation in Unknown Environments. *International Journal of Electrical and Computer Engineering Systems, 1*(2), 1–8.

Mojtaba, N., Alimardani, R., Sharifi, A., & Tabatabaeefar, A. (2009). A Microcontroller-Based Data Logging System for Cone. *Tarim Makinalaji Bilimi Dergisi, 5*(4), 379–384.

Murphy, R. R., Tadokoro, S., Nardi, D., Jacoff, A., Fiorini, P., Choset, H., & Erkmen, A. M. (2008). Search and rescue robotics. In *Springer handbook of robotics* (pp. 1151–1173). Berlin: Springer. doi:10.1007/978-3-540-30301-5_51

Murrieta-Rico, F. N., Hernandez-Balbuena, D., Petranovskii, V., Nieto-Hipolito, J. I., Pestryakov, A., Sergiyenko, O., ... Tyrsa, V. (2014). Acceleration measurement improvement by application of novel frequency measurement technique for FDS based INS. In *2014 IEEE 23rd International Symposium on Industrial Electronics (ISIE)* (pp. 1920-1925). IEEE. 10.1109/ISIE.2014.6864909

Murrieta-Rico, F. N., Hernandez-Balbuena, D., Rodriguez-Quiñonez, J. C., Petranovskii, V., Raymond-Herrera, O., Hipolito, J. I. N., ... Melnyk, V. I. (2015). Instability measurement in time-frequency references used on autonomous navigation systems. In *2015 IEEE 24th International Symposium on Industrial Electronics (ISIE)* (pp. 956-961). IEEE. 10.1109/ISIE.2015.7281600

Murrieta-Rico, F. N., Mercorelli, P., Sergiyenko, O. Y., Petranovskii, V., Hernandez-Balbuena, D., & Tyrsa, V. (2015). Mathematical modelling of molecular adsorption in zeolite coated frequency domain sensors. *IFAC-PapersOnLine, 48*(1), 41–46. doi:10.1016/j.ifacol.2015.05.060

Murrieta-Rico, F. N., Petranovskii, V., Sergiyenko, O. Y., Hernandez-Balbuena, D., & Lindner, L. (2017). A New Approach to Measurement of Frequency Shifts Using the Principle of Rational Approximations. *Metrology and Measurement Systems, 24*(1), 45–56. doi:10.1515/mms-2017-0007

Murrieta-Rico, F. N., Sergiyenko, O. Y., Petranovskii, V., Hernandez-Balbuena, D., Lindner, L., Tyrsa, V., ... Karthashov, V. M. (2016). Pulse width influence in fast frequency measurements using rational approximations. *Measurement, 86*, 67–78. doi:10.1016/j.measurement.2016.02.032

Murrieta-Rico, F. N., Sergiyenko, O. Y., Petranovskii, V., Hernandez-Balbuena, D., Lindner, L., Tyrsa, V., ... Nieto-Hipolito, J. I. (2018). Optimization of pulse width for frequency measurement by the method of rational approximations principle. *Measurement, 125*, 463–470. doi:10.1016/j.measurement.2018.05.008

Nashashibi, F., & Bargeton, A. (2008). Laser-based vehicles tracking and classification using occlusion reasoning and confidence estimation. *IEEE Intelligent Vehicles Symposium*, 847–852. 10.1109/IVS.2008.4621244

Negenborn, R. (2003). *Robot Localization and Kalman Filters: On finding your position in a noisy world* (Thesis). Utrecht University.

Noureldin, A., Karamat, T. B., & Georgy, J. (2012). *Fundamentals of inertial navigation, satellite-based positioning and their integration.* Springer Science & Business Media.

Nurlansa, O., Istiqomah, D. A., Astu, M., & Pawitra, S. (2014). *AGATOR (Automatic Garbage Collector) as Automatic Garbage Collector Robot Model.* doi:10.7763/IJFCC.2014.V3.329

Oh, S.I., & Kang, H.B (2017). Object Detection and Classification by Decision-Level Fusion for Intelligent Vehicle Systems. *Sensors, 17*(1), 1-21.

Ort, T., Paull, L., & Rus, D. (2018). *Autonomous Vehicle Navigation in Rural Environments without Detailed Prior Maps.* MIT CSAIL. doi:10.1109/ICRA.2018.8460519

Özturan, P. M., Bozanta, A., Basarir-Ozel, B., Akar, E., & Coşkun, M. (2015). A roadmap for an integrated university information system based on connectivity issues: Case of Turkey. *The International Journal of Management Science and Information Technology, 17*(17), 1–23. doi:10.14313/JAMRIS

Panovko, G. Y., Savin, S. I., Yatsun, S. F., & Yatsun, A. S. (2016). Simulation of exoskeleton sit-to-stand movement. *Journal of Machinery Manufacture and Reliability, 45*(3), 206–210. doi:10.3103/S1052618816030110

Parker, P. A., & Finley, T. D. (2007). Advancements in aircraft model force and attitude instrumentation by integrating statistical methods. *Journal of Aircraft, 44*(2), 436–443. doi:10.2514/1.23060

Pei, F. J., Liu, X., & Zhu, L. (2014). In-Flight Alignment Using Filter for Strapdown INS on Aircraft. *The Scientific World Journal.* PMID:24511300

Peters, J. F., Ahn, T. C., & Borkowski, M. (2002, October). Obstacle classification by a line-crawling robot: A rough neurocomputing approach. In *International Conference on Rough Sets and Current Trends in Computing* (pp. 594-601). Springer. 10.1007/3-540-45813-1_79

Playter, R., Buehler, M., & Raibert, M. (2006, May). BigDog. In *Unmanned Systems Technology VIII* (Vol. 6230, p. 62302O). International Society for Optics and Photonics. doi:10.1117/12.684087

Poliarus, O., Poliakov, Y., & Lindner, L. (2018). Determination of landmarks by mobile robot's vision system based on detecting abrupt changes of echo signals parameters. *The 44th Annual Conference of the IEEE Industrial Electronics Society,* 3165-3170.

Polytechnic, B. S. P. (n.d.). *Pick and place robot.* Academic Press.

Popirlan, C., & Dupac, M. (2009). An Optimal Path Algorithm for Autonomous Searching Robots. *Annals of University of Craiova. Math. Comp. Sci. Ser., 36*(1), 37–48.

Poulakakis, I., & Grizzle, J. W. (2009, May). Modeling and control of the monopedal robot thumper. In *Robotics and Automation, 2009. ICRA'09. IEEE International Conference on* (pp. 3327-3334). IEEE. 10.1109/ROBOT.2009.5152708

Pradhan, S. S. (2009). *Design and Implementation of Energy Pumping Mechanism and Stabilizing Control on a One-Legged Hopping Robot.* Department of Mechanical EngineeringIndian Institute of Technology, Bombay.

Pratt, G. A. (2000). Legged robots at MIT: What's new since Raibert? *IEEE Robotics & Automation Magazine, 7*(3), 15–19. doi:10.1109/100.876907

Premebida, C., Monteiro, G., Nunes, U., & Peixoto, P. (2007). A Lidar and Vision-based Approach for Pedestrian and Vehicle Detection and Tracking. *IEEE Intelligent Transportation Systems Conference*, 1044–1049. 10.1109/ITSC.2007.4357637

Presicion Motion Controller LM629. (2013, May 6). Retrieved February 27, 2019, from http://www.ti.com/product/LM629?jktype=recommendedresults

Qi, H., & Moore, J. B. (2002). Direct Kalman filtering approach for GPS/INS integration. *IEEE Transactions on Aerospace and Electronic Systems, 38*(2), 687–693. doi:10.1109/TAES.2002.1008998

Radford, N. A., Strawser, P., Hambuchen, K., Mehling, J. S., Verdeyen, W. K., Donnan, A. S., ... Berka, R. (2015). Valkyrie: Nasa's first bipedal humanoid robot. *Journal of Field Robotics, 32*(3), 397–419. doi:10.1002/rob.21560

Raibert, M., Blankespoor, K., Nelson, G., & Playter, R. (2008). Bigdog, the rough-terrain quadruped robot. *IFAC Proceedings Volumes, 41*(2), 10822-10825. 10.3182/20080706-5-KR-1001.01833

Raid, A. M., Khedr, W. M., El-dosuky, M. A., & Ahmed, W. (2014). Jpeg Image Compression Image Compression Using Discrete-Cosine Transform – A Survey. *International Journal of Computer Science & Engineering Survey, 5*(2), 39–47. doi:10.5121/ijcses.2014.5204

Rajyalakshmi, P., & Raju, G. S. N. (2011). Characteristics of Radar Cross Section with Different Objects. *International Journal of Electronics and Communication Engineering, 4*(2), 205-216.

Rakesh, N., A, P. K., & Ajay, S. (2013). *Design And Manufacturing Of Low Cost Pneumatic Pick And Place Robot.* Academic Press.

Rakesh, N., A, P. K., & Ajay, S. (2015). Design And Manufacturing Of Low Cost Pneumatic Pick And Place Robot. *Evolutionary Optimization in Uncertain Environments, 2*(8), 131–133.

Rashkevich, Y.M. (2009). Affine transformations in modifications of the RSA image encryption algorithm. *Automation. Electrotechnical complexes and systems, 2*(24), 59-66.

Rivas-Lopez, M., Sergiyenko, O., & Tyrsa, V. (2008). *Machine Vision: Approaches and Limitations.* InTech. Retrieved from http://cdn.intechopen.com/pdfs/5210/InTech-Machine_vision_approaches_and_limitations.pdf

Sabatini, R., Cappello, F., Ramasamy, S., Gardi, A., & Clothier, R. (2015). An innovative navigation and guidance system for small unmanned aircraft using low-cost sensors. *Aircraft Engineering and Aerospace Technology: An International Journal, 87*(6), 540–545. doi:10.1108/AEAT-06-2014-0081

Sajjad, M., Talpur, H., & Shaikh, M. H. (2012). *Automation of Mobile Pick and Place Robotic System for Small.* Academic Press.

Sakagami, Y., Watanabe, R., Aoyama, C., Matsunaga, S., Higaki, N., & Fujimura, K. (2002). The intelligent ASIMO: System overview and integration. In *IEEE/RSJ international conference on intelligent robots and systems* (Vol. 3, pp. 2478–2483). IEEE. doi:10.1109/IRDS.2002.1041641

Salmador, A., Cid, J. P., & Novelle, I. R. (1989). *Intelligent Garbage Classifier.* Academic Press.

Salomon, D., & Motta, G. (2010). *Handbook of Data Compression* (5th ed.). London: Springer-Verlag. doi:10.1007/978-1-84882-903-9

Salton, J. R. (2010). *Urban Hopper // SPIE Defense.* Orlando, FL: Security and Sensing.

Sankar, A. R., Das, S., & Lahiri, S. K. (2009). Cross-axis sensitivity reduction of a silicon MEMS piezoresistive accelerometer. *Microsystem Technologies, 15*(4), 511–518. doi:10.100700542-008-0740-y

Saravana, G., Sasi, S., Ragavan, R., & Balakrishnan, M. (2016). *Automatic Garbage Separation Robot Using Image Processing Technique.* Academic Press.

Savin, S., & Ivakhnenko, A. (2019). Enhanced Footsteps Generation Method for Walking Robots Based on Convolutional Neural Networks. In Handbook of Research on Deep Learning Innovations and Trends (pp. 16-39). IGI Global. doi:10.4018/978-1-5225-7862-8.ch002

Savin, S., Jatsun, S., & Vorochaeva, L. (2017, November). Modification of Constrained LQR for Control of Walking in-pipe Robots. In *2017 Dynamics of Systems, Mechanisms and Machines (Dynamics)* (pp. 1-6). IEEE.

Savin, S., Khusainov, R., & Klimchik, A. (2019) Admissible region ZMP trajectory generation for bipedal robots walking over uneven terrain. *Zavalishin's Readings.* (forthcoming)

Savin, S., Yatsun, A., & Loktionova, O. (2019) Footstep Planning for Bipedal Robots and Lower Limb Exoskeletons Moving Through Narrow Doors. *Zavalishin's Readings.* (forthcoming)

Savin, S. (2017, June). An algorithm for generating convex obstacle-free regions based on stereographic projection. In *2017 International Siberian Conference on Control and Communications (SIBCON)* (pp. 1-6). IEEE. 10.1109/SIBCON.2017.7998590

Savin, S., Vorochaev, A., & Vorochaeva, L. (2018). Inverse Kinematics for a Walking in-Pipe Robot Based on Linearization of Small Rotations. The Eurasia Proceedings of Science, Technology. *Engineering & Mathematics, 4,* 50–55.

Savin, S., & Vorochaeva, L. (2017, May). Nested quadratic programming-based controller for pipeline robots. In *2017 International Conference on Industrial Engineering, Applications and Manufacturing (ICIEAM)* (pp. 1-6). IEEE. 10.1109/ICIEAM.2017.8076142

SBG Systems. (2019). *Ellipse2*. March, 03, 2019, from SBG Systems Website: https://www.sbg-systems.com/products/ellipse-2-series/

Schapire, R. E. (1999). A Brief Introduction to Boosting. In *International Joint Conference on Artificial Intelligence*. Morgan Kaufmann Publishers Inc.

Schultz, J., & Miner, J. (n.d.). *Wall-E Robot Final Report*. Academic Press.

Schwarz, K. P. (1996). Aircraft position and attitude determination by GPS and INS. *International Archives of Photogrammetry and Remote Sensing*, *31*(B6), 67–73.

Sekiya, H., Kinomoto, T., & Miki, C. (2016). Determination method of bridge rotation angle response using MEMS IMU. *Sensors (Basel)*, *16*(11), 1882. doi:10.339016111882 PMID:27834871

Sergiyenko, O., Hernandez, W., Tyrsa, V., Devia Cruz, L., Starostenko, O., & Pena-Cabrera, M. (2009, May 21). Remote Sensor for Spatial Measurements by Using Optical Scanning. *Sensor*, 5477-5492. doi:10.339090705477

Sergiyenko, O., Tyrsa, V., Basaca-Preciado, L., Rodriguez-Quinonez, J., Hernandez, W., Nieto-Hipolito, J., ... Starostenko, O. (2011). Electromechanical 3D Optoelectronic Scanners: Resolution Constraints and Possible Ways of Improvement. In Optoelectronic Devices and Properties. InTech. doi:10.5772/14263

Sergiyenko, O. Y. (2016). Data transferring model determination in robotic group. *Robotics and Autonomous Systems*, 1–10.

Sergiyenko, O. Y., Balbuena, D. H., Tyrsa, V. V., Mendez, P. L. R., Hernandez, W., Hipolito, J. I. N., ... Lopez, M. R. (2012). Automotive FDS resolution improvement by using the principle of rational approximation. *IEEE Sensors Journal*, *12*(5), 1112–1121. doi:10.1109/JSEN.2011.2166114

Sergiyenko, O., Balbuena, D. H., Tyrsa, V., Mendez, P. L. A. R., Lopez, M. R., Hernandez, W., ... Gurko, A. (2011). Analysis of jitter influence in fast frequency measurements. *Measurement*, *44*(7), 1229–1242.

Serrano, J. M., Peça, J. O., Shahidian, S., Nunes, M. C., Ribeiro, L., & Santos, F. (2011). Development of a Data Acquisition System to optimizing the Agricultural Tractor Performance. *Journal of Agricultural Science and Technology*, 756–766.

Shepherd, R. F., Ilievski, F., Choi, W., Morin, S. A., Adam, A., Mazzeo, A. D., ... Whitesidesb, G. M. (2017). *Multigait soft robot Linked references are available on JSTOR for this article : Multigait soft robot*. doi:10.1073/pnas

Shifrin, J. S. (1971). Statistical antenna theory. Golem Press.

Short, A., & Bandyopadhyay, T. (2018). Legged motion planning in complex three-dimensional environments. *IEEE Robotics and Automation Letters*, *3*(1), 29–36. doi:10.1109/LRA.2017.2728200

Singh, S. K. (2008). *Kinematics Fundamentals*. Houston, TX: Connexions.

Song, X., Seneviratne, L., & Althoefer, K. (2009). *A Vision Based Wheel Slip Estimation Technique for Mining Vehicles*. IFACMMM.

Stoeter, S. A., & Papanikolopoulos, N. (2006). Kinematic motion model for jumping scout robots. *IEEE Transactions on Robotics*, *22*(2), 397–402. doi:10.1109/TRO.2006.862483

Sugiura, R., Fukagawa, T., Noguchi, N., Ishii, K., Shibata, Y., & Toriyama, K. (2003). Field information system using an agricultural helicopter towards precision farming. *Proceedings IEEE/ASME International Conference on Advanced Intelligent Mechatronics*, *2*, 1073–1078. 10.1109/AIM.2003.1225491

Surrécio, A., Nunes, U., & Araújo, R. (2005). *Using Kalman Filters and Augmented System Models for Mobile Robot Navigation*. Dubrovnic, Croatia: IEEE ISIE.

Sysoev, S. (2009). Magnetically controlled, MEMS and multisensory motion sensors of 2009 are more functional, more precise, miniature predecessors. *Component Technology*, *97*, 54–63.

Tanaka, H., Hoshino, K., Matsumoto, K., & Shimoyama, I. (2005, August). Flight dynamics of a butterfly-type ornithopter. In *Intelligent Robots and Systems, 2005.(IROS 2005). 2005 IEEE/RSJ International Conference on* (pp. 2706-2711). IEEE. 10.1109/IROS.2005.1544999

Thompson, C. M., & Raibert, M. H. (1990). Passive dynamic running. In *Experimental Robotics I* (pp. 74–83). Berlin: Springer. doi:10.1007/BFb0042513

Torres-garcía, A., Rodea-aragón, O., Longoria-gandara, O., Sánchez-garcía, F., & González-jiménez, L. E. (2015). *Intelligent Waste Separator*. doi:10.13053/CyS-19-3-2254

Tsagarakis, N. G., Caldwell, D. G., Negrello, F., Choi, W., Baccelliere, L., Loc, V. G., ... Natale, L. (2017). Walk-man: A high-performance humanoid platform for realistic environments. *Journal of Field Robotics*, *34*(7), 1225–1259. doi:10.1002/rob.21702

Tsukagoshi, H., Sasaki, M., Kitagawa, A., & Tanaka, T. (2005, April). Design of a higher jumping rescue robot with the optimized pneumatic drive. In *Robotics and Automation, 2005. ICRA 2005. Proceedings of the 2005 IEEE International Conference on* (pp. 1276-1283). IEEE. 10.1109/ROBOT.2005.1570291

Tyrsa, V. E., & Zenya, A. D. (1983). Analysis of errors in frequency comparison by the pulse coincidence method. *Measurement Techniques*, *26*(7), 576–579.

UNEP. (2016). Waste Management Criteria. *Cleaner Production*, *4*(5), 12–23.

Vaz, D., Serralherio, A., & Gerald, J. (n. d.). *Navigation System for a Mobile Robot Using Kalman Filters*. Academic Press.

Vermeulen, J. (2004). *Trajectory generation for planar hopping and walking robots: An objective parameter and angular momentum approach. Vrije Universiteit Brussel.*

Vernier. (2019). *3-Axis Accelerometer.* March, 03, 2019, from Vernier Website: https://www.vernier.com/products/sensors/accelerometers/3d-bta/

Volkova, L. Y., & Jatsun, S. F. (2013). Studying of regularities of movement of the jumping robot at various positions of a point of fixing of a foot. *Nelineinaya Dinamika, 9*(2), 327–342. doi:10.20537/nd1302009

Volkova, L. Y., & Yatsun, S. F. (2013). Simulation of motion of a multilink jumping robot and investigation of its characteristics. *Journal of Computer and Systems Sciences International, 52*(4), 637–649. doi:10.1134/S1064230713030155

Vorochaeva, L. Y., Malchikov, A. V., & Postol'niy, A. A. (2018). Approaches to designing wheeled jumping robot. *Extreme robotics and conversion tendencie*s, 308-316.

Vorochaeva, L. Y., Efimov, S. V., Loktionova, O. G., & Yatsun, S. F. (2018). Motion Study of the Ornithopter with Periodic Wing Oscillations. *Journal of Computer and Systems Sciences International, 57*(4), 672–687. doi:10.1134/S1064230718040147

Vorochaeva, L. Y., Panovko, G. Y., Savin, S. I., & Yatsun, A. S. (2017). Movement Simulation of a Five-Link Crawling Robot with Controlled Friction Forces. *Journal of Machinery Manufacture and Reliability, 46*(6), 527–535. doi:10.3103/S1052618817060152

Vorochaeva, L. Y., & Yatsun, S. F. (2015). Mathematical simulation of the controlled motion of the five-link wheeled jumping robot. *Journal of Computer and Systems Sciences International, 54*(4), 567–592. doi:10.1134/S1064230715030168

Vukobratović, M., & Borovac, B. (2004). Zero-moment point—thirty five years of its life. *International Journal of Humanoid Robotics, 1*(1), 157-173.

Wang, H., Wang, C., Luo, H., Li, P., Cheng, M., Wen, C., & Li, J. (2014). Object Detection in Terrestrial Laser Scanning Point Clouds Based on Hough Forest. IEEE Geoscience and Remote Sensing Letters, 11, 1807–1811.

Wang, T., Zheng, N., Xin, J., & Ma, Z. (2011). Integrating Millimeter Wave Radar with a Monocular Vision Sensor for On-Road Obstacle Detection Applications. *Sensors (Basel), 11*(9), 8992–9008. doi:10.3390110908992 PMID:22164117

Watanasophon, S., & Ouitrakul, S. (2014). *Garbage Collection Robot on the Beach using Wireless Communications.* . doi:10.7763/IPCBEE

Weiland, M., Boekhoff, A., & Staloch-schultz, T. (2017). n *Strategies.* Academic Press.

Wescott, T. (2006). *Applied Control Theory for Embedded Systems.* Newnes.

Xiang, S., Chen, S., Wu, X., Xiao, D., & Zheng, X. (2010, February). Study on fast linear scanning for a new laser scanner. *Optics & Laser Technology, 42*(1), 42–46. doi:10.1016/j.optlastec.2009.04.019

Xia, Z., Xiong, J., & Chen, K. (2011). Global navigation for humanoid robots using sampling-based footstep planners. *IEEE/ASME Transactions on Mechatronics, 16*(4), 716–723. doi:10.1109/TMECH.2010.2051679

Yahya, A. (2000). *Tractor with Built-in DGPS for Mapping Power and Energy Demand of Agricultural Field Operations in Malaysia.* UPM Research Report 1997-2000, II/2, 129–131.

Yahya, A., Zohadie, M., Kheiralla, A. F., Gew, S. K., Wee, B. S., & Ng, E. B. (2004). Dewe-2000. Precision system for mapping terrain trafficability, tractor-implement performance and tillage quality. *Proceedings of the 7th International Conference on Precision Agriculture and Other Precision Resources Management,* 23–41.

Yemets, V. (2003). Modern cryptography. Lviv: Baku.

Yi, S. J., McGill, S., Vadakedathu, L., He, Q., Ha, I., Rouleau, M., . . . Lee, D. D. (2014, September). Modular low-cost humanoid platform for disaster response. In *Intelligent Robots and Systems (IROS 2014), 2014 IEEE/RSJ International Conference on* (pp. 965-972). IEEE. 10.1109/IROS.2014.6942676

Yi, S. J., McGill, S. G., Vadakedathu, L., He, Q., Ha, I., Han, J., ... Yim, M. (2015). Team thor's entry in the darpa robotics challenge trials 2013. *Journal of Field Robotics, 32*(3), 315–335. doi:10.1002/rob.21555

Yu, L., Santoni-Bottai, G., Xu, B., Liu, W., & Giurgiutiu, V. (2008). Piezoelectric wafer active sensors for in situ ultrasonic-guided wave SHM. *Fatigue & Fracture of Engineering Materials & Structures, 31*(8), 611–628. doi:10.1111/j.1460-2695.2008.01256.x

Yu, Y., Li, J., Guan, H., Jia, F., & Wang, C. (2015). Learning Hierarchical Features for Automated Extraction of Road Markings From 3-D Mobile LiDAR Point Clouds (2015). *IEEE Journal of Selected Topics in Applied Earth Observations and Remote Sensing, 8*(2), 709–726. doi:10.1109/JSTARS.2014.2347276

Zeglin, G. J. (1991). *Uniroo--a one legged dynamic hopping robot* (Doctoral dissertation). Massachusetts Institute of Technology.

Zewei, X., Jie, W., & Xianqiao, C. (2015). Vehicle recognition and classification method based on laser scanning point cloud data. *International Conference on Transportation Information and Safety,* 44–49. 10.1109/ICTIS.2015.7232078

Zhang, F., & Knoll, A. (2016). Vehicle Detection Based on Probability Hypothesis Density Filter. *Sensors, 16*(4), 1-13.

Zhao, J., Yang, R., Xi, N., Gao, B., Fan, X., Mutka, M. W., & Xiao, L. (2009, October). Development of a miniature self-stabilization jumping robot. In *Intelligent Robots and Systems, 2009. IROS 2009. IEEE/RSJ International Conference on* (pp. 2217-2222). IEEE. 10.1109/IROS.2009.5353949

Zhao, Y., Horemuz, M., & Sjöberg, L. E. (2011). Stochastic modelling and analysis of IMU sensor errors. *Archiwum Fotogrametrii, Kartografii i Teledetekcji, 22*.

Zhao, X., Gao, H., Zhang, G., Ayhan, B., Yan, F., Kwan, C., & Rose, J. L. (2007). Active health monitoring of an aircraft wing with embedded piezoelectric sensor/actuator network: I. Defect detection, localization and growth monitoring. *Smart Materials and Structures, 16*(4), 1208–1217. doi:10.1088/0964-1726/16/4/032

Zheng, Y., Lin, M. C., Manocha, D., Adiwahono, A. H., & Chew, C. M. (2010, October). A walking pattern generator for biped robots on uneven terrains. In *2010 IEEE/RSJ International Conference on Intelligent Robots and Systems* (pp. 4483-4488). IEEE. 10.1109/IROS.2010.5653079

Zhongdong, Y., Peng, W., Xiaohui, L., & Changku, S. (2014, March). 3D laser scanner system using high dynamic range imaging. *Optics and Lasers in Engineering, 54*, 31–41. doi:10.1016/j. optlaseng.2013.09.003

Zimbabwe National Statistics Agency. (2016). Zimbabwe Demographic and Health Survey. *Population of Zimbabwe Urban Areas, 3*(17), 13–24.

Zunaidi, I., Norihiko, K., Yoshihiko, N., & Hirokazu, M. (2019). Positioning System for 4Wheel Mobile Robot: Encoder, Gyro and Accelerometer Data Fusion with Error Model Method. *CMU. Journal, 5*(1), 1–14.

About the Contributors

Oleg Sergiyenko received the B.S., and M.S., degrees in Kharkiv National University of Automobiles and Highways, Kharkiv, Ukraine, in 1991, 1993, respectively. He received the Ph.D. degree in Kharkiv National Polytechnic University on specialty "Tools and methods of non-destructive control" in 1997. He has editor of 1 book, written 8 book chapters, 87 papers and holds 1 patent of Ukraine. Since 1994 till the present time he was represented by his research works in several International Congresses of IEEE, ICROS, SICE, IMEKO in USA, England, Japan, Italy, Austria, Ukraine, and Mexico. Dr.Sergiyenko in December 2004 was invited by Engineering Institute of Baja California Autonomous University for researcher position. He is currently Head of Applied Physics Department of Engineering Institute of Baja California Autonomous University, Mexico, director of several Master's and Doctorate thesis. He was a member of Program Committees of various international and local conferences. He is member of Scientific Council on Electric specialties in Engineering Faculty of Autonomous University of Baja California and Academy of Engineering. Included in the 2010-2015 Edition of Marquis' Who's Who in the World.

Moisés Rivas-López was born in June, 1, 1960. He received the B.S. and M.S. degrees in Autonomous University of Baja California, México, in 1985 and 1991, respectively and the PhD degree in Applied Physics, in the same university, in 2010. He is editor of a book, has written 38 papers and 6 book chapters in optical scanning, 3D coordinates measurement, and structural health monitoring applications. He holds a patent and has presented different works in several international congresses, of IEEE, ICROS, SICE, AMMAC in America and Europe. He was Dean of Engineering Institute of Autonomous University Baja California (1997-2005) and Rector of Polytechnic University of Baja California (2006 -2010). He is member of National Researcher System Dr. Rivas was Head of Engineering Institute of Baja California Autonomous University Since 1997 to 2005; was Rector of Baja California Polytechnic University Since 2006 to 2010 and now is full researcher and the head of physic engineering department, of Engineering Institute of UABC, Mexico.

Wendy Flores-Fuentes received the master's degree in engineering from Technological Institute of Mexicali in 2010, and the Ph.D. degree in science, applied physics, with emphasis on Optoelectronic Scanning Systems for SHM, from Autonomous University of Baja California in June 2014. Until now she is the author of 24 journal articles in Elsevier, IEEE Emerald and Springer, 13 book chapters and 5 books in Intech, IGI global and Springer, 27 proceedings articles in IEEE ISIE 2017-2017, 2019, IECON 2018-2019, the World Congress on Engineering and Computer Science (IAENG 2013), IEEE Section Mexico IEEE ROCC2011, and the VII International Conference on Industrial Engineering ARGOS 2014. Recently, she has organized and participated as Chair of Special Session on "Machine Vision, Control and Navigation" at IEEE ISIE 2015, 2017 and 2019. She has been a reviewer of several articles in Taylor and Francis, IEEE, Elsevier, and EEMJ (Gh. Asachi Technical University of Iasi. Currently, she is a full-time professor-researcher at Universidad Autónoma de Baja California, at the Faculty of Engineering.

Julio C. Rodríguez-Quiñonez received the Ph.D. degree from Baja California Autonomous University, México, in 2013. He is currently Professor of Electronic Topics with the Engineering Faculty, Autonomous University of Baja California. His current research interests include automated metrology, stereo vision systems, control systems, robot navigation and 3D laser scanners. He has written over 50 papers, 3 Book Chapters, has been guest editor of Journal of sensors, book editor, and has been reviewer for IEEE Sensors Journal, Optics and Lasers in Engineering, IEEE Transaction on Mechatronics and Neural Computing and Applications of Springer, he participated as a reviewer and Section Chair of IEEE conferences in 2014, 2015, 2016 and 2017. He is involved in the development of optical scanning prototype in the Applied Physics Department and is currently research head in the development of a new stereo vision system prototype.

Lars Lindner was born on July 20th 1981 in Dresden, Germany. He received his M.S. degree in mechatronics engineering from the TU Dresden University in January 2009. He was working as graduate assistant during his studies at the Fraunhofer Institute for Integrated Circuits EAS in Dresden and also made his master thesis there. After finishing his career, he moved to Mexico and started teaching engineering classes at different universities in Mexicali. Since August 2013 he began his PhD studies at the Engineering Institute of Autonomous University of Baja California in Mexicali with the topic "Theoretical Method to Increase the Speed of Continuous Mapping of a Three-dimensional Laser Scanner using Servomotor Control", in which he worked in the development of an optoelectronic prototype for the measurement of 3D coordinates, using laser dynamic triangulation. Its academic products include 13 original research articles, 3 articles in national congresses, 20 articles in international congresses and 9 book chapters. In September 2017, he was appointed as a Level 1 National Researcher by the National System of Researchers CONACYT for the period 2018-2020. Right now he is working as a technician assistant at the Engineering Institute of Autonomous University of Baja California for the department of applied physics.

* * *

Roman Antoshchenkov was born on September 14, 1982 in Kharkov, Ukraine. In 2006, he graduated from the Kharkiv Petro Vasylenko National Technical University of Agriculture (Kharkov, Ukraine) and obtained the qualification of an agricultural mechanical engineer. The degree of Candidate of Technical Sciences at the Kharkiv Petro Vasylenko National Technical University of Agriculture (Kharkov, Ukraine) in the specialty "Machines and means of agricultural mechanization" in 2010 The degree of doctor of (technical) sciences was received at the Kharkiv Petro Vasylenko National Technical University of Agriculture (Kharkov, Ukraine) with a degree in Machinery and Means of Agricultural Production Mechanization in 2018. He is the author of a monograph, 2 textbooks, 75 articles, 5 patents of Ukraine. Since 2006, he has participated in international scientific conferences in Bulgaria, Belarus, Russia and Ukraine. He began his scientific activity in 2006 as a junior researcher at the Kharkiv Petro Vasylenko National Technical University of Agriculture (Kharkiv, Ukraine). Currently, he is the head of the department of mechatronics and machine parts of the Kharkiv Petro Vasylenko National Technical University of Agriculture. Under his leadership, defended dozens of master's works. He leads the work of a graduate student.

Viktor Antoshchenkov was born on March 24, 1957. In 1985 he graduated from the Kharkov Institute of Mechanization and Electrification of Agriculture (Kharkov, Ukraine) and received the qualification of an engineer-mechanic. He obtained the degree of PhD at the Kharkov Institute of Mechanization and Electrification of Agriculture (Kharkov, Ukraine) in the specialty "Mechanization of Agricultural Production" in 1991. He is the author of 5 textbooks, 53 articles, 3 copyright certificates of the USSR and 5 patents of Ukraine. He is a member of the organizing committee of the International Youth Forum "Youth and Agricultural Machinery in the XXI Century"; scientific and practical workshop; scientific-practical conference "Technical progress in the agricultural sector"; scientific-practical conference "Mechanization of Agriculture"; Interuniversity student seminar "Tractor Energy". He began his scientific activity in 1978 as a senior technician in the department of tractors and automobiles, as an engineer, junior researcher, assistant, associate professor, and from 2014 a professor in this department. From 2007 to 2015, he held the position of Deputy Director, and from 2015 to 2017, Acting Director of the Educational and Scientific Institute of Mechatronics and Management Systems. Since 2017, he holds the position of Professor at the Department of Tractors and Automobiles of the Kharkiv Petro Vasylenko National Technical University of Agriculture. Under his leadership several dozens of master's works and one dissertation of the candidate of technical sciences were defended.

Joel Antúnez-García was born in Ensenada B. C., México, in 1975. He received the B. Sc. degree in Physics from Universidad Autónoma de Baja California (UABC), México, in 1999. The M. Sc. from Centro de Investigación Científica y de Educación Superior de Ensenada (CICESE), México, in 2004. The Ph. D. in Physical-Industrial Engineering from Universidad Autónoma de Nuevo Léon (UANL), Méxio, in 2010. From 2012 to 2013 he did a postdoctoral stay at Centro de Nanociencias y Nanotecnología at UNAM working on DFT calculations to obtain the electronic properties of different zeolites. From 2013-2015 he was working as professor at Centro de Enseñanza Técnica y Superior (CETYS university). From 2016 to date, he has been involved in the theoretical study of bi-and tri-metallic catalysts based on MoS2 compounds.

Danilo Cáceres Hernández received the Bachelor's degree in electrical and electronic engineering from the Universidad Tecnológica de Panamá, Panamá City, Panama, the Master of Science in electrical engineering, and the Ph.D. in electrical engineering from the University of Ulsan, Ulsan, South Korea, in 2004, 2011, and 2017, respectively. He is currently a Full-Time Professor in the Electrical Department at the Universidad Tecnológica de Panamá.

Wilmar Hernandez graduated in 1992 with an Electronics Engineering degree from the Instituto Superior Politécnico Jose Antonio Echeverria (ISPJAE), Havana, Cuba, and received a Specialist degree in Microelectronics from the ISPJAE in 1994. Also, he received a M.S. degree in Signal Treatment and a Ph.D. degree in Electronic Engineering from Enginyeria La Salle at the Universitat Ramon Llull, Barcelona, Spain, in 1997 and 1999, respectively. From 1992 to 1995, he was a lecturer in the Electrical Engineering Faculty at the ISPJAE and a researcher in the Microelectronic Research Center at the same university. From 1999 to 2003, he was with the Department of Electronics and Instrumentation in the University Institute for the Automobile Research at the Universidad Politécnica de Madrid (UPM), Spain, where he was the Technical Director of such a department from January 2003 to January 2004. From January 2004 to March 2013 he was an Associate Professor of Circuits and Systems in the Department of Circuits and Systems in the EUIT de Telecomunicación at the UPM. From September 2014 to September 2015 he was a researcher of SENESCYT, Ecuador, under the Prometeo fellowship program. From December 2015 to November 2017 he was a professor at the Universidad Técnica Particular de Loja, Ecuador, and currently he is a professor at the Universidad de Las Américas, Ecuador.

Juan Ivan Nieto Hipólito received his M.Sc. degree from CICESE Research Center in 1994 (México). His PhD degree from Computer Architecture Department at Polytechnic University of Catalonia (UPC, Spain) in 2005. Since august 1994 he is full professor at the Autonomous University of Baja California (UABC, México), where he was the leader of the Telematic research group from 2007 to 2012. His research interest is in applications of ICT, mainly wireless, mac and routing protocols for e-health.

Kang-Hyun Jo received the Ph.D. degree in Computer Controlled Machinery from Osaka University, Japan, in 1997. After a year of experience at ETRI as a postdoctoral research fellow, he joined the School of Electrical Engineering, University of Ulsan, Ulsan, Korea. He has served as a director or an AdCom member of Institute of Control, Robotics and Systems, The Society of Instrument and Control Engineers, and IEEE IES Technical Committee on Human Factors Chair. Currently, he is serving as AdCom member, and from 2018, as the Secretary, of the IEEE IES. He has also been involved in organizing many international conferences such as International Workshop on Frontiers of Computer Vision, International Conference on Intelligent Computation, International Conference on Industrial Technology, International Conference on Human System Interactions, and Annual Conference of the IEEE Industrial Electronics Society. At present, he is an Editorial Board Member for international journals, such as the International Journal of Control, Automation, and Systems and the Transactions on Computational Collective Intelligence. His research interests include computer vision, robotics, autonomous vehicle, and ambient intelligence.

Vladimir Krasilenko was born 20 July 1953, Vinnitsa Region, Ukraine. Education: Radio Engineers Diploma, 1975, Candidate of Sciences Degree (PhD), 1988, Information systems, Vinnitsa State Technical University. Engineering positions and head of the department of research institutes and enterprises - 1975-1982; PhD student and Lecturer Assistant, 1982-88; Senior, Leading Scientific Researcher, Head of Special Design and Technology Bureau of Vinnitsa National Technical University, Leading Scientist, Enterprise «Injector», Science Research Institute of Videotechnic, 1988-2001; Associated Professor, Professor of Information Technology Department, Vinnitsa Social Economy Institute of Univ. "Ukraine", 2001-2015. He has authored/co-authored more 400 scientific works, including 188 inventions, about 80 articles in scientific journals and Proc. SPIE, 2 chapter published by InTech, 5 tutorial books. Krasilenko named best Young Inventor of Ukraine in 1985, Member of SPIE with 1995 and Senior Member of SPIE with 2012. Research interests: optoelectronic devices and multifunctional logic elements for parallel image processing and for computing, neural networks, multi-valued, continuous matrix logic, recognition, cryptography, information protection, analog-to-digital transformations.

Laksono Kurnianggoro received his bachelor of engineering from the University of Gadjah Mada, Indonesia, in 2010. He is currently a Ph.D. student at the Graduate School of Electrical Engineering, University of Ulsan, Ulsan, Korea. He is actively participating as a member of societies such as IEEE. His research interest include stereo vision, 3D image processing, computer vision, and machine learning. He has scientific publications in some publishers such as IEEE, Springer, and Elsevier. He also involved in several projects including development of autonomous vehicle system, advanced car washer system, low-cost 3D scanner, autonomous robot, and many mores. He is also an active contributor of the popular computer vision library, like OpenCV.

Alexander Lazarev is an assistant professor at the Vinnytsia National Technical University (VNTU), Ukraine. He completed his graduate studies with a PhD from the VNTU in 2003 and his undergraduate studies at the VNTU with a MSc in Electronics Engineering in 1998. He has authored/co-authored more than 250 publications.

Andrei Malchikov is a candidate of technical sciences Academic title: assistant Professor From 2004 to 2010 studied mechatronics and robotics at Southwest State University and graduated with degrees at bachelors and masters level. In 2013 he defended his candidate thesis: "Dynamics of controlled motion of a six-link in-pipe robot". He published more than 50 publications and has more than 10 patents on inventions.

Andres Santiago Martinez Leon, 26 years, born in Ecuador, received a B.Sc. (with honors) in 2015 and a M.Sc. (with honors) in 2017 in Mechatronic and Robotic Engineering at Southwest State University, Kursk, Russian Federation. At present, pursuing a Ph.D. in Mechanical Engineering at Southwest State University, Kursk, Russian Federation. Since 2017 Associated Teacher at Escuela Politécnica Nacional, Quito, Ecuador and Universidad Estatal Amazónica, Puyo, Ecuador. Author and coauthor of several international scientific papers and also 1 RU patent and 2 pending patents. The current research involves the development of Unmanned Aerial Vehicles (UAV).

Viktor I. Melnyk was born January 8, 1958. In 1980 was graduated from Poltava Agricultural Institute (Poltava, Ukraine) and was qualified as a mechanical engineer in agriculture. PhD degree obtained at the Kharkov State Technical University of Radio Electronics (Kharkiv, Ukraine) on specialty "Technology, equipment and production of electronic devices" in 2000. He get the Degree of Doctor of Sciences (in Technics) at the Kharkov National Technical University of Agriculture named after Peter Vasilenko (Kharkiv, Ukraine) on specialty "Machinery and mechanization of agricultural production" in 2011. He is the author of 4 books, 261 articles, 39 patents of the Russian Federation and 19 patents of Ukraine. Since 1986 participated in the international scientific conferences in the United States, Belarus, Russia and Ukraine. He is currently a professor in the Department of technologic systems optimization, the Head of the research laboratory «Engineering of nature management», as well as the deputy director for scientific activities in Institute of Mechatronics and management systems at the Kharkov National Technical University of Agriculture. He was tutor of several masters' and two PhD thesis. He is deputy chairman of the specialized scientific council for doctoral dissertations on specialty "Machinery and mechanization of agricultural production" at the Kharkov National Technical University of Agriculture. Since 2014 he is the Editor-in-Chief of the scientific journal "Engineering of nature management".

Alfredo Méndez Alonso was born in Madrid, Spain, in June 6, 1958. He graduated with a degree in Mathematical Sciences (Section of Fundamental Mathematics) in 1981, from the Universidad Complutense de Madrid (UCM). Also, he received a M.S. degree in Mathematical Sciences in 1987 and a Ph.D. degree in Mathematical Sciences (Section of Statistics and Operational Research) in 1995 from the UCM. From 1983 to 1993, he was a lecturer at the EUIT Agrícola at the Universidad Politécnica de Madrid (UPM), Spain. Since 1993 he has been professor of Mathematics in the Department of Applied Mathematics to Information and Communication Technologies, of the ETS Ingeniería y Sistemas de Telecomunicación at the UPM, where he was the director of the department from May 2004 to May 2012.

Fabian N. Murrieta-Rico was born in September the 7th of 1986. He received the B.Eng. and M.Eng. degrees from Instituto Tecnológico de Mexicali (ITM) in 2004, and 2013 respectively. In 2017 he received his Ph.D. in Materials Physics from Centro de Investigación Científica y Educación Superior de Ensenada (CICESE). He has worked as automation engineer, systems designer, and as a university professor. His research has been published in different journals, and presented in international conferences since 2009. His research interests are in the field of time and frequency metrology, wireless sensor networks design, automated systems, and highly sensitive chemical detectors. Currently he is involved in development of new frequency measurement systems, and highly sensitive sensors for detection of chemical compounds.

Tawanda Mushiri is a holder of BSc Mechanical Engineering (UZ), Master of Science in Manufacturing Systems and Operations Management (MSOM) (UZ) a PhD in Automation, Robotics and Artificial Intelligence (U.J) of machinery monitoring systems. He is currently a Senior Lecturer at the University of Zimbabwe teaching Machine Dynamics, Robotics, Solid Mechanics and Finite Element Analysis. He is also the coordinator of Undergraduate projects and Master of Science in Manufacturing Systems and Operations Management (MSOM). Tawanda has supervised more than 100 students' undergraduate projects and 1 Masters Student to completion. He has published 2 Academic Textbooks, 3 Chapters in a book, 14 Journals and 84 Conference Papers plus a Patents in highly accredited publishers. He has done a lot of commercial projects at the University of Zimbabwe. He is a reviewer of 4 journals highly accredited. He has been invited as a keynote speaker in workshops and seminars. Beyond work and at a personal level, Tawanda enjoys spending time with family, travelling and watching soccer.

Diana V. Nikitovich i an dispatcher of Faculty for Radio Engineering, Tele-communication and Electronic Instrument Engineering at the Vinnytsia National Technical University (VNTU), Ukraine. She graduated from the magistracy from Vinnitsa Social and Economic Institute in 2012 and its MSc and his bachelor in document management and information activities. She is the author / coauthor of about 50 publications. Research interests: neural networks, devices and logic elements for image processing, continuous matrix logic, recognition, information protection, cryptography, analog-to-digital transformations.

Vitalii Petranovskii received his PhD in Physical chemistry from Moscow Institute of Crystallography in 1988. In 1993-1994 he worked as invited scientist at National Institute of Materials and Chemical Research, Japan. Since 1995, he is working at "Centro de Nanociencias y Nanotecnología, Universidad Nacional Autónoma de México" (2006-2014 – as the Nanocatalysis department chair). His research interests include synthesis and properties of nanoparticles supported over zeolite matrices. He is a member of Mexican Academy of Sciences, International Zeolite Association, and Mendeleev Russian Chemical Society. He has published over 140 papers in peer-reviewed journals and 5 invited book chapters. Also he is co-author of monograph "Clusters and matrix isolated cluster superstructures", SPb, 1995.

Oleksandr Poliarus was born on February 18, 1950 in Gadyach town of Poltava region (Ukraine). From 1967 to 1973 he studied at the Moscow Higher Technical School named after M. E. Bauman, after which he worked as an engineer at a research institute for one year. From 1974 to 1999 he served in the Armed Forces of the USSR and Ukraine. In 1980 graduated from the Military Radio Engineering Academy of Air Defense in Kharkiv. He was in teaching positions of Academy, and since 1996 - the head of the department of antenna-feeder devices. In 1994 he defended the doctoral thesis. Since September 2007 he is the head of the Department of Metrology and Life Safety of the Kharkiv National Automobile and Highway University.

Yevhen Polyakov was born in 1985 in Pavlograd, Dnipropetrovsk region. In 2008 he graduated from Kharkiv National Automobile and Highway University and received a full higher education in the specialty "Automated Control of Technological Processes". From 2008 to 2012, graduate student of the Department of Metrology and Life Safety of Kharkiv National Automobile and Highway University. From 2012 he works at the Kharkiv National Automobile and Highway University as Associate Professor of the Department of Metrology and Life Safety. In 2014 he defended his Ph.D. thesis on the theme "Improvement of methods of sensors dynamic errors decrease".

Miguel Reyes-Garcia was born in Hidalgo state, Mexico, September, 29, 1989. He is a graduated mechatronics engineer of the Universidad Politécnica de Baja California (UPBC) in 2014. Currently, He is studying his Master degree with his thesis named: Theoretical method to reduce the positioning error in a scanning laser system, using an embedded digital controller and direct current motors. Working in the Optoelectronics and Automated Measurement Laboratory, Engineering Institute of the Universidad Autónoma of Baja California (UABC), Mexicali, B.C., Mexico.

Juan de Dios Sanchez Lopez received the B. Eng. (Electrical engineering) degree from the Mechanical and Electrical Department of the Technological Institute of Madero City in 1988. He received his M.Sc. and Ph.D. degrees from the CICESE Research Center in 1999 and 2009 respectively. His research interest include wireless communications, optical communications, analog-digital signal processing, electronic instrumentation and bio-optical signals.

Sergei Savin graduated from Southwest State University (SWSU) in 2011, and received PhD in 2014. From 2013 worked at the department of Mechanics, Mechatronics and Robotics at SWSU (senior researcher and docent), from 2018 is a senior research at the Innopolis University. Research interests include application of optimal control, convex optimization and computational geometry to mobile robotics. The areas of research include legged locomotion in general, bipedal walking robots, exoskeletons, in-pipe robots and multi-link mobile robots, machine learning in robotics. He is author of 4 books, more than 50 scientific papers and over 20 patents.

Cesar Sepúlveda-Valdez was born in Mexico in 1994, he obtained his degree in mechatronical engineering from the Autonomous University of Baja California in 2016. He is currently in the master and doctorate program of the UABC Engineering Institute to acquire his master degree in engineering in the applied physics department to which he joined in 2018. He participated in ISIE Vancouver 2019 Congress, obtaining a certificate for the best session presentation. He has written articles in the machine vision area which is the topic of his thesis degree.

Vera (Vira) V. Tyrsa was born on July 26, 1971. She received the B. S. and M. S. degrees in Kharkov National University of Automobiles and Highways, Kharkov, Ukraine, in 1991, 1993, respectively (Honoris Causa). She received the Ph.D. degree in Kharkov National Polytechnic University on specialty "Electric machines, systems and networks, elements and devices of computer technics" in 1996. She has written 1 book, 7 book chapters, and more than 50 papers. She holds one patent of Ukraine and one patent of Mexico. From 1994 till the present time, she is represented by her research works in international congresses in USA, England, Italy, Japan, Ukraine, and Mexico. In April 1996, she joined the Kharkov National University of Automobiles and Highways, where she holds the position of associated professor of Electrical Engineering Department (1998–2006). In 2006–2011, she was invited by Polytechnic University of Baja California, Mexico for professor and researcher position. Currently, she is a professor of electronic topics with the Engineering Faculty, Autonomous University of Baja California. Her current research interests include automated metrology, machine vision systems, fast electrical measurements, control systems, robot navigation and 3D laser scanners. Now she works at the Autonomous University of Baja California.

Mabel Vazquez-Briseño received her Ph.D in Computer Science from Telecom SudParis (ex INT) and Pierre et Marie Curie University, France in 2008. She received the M.Sc degree in Electronics and Telecommunications from CICESE Research Center, Mexico, in 2001. She is now researcher-professor at the Autonomous University of Baja California (UABC), where she is a member of the Telematics research group. Her research interests include computer networks, mobile computing, sensor networks and protocols.

Rosario Isidro Yocupicio Gaxiola was born in Los Mochis, Sinaloa, Mexico. He studied at the Instituto Tecnológico de Los Mochis and completed a PhD at Centro de Nanociencias y Nanotecnología of Universidad Nacional Autónoma de México (CNyN.UNAM) in 2017. Actually, he is a posdoctoral researcher at CNyN-UNAM. His research interests lie in the synthesis, analysis and applications of porous solids, particularly the study of zeolites as microporous and mesoporous systems. His research is focused in the use of microporous materials as catalysts, supports of catalysts, sorbents and opticals devices.

Index

A

Acceleration 3, 5, 9, 11-13, 15, 17, 19, 22-23, 27-29, 33-36, 39-40, 46, 48, 51-55, 58-70, 73, 75-76, 78-81, 83-86, 89, 121, 221-222, 229, 232, 235, 264-265, 271

Acceleration Module 52-54, 58-65, 67-70, 75-76, 79-81, 83-84

Accelerometer 1, 3, 5, 7-9, 11-12, 17-18, 22-23, 25-30, 33-34, 46-47, 49-51

Active Node 265

Angle of Rotation in Flight 89

Angular Velocity 5, 9, 17-18, 21, 23, 60, 82, 103, 232, 234

Anthropomorphic Robot 285

Arduino 33, 90, 108, 113, 143, 146-149, 165, 217-218, 227, 229-230, 237, 247-248, 250

B

Bipedal Robot 275, 277, 285

C

Center of Mass Trajectory Generation 268-269, 274, 278

Compression Rate 313, 315-316, 319, 321-325

Control System 24, 52-53, 58, 67, 82-84, 123, 143, 217-218, 223-224, 227-228, 234, 236-239, 247-248, 250, 265, 275

Cryptographic Image Transformations 170, 211-212

D

Dead Reckoning 28, 51

Decryption 176, 196, 198, 212-213

Diffuse Reflection 265

Dynamics 1-3, 5-13, 15, 17-19, 21-23, 47, 51, 85, 87, 267-268, 271-272, 277, 281, 283-284

E

Encryption 174, 176, 180, 183, 188, 196, 198, 212-214

F

Filter 1, 6, 12-14, 17, 19, 21-24, 43, 49, 277, 288, 329, 353

Flight 54-55, 57, 60, 68-69, 73, 75, 79, 81-82, 84-85, 87, 89

Frequency Measurement 27-29, 31-33, 43, 45-49

Frequency Output 26-27, 29

Frequency Shift 40, 43, 47

G

Garbage Collection Robot 90, 92, 108, 115, 117, 127, 133-134, 137, 141, 143, 159-160, 162-165, 167

H

Height of the Jump 89

Hybrid Wheeled Jumping Robot 52

Ensure Quality Research is Introduced to the Academic Community

Become an IGI Global Reviewer for Authored Book Projects

Premier Reference Source

Emerging GIS Applications for Emergency and Disaster Management

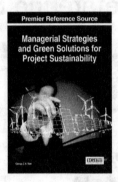

Premier Reference Source

Managerial Strategies and Green Solutions for Project Sustainability

Premier Reference Source

Comparative Approaches to Using R and Python for Statistical Data Analysis

Premier Reference Source

Solutions for High-Touch Communications in a High-Tech World

The overall success of an authored book project is dependent on quality and timely reviews.

In this competitive age of scholarly publishing, constructive and timely feedback significantly expedites the turnaround time of manuscripts from submission to acceptance, allowing the publication and discovery of forward-thinking research at a much more expeditious rate. Several IGI Global authored book projects are currently seeking highly-qualified experts in the field to fill vacancies on their respective editorial review boards:

Applications and Inquiries may be sent to:
development@igi-global.com

Applicants must have a doctorate (or an equivalent degree) as well as publishing and reviewing experience. Reviewers are asked to complete the open-ended evaluation questions with as much detail as possible in a timely, collegial, and constructive manner. All reviewers' tenures run for one-year terms on the editorial review boards and are expected to complete at least three reviews per term. Upon successful completion of this term, reviewers can be considered for an additional term.

If you have a colleague that may be interested in this opportunity, we encourage you to share this information with them.

Celebrating Over 30 Years of Scholarly Knowledge Creation & Dissemination

InfoSci®-Books

A Database of Over 5,300+ Reference Books Containing Over 100,000+ Chapters Focusing on Emerging Research

GAIN ACCESS TO **THOUSANDS** OF REFERENCE BOOKS AT **A FRACTION** OF THEIR INDIVIDUAL LIST **PRICE**.

InfoSci®-Books Database

The **InfoSci®-Books** database is a collection of over 5,300+ IGI Global single and multi-volume reference books, handbooks of research, and encyclopedias, encompassing groundbreaking research from prominent experts worldwide that span over 350+ topics in 11 core subject areas including business, computer science, education, science and engineering, social sciences and more.

Open Access Fee Waiver (Offset Model) Initiative

For any library that invests in IGI Global's InfoSci-Journals and/or InfoSci-Books databases, IGI Global will match the library's investment with a fund of equal value to go toward **subsidizing the OA article processing charges (APCs) for their students, faculty, and staff** at that institution when their work is submitted and accepted under OA into an IGI Global journal.*

INFOSCI® PLATFORM FEATURES

* No DRM
* No Set-Up or Maintenance Fees
* A Guarantee of No More Than a 5% Annual Increase
* Full-Text HTML and PDF Viewing Options
* Downloadable MARC Records
* Unlimited Simultaneous Access
* COUNTER 5 Compliant Reports
* Formatted Citations With Ability to Export to RefWorks and EasyBib
* No Embargo of Content (Research is Available Months in Advance of the Print Release)

*The fund will be offered on an annual basis and expire at the end of the subscription period. The fund would renew as the subscription is renewed for each year thereafter. The open access fees will be waived after the student, faculty, or staff's paper has been vetted and accepted into an IGI Global journal and the fund can only be used toward publishing OA in an IGI Global journal. Libraries in developing countries will have the match on their investment doubled.

To Learn More or To Purchase This Database:
www.igi-global.com/infosci-books

eresources@igi-global.com • Toll Free: 1-866-342-6657 ext. 100 • Phone: 717-533-8845 x100

Printed in the United ...
in Singapore ...

Printed in the United States
By Bookmasters